MEANT TO BE WILD

MEANT TO BE WILD

*The Struggle to Save Endangered Species
Through Captive Breeding*

Jan DeBlieu

Fulcrum Publishing
Golden, Colorado

Portions of Chapter I, "Return to the East," and Chapter IV, "A New Freedom," appeared in slightly different form in *Outer Banks Magazine,* under the title, "Red Wolves Roaming." Portions of Chapter IX, "The Essence of Wildness," appeared in *Orion Magazine* under the title, "The Polar Bear's Dance."

Library of Congress Cataloging-in-Publication Data

DeBlieu, Jan.
    Meant to be wild : the struggle to save endangered species through captive breeding / Jan DeBlieu.
        p. cm.
    Includes bibliographical references and index.
    ISBN 1-55591-166-8
    1. Wildlife reintroduction—North America. 2. Captive wild animals—North America—Breeding. 3. Endangered species—North America. 4. Wildlife conservation—North America. I. Title.
QL83.4.D43 1991
639.9—dc20   91-71365
                                                     CIP

Printed in the United States of America

0 9 8 7 6 5 4 3 2 1

Fulcrum Publishing
350 Indiana Street, Suite 350
Golden, Colorado 80401

For my brother Ken

TABLE OF CONTENTS

THE WOLVES OF ALLIGATOR RIVER

I	RETURN TO THE EAST	3
II	THE FIRST SPRING	15
III	RESCUE MISSION	29
IV	A NEW FREEDOM	53
V	A CHORUS OF WOLVES	67

BIRDS IN THE HAND

VI	A RUINOUS LEGACY	91
VII	THE NEED FOR COMPASSION The Whooping Crane and the Peregrine Falcon	105
VIII	LEAVING THE ARK The Arabian Oryx and the Golden Lion Tamarin	129
IX	THE ESSENCE OF WILDNESS	149

MEANT TO BE WILD

X	A SINGLE STRUGGLING FLOCK The Puerto Rican Parrot	167
XI	THE SOUL OF THE CONDOR	191
XII	STEPPING BACK FROM EXTINCTION The Black-footed Ferret	225
XIII	THE PANTHER VERSUS FLORIDA	249
XIV	EPILOGUE	275
	BIBLIOGRAPHY	287
	INDEX	295

AUTHOR'S NOTE

I would like to express my deep appreciation to the scientists and wildlife conservationists who appear within these pages. Without exception they were extremely gracious and generous with their time, their resources, and their expertise. They have devoted themselves to preserving some of the most beautiful forms of life on earth. We owe them a great debt.

I have included a bibliography at the end of the text for readers who would like more detailed information than presented here.

We were flying over a range of snowpeaked mountains
And throwing dice for the soul of the condor.
—Should we grant reprieve to the condor?
—No, we won't grant reprieve to the condor.
It didn't eat from the tree of Knowledge and so it must perish.

From *Diary of a Naturalist*
by Czeslaw Milosz

Where there is no vision
The people perish.

Proverbs 29:18

THE WOLVES
OF ALLIGATOR RIVER

I

RETURN TO THE EAST

On a mid-November morning in 1986, a day as bleak and gray as gun metal, a dozen reporters huddled loosely together in the parking lot of a small airport on the North Carolina coast. A light west wind blew in from the flat waters of Croatan Sound, bringing the damp, salty smell of marshland and rain. Earlier, the clouds had built almost to darkness, but in the past hour they had thinned enough to let through a harsh glare. The reporters and camera crews, bored by unexpected delays, lounged grumpily against their cars.

The day had started with a gust of activity as the national media arrived on Roanoke Island, an oasis of quiet neighborhoods and piney forests just west of the sandy reefs known as the Outer Banks. The press contingent included reporters from the *New York Times*, *Newsweek*, *Smithsonian*, two national television networks, and several regional papers. They had come to witness the close of a long and painful era in natural history and the heady first moments of what was being billed as a new age. Sometime that morning a helicopter would deliver a cargo of eight red wolves that were to be set free on a nearby wildlife refuge. Their release would mark the first time in more than a century that wolves had roamed free along the eastern seaboard of the United States.

It was a symbolic event, the kind the media chases with a terrible voraciousness. The reporters and photographers arrived dressed as instructed for rain and mud, several of them in shiny duck shoes and Goretex jackets that looked as if they had been purchased for the occasion. Impatient and professionally aloof, they seemed out-of-place among the local residents, who showed up in the usual winter garb of the Outer Banks—in jeans and wool shirts and in no hurry at all.

On the edge of the main tarmac a slim, graying man named Warren Parker scanned the sky and kicked distractedly at a low fence. I had met Parker only briefly, but from my position thirty yards away I could see that he was nervous. As the leader of the U.S. Fish and Wildlife Service's red wolf reintroduction project, Parker had worked for three years to find a place where the animals could be freed. For most of the morning he had been followed by reporters, pulled aside for interviews, taped and photographed, until, to his dismay, the reporters had run out of questions. Now all he could do was wait for the Coast Guard helicopter that was supposed to have arrived an hour before.

As the morning wore on, Parker's face had grown pinched with worry. Tension showed in his rigid stance, in the way he kept his fists shoved forward into his pockets like stones. He had no reason to think anything had gone wrong, but he had the jitters. At the sound of an engine he cocked his head, then shook it, disappointed. A second later he cocked it again. A small plane had landed and coasted to a stop, but behind it was another noise, a thin throbbing, the sound of rotors slicing through humid air. The reporters around me also looked up; one or two wandered over to the fence where Parker waited. Finally the helicopter appeared in the west, and Parker broke into a broad smile. Heat played in waves across the long body of the craft, rippling the orange and blue emblem on its side. As it settled onto the tarmac, a knot of people burst past the fence and sprinted across the wind-blown cement.

History had arrived in the dismal midday light, history in the form of eight rangy, copper-colored wolves. Behind the aircraft, two biologists in brown Fish and Wildlife Service jackets stepped forward and sprang the hatch to the cargo bin. Parker hovered behind them, relief and joy in his smile. Together the men brought out a large vinyl shipping kennel, lifting it high enough for onlookers to glimpse the silhouette of two pointed ears through the door. Reporters and sound crews, nudged forward by others straining to see, pressed close around, holding booms and microphones aloft. As a pickup truck backed to within a few feet of the helicopter, the men slid the kennel into the bed and reached through the cargo hatch for another.

For the next five years Parker and his colleagues would be responsible for running one of the most significant wildlife research projects ever undertaken in the United States. The red wolf was extinct in the wild; only seventy-four remained, and all had been raised in pens. No one knew whether the wolves arriving that day retained enough fear of humans to fend for themselves and stay out of the small communities near the release site. The success or failure of the project would have direct bearing on plans to return wolves to other areas of the country, including Yellowstone National Park. And as one of the first attempts to rebuild a wild population from captive animals, the red wolf project, whatever its outcome, would serve as a model for efforts to save endangered species around the world.

The project had another, more symbolic importance. By releasing wolves in a North Carolina swamp, the Fish and Wildlife Service was attempting to force a break in the sad march of American natural history. It was trying to create a new era of tolerance, to reclaim some of the nation's biological richness while reshaping the attitudes that had forced wolves and other predators to the edge of extinction. At the Alligator River National Wildlife Refuge, biologists hoped to show that wolves and people could comfortably coexist, as long as the wolves were carefully managed and the people were willing to yield the animals some ground.

The landscape to which the species was being returned had been beset by changes; so had the nature of the people the wolves were likely to encounter. Two hundred years before, when wolves were still thick in the eastern forests, the coastal swamps had been untouched by development. Now even remote areas were crossed by canals and dirt roads. Two hundred years before, trappers and hunters had caught only fleeting glimpses of wolves. Now biologists could use electronics to monitor where and when each wolf moved, and with whom it bred. Although none of us on the tarmac that day had any way of knowing it, in coming years the red wolf project would unfold as a story of early frustrations and, in the end, surprising fulfillment. Finally, it would prove to be a story of how wildness—the silent wit and ken of free-ranging animals—surpasses the scope of human understanding.

Within minutes the kennels had been loaded by pairs into four mud-encrusted trucks, each of which would be driven to a pen deep in the woods of the Alligator River National Wildlife Refuge. The trucks idled quietly as I jogged back to the parking lot, past several photographers who were hurrying to cram their equipment into the back of an overloaded station wagon. After engines turned over, choked, sputtered, and caught, we pulled into a convoy heading west across the Croatan Sound bridge to the mainland. We curled through Manns Harbor, a community of modest houses and small businesses set on a piney ridge. To the west of town the highway straightened and followed two silent black canals carved through a marsh of slender reeds. The convoy drove on, entering a swamp of pine, cedar, sawgrass, and sulfurous soil that collapsed beneath any weight. A swamp where red wolves were soon to roam.

If wolves are animals of savage and demonic qualities, as myth and folklore portray them, then red wolves have been doubly damned. They are despised, on the one hand, by people who think of wolves as bloodthirsty and sinister, yet they are often overlooked by those who might be expected to rush to their defense. A shy animal that once roamed widely through southeastern marshes and forests, the red wolf, *Canis rufus*, is smaller, more secretive, and more solitary than the gray wolf, *Canis lupus*, of Alaska, Canada, and the northern United States. Perhaps because of this, it has not received the attention that has made the gray wolf a symbol of thrilling beauty to people concerned about the degradation of the nation's wilderness. Where conservationists might describe the gray wolf as regal, intelligent, and socially complex, many have never heard of its red cousin.

Physically, the red wolf so resembles a cross between the coyote and the gray wolf that its classification as a separate species has long been a point of debate. Generally weighing between seventy and eighty pounds, the male red wolf is lankier than the gray and not as broad through the shoulders, almost like an adolescent who is still waiting to fill out. Its coat is not really red but tan or cinnamon, filled with gold and auburn highlights and streaked with black. Whether it is a true species or a coyote-gray wolf mix was of little concern to the settlers of the Southeast, who killed all predators with equal fervor. Through the mid-1960s the red wolf was trapped by government hunters and ranchers and crowded by development into a diminishing range, until finally it subsisted only along the Texas-Louisiana coast. Even then, predator control officers routinely killed any animal that dared linger near ranch fields and towns.

But while historically the red wolf has been counted among the pariahs of wildlife, since 1973 it has been counted among the chosen, at least in the graces of the federal government. Shortly after Congress passed the Endangered Species Act, the U.S. Fish and Wildlife Service began trying to save the red wolf, first through a small field program and, later, when biologists realized the species was about to perish, by trapping red wolves and shipping them to a zoo near Tacoma, Washington, for breeding.

It was the first time the agency had taken such radical measures to rescue a dying species. The red wolf was so close to extinction that it seemed more sensible to gather the few animals that remained and rebuild the species artificially than to try to preserve it in its original habitat. By relying wholly on captive breeding, however, government biologists were gambling that the red wolf could be frozen in time. Its imprisonment would be too short (the biologists hoped) to affect either its temperament or its evolution. As soon as conditions were right, *Canis rufus* would be returned to its native lands. Whatever small changes occurred in its behavior, or its genetic makeup, or the way it perceived its surroundings would not, the biologists trusted, have much effect on its well-being.

To justify eliminating the red wolf as a free-ranging species, the men and women of the Fish and Wildlife Service made at least two significant assumptions about the character of wild animals and the factors that propel them toward extinction. They agreed, first, that the red wolf was probably in too precarious a state to survive on its own, or with the help of conventional wildlife management techniques. In captivity the red wolf could be vaccinated against disease and fed a balanced diet. It could be treated for worms and mange, fenced off from poachers, and protected from the threats, human and natural, it would otherwise face. The second assumption required a greater leap of faith. The biologists agreed there would probably not be much difference between a wolf reared in the wild and one raised in a pen, though at the time no one knew for certain that red wolves would breed in captivity. Without immediate help the species seemed doomed, and a breeding program appeared to be the quickest, most practical way to intervene.

This same rationale has been used many times in endangered species conservation projects in the United States and abroad. Since the 1970s the California condor, the Puerto Rican parrot, the whooping crane, the masked bobwhite, and the black-footed ferret all have been the subjects of breeding experiments run by the Fish and Wildlife Service. Wildlife scientists from the Smithsonian Institution, the National Zoological Park, the San Diego Wild Animal Park, the New York Zoological Society, and other conservation organizations have begun raising critically endangered species from around the world. Some of these animals no longer live in the wild at all. Eventually scientists hope to free the offspring of endangered captives, although the problems that have stricken their populations tend to grow from the stress of having too little suitable habitat.

In some ways the philosophy behind captive breeding resembles the old saw about the comparative value of a bird in the hand. Wild animals live in constant danger of death, and the daily uncertainty of their survival increases as they become more rare. Many scientists believe it is more important to maintain a good genetic stock of critically endangered animals in zoos and breeding centers than to preserve inbred populations in the wild. In this sense, any rare animal in the hand is worth two in the bush. Yet a number of wildlife researchers and conservationists have come to view the trend toward captive breeding with alarm. Some complain that so much money has been diverted for breeding programs that field study—once the crux of endangered species work—is being seriously neglected. Others worry about the subtle physical and psychological changes that may beset wild animals that are confined to cages, even when their contact with people is limited. Their muscles may degenerate, for example, or their eyesight grow lazy from staring at the same unchanging scenery. Protected from predators for generations, they may lose the intuition they need to avoid danger, and the desire to teach their young to be constantly on guard.

However wildlife scientists may view captive breeding, they agree that it is no more than a palliative. It provides no lasting solution to the problem of vanishing species, no means of stemming the environmental destruction that threatens to bleed the world of most of its natural diversity. It offers hope that someday human attitudes will change, political turmoil will cease, and wild landscapes will be restored; but it does not address the deep cultural and religious beliefs that encourage people to kill wildlife and destroy natural areas. Nor do breeding programs preserve the complex social structures and behavioral patterns of the animals, although the best programs may try.

When the time comes for the rarest captives to be set free, they will be released into a world completely devoid of their kind. It will be as if a holocaust has destroyed everyone but them. Each will have to learn the tricks of survival—where to find food, how to make shelter, when to hide from danger. And each population will have to establish its own new social structure, which, in the case of complex, intelligent animals like wolves, may require many generations.

How quickly and easily a reintroduced species adjusts to its new existence depends on how deeply it can draw from the well of wild behavior that nourished it before its capture. Most animals, even domestic cats and dogs, retain a few wild impulses. It is not possible, however, to know when the source of wild behavior begins to grow polluted; it is not possible to know what subtle characteristics may be altered as a species spends a lengthening time in confinement.

The wolves to be released in North Carolina had been in captivity for only a few generations. It seemed likely that they would hunt skillfully enough to feed themselves and would shun any contact with humans. If need be, food supplements could be set out during the first months after they were freed. Since the animals were to be fitted with radio collars, their movements could be carefully watched. Trackers could drive off any that wandered too close to nearby villages. The wolves would be managed too much to be considered truly wild, of course. But, under the circumstances, a managed population of red wolves seemed better than none at all.

It was this line of thought that had brought the Fish and Wildlife Service to the North Carolina coast for a grand experiment on how an extinct predator might be resurrected in twentieth-century America. As Warren Parker and his colleagues stood on the tarmac and hailed the arrival of the wolves, they had an unspoken but pervasive hope that the full recovery of the species was only five or six years away. By releasing even a few of the animals at Alligator River they would lay a cornerstone of sorts, and from there they could continue to build in orderly fashion, until the red wolf was once again on safe, unshakable ground.

In many respects the red wolf exemplifies the story of American wildlife, for it was nearly destroyed before government scientists began piecing together its history and wondering what it needed to survive. Scattered fossils indicate that *Canis rufus* was once common through the Deep South, and that it ranged as far north as Pennsylvania and as far west as Illinois, Missouri, Oklahoma, and central Texas—in other words, to where eastern forests gave way to western grasslands. The species existed nowhere else on earth.

Biologists cannot say in what areas the red wolf prospered most, since the writings of early settlers make no distinction between it and the gray wolf. The red wolf probably preferred to live deep within forests and coastal marshes, where it hunted in groups of two or three for deer, rabbits, rodents, muskrats, and other small prey. No one really knows, though, because no detailed studies were made of its behavior until the late 1960s. By then the small bands of wolves that remained had been crowded into a few pockets of coastal Texas and Louisiana, where they lived in low, sodden marshes, squeezed between ranches, oil fields, and the sprawling refineries of the petrochemical industry. Hookworms killed many of the puppies. Those that lived to adulthood suffered from heartworms and a mange that destroyed their fur and left their exposed skin grayish-blue. Under such conditions it is unlikely that the wolves bore much resemblance, behaviorally, to their ancestors. The species had certainly changed one notable aspect of its social habits: It had begun to interbreed with coyotes, a clear sign that the population had fallen too low to sustain itself.

Alarmed that the existence of the red wolf was threatened by hybridization, the Fish and Wildlife Service set out in the 1970s to capture the remaining individual animals and put them in breeding centers. Red wolves had never been raised in pens. But with the proliferation of coyote-wolf hybrids, and with so many wolf pups dying from parasites each spring, the wildlife managers assigned to the project believed they had no choice. Between 1973 and 1980, government biologists located more than four hundred coyotes and wolflike hybrids, but fewer than twenty full-blooded red wolves. They retrieved the last purebred animals in 1980 from a marsh near an industrialized section of Galveston, Texas. All were taken to a breeding facility run by the Point Defiance Zoo and Aquarium outside Tacoma, Washington, nearly two thousand miles from the swampy coast where the species had nearly perished.

With so few surviving animals, the odds of recovery did not seem promising. Kept in one facility, the animals were unusually susceptible to disease; an outbreak of distemper, for example, might kill them all. In addition, any number might refuse to breed under the stress of captivity. Even if the wolves did survive and begin to reproduce, much of their original habitat had been lost to logging, ranching, and urban development. It seemed far-fetched to think a release site could be found in the populous Southeast. Yet, as it turned out, captivity did offer the red wolf another chance at life. After a slow start, the animals bred well, and in the spring of 1986 their numbers stood at seventy-four. The problem of finding a new home for the species proved more troublesome. Wolves had been eradicated in the East because they were feared, and because they preyed on the same animals humans hunted and raised for food. As the public became more concerned about vanishing wildlife, the fear of wolves slackened but never completely disappeared. In 1983, when the Fish and Wildlife Service began to talk of releasing red wolves in a public recreation area on the Tennessee-Kentucky border, hunters and farmers raised vehement objections. The service scrapped the proposal after wildlife commissioners in both states voted against it.

Warren Parker was then in charge of the southeast region's endangered species program. A southerner with a mellifluent drawl, Parker cut an impressive figure when he spoke before hunters and state wildlife boards. He was amiable and unpretentious, and he counted hunting and fishing among his favorite pleasures.

The failure of the reintroduction proposal in Kentucky and Tennessee had left Parker with a major problem. Red wolves had already been in captivity for nearly a decade. They needed to be released soon, before their natural behavior and wariness toward humans began to break down. But the residents near the recreation area had argued so bitterly against a wolf release that Parker doubted the species could ever be freed without creating an uproar. Any reintroduction site would have to be large and isolated, preferably with no dairy or livestock farms nearby. It would also have to be accessible enough for biologists to radio track the wolves and, if necessary, trap them. He knew of no such place in the Southeast.

Then, in early 1984, the Fish and Wildlife Service suddenly received a gift of 118,000 acres of scrub forest and freshwater bog on a peninsula just west of the North Carolina Outer Banks. To Parker the gift seemed providential. The property was thirty miles from any sizable town and was surrounded on three sides by bodies of water too wide for wolves to swim. Swampy and thick with brier, the new Alligator River National Wildlife Refuge looked—as far as biologists could tell—like good habitat for red wolves. Certainly it was no less hospitable than the Texas swamps where the last wild wolves had been caught.

What were the chances, Parker wondered, that the people who lived near the new refuge would agree to a release of red wolves? In retrospect, it seemed that the reintroduction in Kentucky and Tennessee had been sabotaged by poor planning. For one thing, many of the residents around the recreation area had voiced great animosity toward the Tennessee Valley Authority, the owner of the property. Some people had probably opposed the project simply because it was backed by TVA. The Fish and Wildlife Service had never done much to inform local residents about the reclusive nature of the wolves they hoped to reintroduce. If the biologists were more open about their plans in North Carolina, if they tried to answer all questions and assuage all concerns, perhaps the public would be more amenable to the release.

Even so, the top administrators in the southeast regional office were not sure they could win public acquiescence without some innovative tactics. The possibility of failure troubled them, until after some debate they added a new component to their plans, a secret weapon of sorts. Before its release, each wolf would be outfitted with a special radio "capture collar" that contained a tiny computer and two tranquilizer darts. By sending coded radio commands to the computer, biologists would be able to fire the darts, drugging the animal by remote control. The device would allow them to retrieve a wolf virtually on a whim.

Nothing so sophisticated had ever been used in field biology. The capture collars were being designed by the 3M Corporation, and if they worked—something several people involved with the red wolf project questioned—they could greatly reduce the danger and hardship of handling predators. They could also give biologists an unprecedented degree of control; the wolves would be free-ranging, but only as long as they stayed within the boundaries of the Alligator River refuge.

In February 1986 Parker traveled to Dare County, North Carolina, a countryside of extremes. The barrier islands of the Outer Banks are crowded with affluent resorts that strain against the boundaries of a beautiful and popular national seashore. Just to the west, boggy, desolate forests spread across the thumb-shaped Alligator River peninsula. The darkness of the woods is broken only by a few farm fields and small, insular towns populated by laborers and fishermen.

Parker was prepared to field hostile questions at meetings in East Lake, Manns Harbor, and Stumpy Point, the three communities bordering the new wildlife refuge. He was pleasantly surprised. The residents liked solitude and were used to seeing black bear. If they were bothered by the notion of living near the small, shy wolves Parker described, they said little about it. But they questioned service officials sharply about the management plan that had been written for the refuge. Several men complained that the plan would no longer allow them to hunt deer with teams of trained dogs, as they had done all their lives. They seemed mollified when told that the property would still be open for traditional hunting.

Despite the reassurances he had offered residents, despite the friendliness they had shown in return, Parker was apprehensive. In August 1986, seven months after the public hearings, the Fish and Wildlife Service adopted a final management plan for the refuge. The hunters would still be able to use the land, but not all of it, and not in the summer, the season when they had normally trained their dogs. The restrictions brought cries of outrage from men in East Lake and Manns Harbor. If a couple of hunters decided to seek revenge by shooting red wolves, Parker knew they might very well succeed.

The convoy of reporters that had come to welcome the wolves to North Carolina drove west into one of the last vestiges of wilderness along the Atlantic Coast, into woods that crowded against the road, dense, unruly, bereft of light, black with rain. On the Alligator River National Wildlife Refuge, tree trunks grow canted beneath the weight of vines. Limbs twist and splay, pushing outward in struggle for space, upward in struggle for sun. Dense stands of cane grass obscure the soil. Pines and gums erupt from a treacherous foundation of sucking mud.

Two hundred years ago much of the Dare County peninsula was cypress swamp, but it has been logged, and logged again, until most of the trees that remain are of no commercial value. At one time a farming conglomerate planned to drain the swamps and plant them, but the soil proved to be so peaty and full of submerged logs that the company donated its holdings to the federal government. To many people the peninsula is without beauty or value. From its desolate forests the government has carved a military bombing range and, encircling the range, the Alligator River refuge.

The line of reporters drove slowly down muddy roads, through woodlands of pine, myrtle, and bay, heading for a wolf pen on the west side of the refuge. At one point the woods fell away, and we passed a broad field covered with more ferns than I had ever seen in a single place. For a half mile in three directions their upturned fronds lit the field like small greenish-yellow flames. Later I would learn that Navy pilots, flying F-14s equipped with smoke bombs and lasers, use the field as a target for honing their combat skills.

We drove and drove, following the wide black canals. Near a slippery intersection the caravan halted. Reporters and photographers stepped out in the mud and transferred to four-wheel-drive trucks for the final leg of the trip. The road narrowed almost to the width of a cart path; branches screeched across windshields and thumped against fenders. A sharp turn to the west, another to the north, and the trucks stopped to dispatch us. We had reached one of four pen sites, where we would be allowed to see a pair of wolves fitted with the conventional radio collars they would wear during the acclimation period.

Around us the woods swelled with the sounds of birds. Narrow, opaque canals paralleled the road on either side, with dull-green lily pads bobbing on the surface. Beyond, the ground appeared solid, although littered with half-decayed logs. It occurred to me that in the coming months the wolves, for the first time, would be exposed to the pests of the South, the unavoidable mosquitoes, ticks, and chiggers. The summers would be more humid than in Washington, and the winter winds more harsh. When they were freed they would have to learn to negotiate a labyrinth of briers and fallen trees, chasing through muck for prey, avoiding cottonmouths, copperheads, and rattlesnakes. Would life for them, besides being more perilous, become richer when they were released? I wanted to believe it would.

I wandered over to the pen. Twenty feet wide and eighty feet long, it stood on a bed of sandy soil that had been piled up by logging companies forty years before, when roads were built to haul out timber. Chain-link fencing reached straight up for eight feet, then pitched in at the top. Three strands of electrically charged cable encircled the pen at knee level, partly to keep away bears but also to discourage unauthorized visitors. Inside, the ground was bare of vegetation except for scattered weeds and a few scraggly pines. Two small plywood houses would provide the wolves with cover from rain and wind.

The animals were to remain penned for six months so they would have time to adjust to their new surroundings. This acclimation period was considered essential to the success of the reintroduction. In 1976 the Fish and Wildlife Service had freed a pair of wild red wolves on Bulls Island, a barrier island off the South Carolina coast. The wolves had been kept on the island a little over a month. A week after their release, the female bolted for the mainland, crossing three miles of salt marsh before being recaptured. Two years later biologists repeated the experiment with another pair of wolves, but this time they held the wolves on the island for six months. After the door to the pen was tied open, the animals explored the island and showed no inclination to stray to the mainland. They were recaptured the following year. The same procedure was to be followed at Alligator River; the wolves would not be freed until May 1987. In the interim they would be given live prey so they could learn to kill.

The last of the press arrived at the pen. The sky grew darker, threatening more rain. I was glancing south toward some gathering clouds when a Navy jet split the silence and banked into a deep turn, the roar of its boosters causing everything in our immediate world to vibrate—the ground, the trees, the blood rushing through my head.

Several seconds passed before two men standing near me recovered enough to unclench their jaws and let their hands fall from their ears. "Whew," someone gasped. The voices around me sounded toneless and faint. Other jets had passed overhead that day, but none had so rattled our senses or come with so little warning. The change in scenery would not be the only thing the wolves might find shocking, I thought, as a mud-splattered pickup turned slowly onto the road to deliver wolves 140 and 231.

Reporters scrambled to find positions along the fence, shoving each other and muttering terse apologies, as refuge staff members carried the shipping kennels from the truck to an open area in the pen. None of the spectators spoke above a whisper; the loudest sounds were the clicks and whirrs of cameras.

A cheerful-looking woman with strawberry-blond hair had also arrived, and now she tipped one of the kennels on its end. This was Sue Behrns, an animal handler who raised red wolves at the breeding center in Washington State. "Let's do the female first," she said, obviously in charge. She sprang the catch to the door of the upended kennel. Inside, wolf 231, a three-year-old female two generations removed from the wild, could be heard bumping roughly against the vinyl sides.

One of the refuge workers braced the cage door open several inches. A second handed Behrns a six-foot aluminum pole with a loop of cable on one end to be fitted around the neck of the wolf. The pole was something like a rigid leash; the loop could be tightened from its far end, allowing the handler to control the animal from two yards away. Behrns pushed the pole into the kennel, scraping it against the sides as she worked the loop over the wolf's head. With a quick jerk she pulled a length of cable from its free end, tightening the noose.

The refuge workers pulled the animal out of the kennel by her shoulders, just far enough for Behrns to muzzle her by winding a piece of gauze around her snout. Then they lifted her out and pushed her to the ground, holding her flattened on her belly. A murmur ran among the spectators; this was what they had come to see, a rare red wolf, one of only seventy-four alive. Almost like a German shepherd, really, but with rich reddish-tan fur. Wolf 231's eyes cast about wildly. Another Navy jet passed to the south, but no one seemed to notice.

A few yards away, John Taylor, the manager of the refuge, hammered an awl through the radio collar to make the holes that would fasten it in place. Taylor was a career service man, normally relaxed and confident, but now he struck the awl slowly and with exaggerated care. "We want to make sure we get this on her snugly, but not too tight," he said to no one in particular. As he slipped the collar around the wolf's neck, Behrns smiled briefly, affectionately, and rubbed the animal's haunch. She had raised these wolves without collars, but today they would begin a new phase of their lives. "Let's put her back in the kennel while we do the male," she said.

Within minutes the female had been lifted into the kennel and freed of the gauze muzzle. The male, 140, was six years old, the offspring of a captive female and a wild male trapped in the swamps of Louisiana. Once he had been fitted with a collar, the pair would be released into the pen.

Behrns tipped up the kennel and began her patient fishing with the pole. Wolf 140 had been handled dozens of times; probably he knew what to expect. If he tried to duck away, though, he was no match for Behrns, who tightened the noose with a single jerk on the end of the pole. With smooth, sure movements she handed the pole to Taylor and stretched the strip of gauze between her fingers, winding it around the wolf's jaws.

The refuge staff members lowered the animal to the ground, pulled his legs out from under him, and rolled him onto his side. All had gone well. With the collaring almost complete, they had begun to relax and smile among themselves. But just as they set the animal down a jet passed close overhead. The wolf began to thrash, shoulders quivering, yellow eyes rolling back, lips pulling away from teeth. Quickly the workers spread their hands over him and pressed him to the earth, pinning him firmly, bading him silently to be still.

II

THE FIRST SPRING

It would not take long for the wolves at the Alligator River refuge to adjust to the hellish rasp of military aircraft overhead. To all appearances the animals learned to ignore the passing jets within a few weeks, although no one could measure the effect of the din on their nerves. For me, however, spending time with the wolf biologists in the humid forests and windy swamps, the jets became potent symbols. Each served to remind me of how difficult it is to escape the white noise of civilization—not just the sounds but the traffic, the pollution, the roadside trash, the glow of distant city lights.

The red wolf was being granted only a provisional existence; the Fish and Wildlife Service was not willing, at least initially, to ask the public for anything more. The species was being reintroduced to a swamp with no apparent economic value, except as a place where the military could fire its lasers and scatter its bombs. Flat and desolate, with a water table that brimmed a few inches below the surface, the land seemed more like a tattered membrane than like solid earth. The Alligator River peninsula was large enough for the animals to avoid the bombing fields without difficulty. And if the release went smoothly, eventually they might be allowed to venture onto the well-drained farmlands south of the wildlife refuge. First, however, it would have to be shown that they could be easily and consistently managed.

In planning the reintroduction, Parker and his colleagues had struggled to be practical, given the political constraints of working with predators. Their cautious attitude was typical of endangered species recovery programs, which are almost always designed to be convenient for people. The predominant goal of such programs is not to restore animal populations to their original conditions but to reshape them so they can exist in a thickly populated, heavily developed, economically expanding nation.

The repercussions of this policy are vast, more vast than one can easily grasp: The animals will be given the territory they need to prosper only if they do not come into direct conflict with development or special interests. For example, the black-footed ferret, a sleek nocturnal predator that once ranged widely through western grasslands, will be able to breed and hunt only on the scattered oases where the government has outlawed the poisoning of prairie dogs, the ferret's natural prey. Many biologists believe that the energetic burrowing of the prairie dog stimulates the growth of range grasses, but western ranchers consider the industrious mammals to be pests and routinely kill them. The current policy also means that the reintroduction of the gray wolf to Yellowstone National Park might be delayed for years by the maneuvering of ranchers. If gray wolves are ever released in Yellowstone, any that stray from the park toward cattle and sheep pastures will probably be trapped and relocated or destroyed.

It is no surprise to most people that endangered species projects can be stymied by special-interest groups. Americans are accustomed to watching government policies be shaped by such lobbies as the National Rifle Association, the American Petroleum Institute, and the beef industry. Nevertheless, we tend to think of ourselves as capable of solving most of our environmental problems. There is a faith in this country that no matter how polluted our air and water become, no matter how much of our forests are cut, no matter how many of our indigenous species disappear, we will ultimately be able to fix things, to repair the intricate interworkings of natural systems by replanting deforested areas and reconstructing populations of wild animals. We also seem to believe that our countrysides could never be as badly damaged or as stripped of natural integrity as those of Latin America, Africa, and other parts of the undeveloped world. A tragedy such as the rampant destruction of tropical rain forests could never occur here; with their wealth, education, and sophisticated understanding, Americans would simply not allow it.

This shows a disturbing gullibility. In the past century U.S. timber companies have destroyed one of the world's most beautiful rain forests in the Pacific Northwest; conservationists and loggers are still fighting, viciously, over the fate of the last meager stands. (Only about ten percent of the original old-growth forests remain.) And this is not the only North American wilderness under siege. In much of the West, cattle grazing has altered the ecology of grasslands, not only on private ranches but in publicly owned forests and ranges. Mineral exploration has opened unpenetrated forests to off-road vehicles, and to oil and gas drilling. In the Rocky Mountains and the Sierra Nevadas, where millions of acres have been logged and developed, many animal populations are confined to isolated islands of uncut woods. Biologists fear that dozens of mammal species may become endangered if their habitat continues to be fragmented.

The bootprints of ranching, logging, and mineral development have trod most deeply into the American West and the industrialized coastline of the Gulf of Mexico. The mark of another special interest group, the hunting lobby, is more universal. In most states, ninety percent or more of the money allocated for wildlife goes to manage game species—deer, trout, elk, waterfowl. The remaining ten percent of money and staff time must be split among all other species, including those that are endangered. Where the demands of hunters and the needs of rare animals clash, the hunters often prevail. In the mid-1980s, for example, biologists began to suspect that teams of hunting dogs, used to chase deer and hogs, were causing serious stress to the endangered panthers that inhabit the swamps of south Florida. Yet not until 1990 did state and federal officials close the Big Cypress National Preserve, a key part of the panther's range, to hunting with dogs.

In extreme cases where the needs of rare species collide with the desires of many special interest groups, biologists sometimes find it more expedient to change the behavior of the animals than to attempt to preserve them in their natural surroundings. One of the most controversial examples involves the California condor. A century ago condors foraged for food over hundreds of miles along the Pacific Coast, but the last birds were removed from the wild for captive breeding in 1986 and 1987. In 1989, convinced that the condor could no longer survive as a free-ranging species, biologists began training young condors to feed only within the bounds of a wildlife refuge outside Los Angeles. By setting out a steady supply of carrion, the scientists hoped to restrict the condors' movements and reduce the likelihood that the birds would be poisoned by lead shot or other toxins.

The pressure from special interests, coupled with the incremental pace of scientific inquiry, have made it difficult for biologists to work quickly enough to rescue species in imminent danger of extinction. As a result, they have come to depend heavily on captive breeding as a way of buying time. Usually, at least a small group of animals is allowed to remain in the wild. But in extreme situations—as with the red wolf, the condor, and the black-footed ferret—every individual of the species must be captured to salvage the needed genetic diversity.

To understand the immense appeal of captive breeding, one must know something of the dilemma that faces conservation scientists. Although accurate counts are not available, biologists estimate that thousands of plants and animals go extinct every year in tropical forests alone. The rate in temperate climates, which hold fewer species, is somewhat slower. Using various mathematical models, scientists have calculated the annual rate of extinction worldwide from as low as 365 species—one a day—to as high as 17,500 species. Many biologists believe the world has entered an era of upheaval more severe than the Cretaceous Period, the epoch of mass extinctions and wide ecological collapse that accompanied the disappearance of the dinosaurs. But where the dinosaur extinction may have been the result of meteor impact or some other natural event, the current crisis is clearly being caused by overpopulation, pollution, habitat destruction, and other consequences of human activity.

No one fully understands the implications of this trend. Without question we will lose dozens of species that would have been valuable to us as potential sources of medicine or food. In many ecological systems the process of evolution will be disrupted or destroyed. Beyond that, however, it is impossible to predict how the loss of diversity will affect either the natural world or the one species that seems bent on overrunning it.

The argument has been made that people are part of nature, and that our influence over other animals, our implicit demand that they accommodate themselves to our presence, is simply part of the march of evolution. Throughout time species have been forced to adjust to sudden, drastic change. Yet the current rate of extinction is so high that many scientists and conservationists are justifiably frightened. A man-made flood of problems threatens to wash the earth clean of its astounding natural diversity.

Captive breeding is a way of trying, however weakly, to slow the torrent. But the need to keep animals caged for generations, and to change their behavior once they are released, raises a difficult philosophical question: To what extent are we altering the fundamental nature of wild creatures to keep them alive?

Since the 1970s a small cadre of wildlife researchers have begun to wonder how they can protect the more ephemeral traits of wildness, traits that determine how rare animals perceive their world and move through it. These men and women tend to practice science with an artful hand. They have found that when they are patient and observant, they may be rewarded with insights into an animal's ability to learn, its bonds with its young, or other qualities that seem familiar within the constraints of human experience. It is clear at such times that there is more to the spirit of wild creatures than can be codified by the strict reductionist methods of Cartesian science.

The reintroduction of the red wolf would bring the disparity between these two problems—maintaining an animal physically and preserving its wild spirit—into sharp relief. Because the project involved a socially complex predator about which little was known, those charged with managing the wolves of Alligator River would spend much of their time wondering how to proceed. In the end, the survival of the wolf in North Carolina would depend heavily on the intelligence and sensitivity of a few key players—the Fish and Wildlife Service administrators, yes, but, perhaps most critically, the staff biologists who were to track the animals and study them in the wild.

All through the day of the red wolves' arrival in North Carolina, through the delay at the airport and the collaring at the wolf pen, a group of young biologists stood quietly on the edge of the festivities, watching with excitement and twinges of nervousness. Once the press had started for town, three of the biologists were taken into the woods and left alone in the gathering dark. For the next six months they would live at isolated camps near the wolves, guarding the animals, feeding them, and warning the public away. As long as the wolves remained in pens they were easy targets for harassment. The continual presence of the biologist-caretakers was meant to discourage curiosity seekers and poachers.

One man was left at a small trailer near the pen where the press had gathered; he was also to tend a pair of wolves being kept a mile to the west. A woman was dropped off at a trailer in a windy freshwater bog almost twenty miles to the east. Another man was taken north to a lake and given the keys to a launch that would be his only transportation to and from a tiny houseboat. Each would stay in the field for three weeks, then take a week off.

Chris Lucash spent his first shift as a caretaker moored in a narrow cove in South Lake, a finger of tannic water that spilled sloppily against the shoreline of the refuge, flooding groves of sawgrass and gum. Here and there, cypress poked through the messy canopy, their trunks and limbs rising like skeletons through a heavy cover of shrubs. A hundred yards into the forest, a large pen held male 211 and female 196, the prettiest, friskiest, and most appealing wolves on the refuge.

Lucash was twenty-five and lean, with curly dark hair, a strong jaw, and an olive complexion that gave him a slightly Mediterranean look. Once a day he pulled on hip waders, drove the launch ashore, and picked his way up a marshy trail to the pen. He passed first through a border of dull-green sawgrass with slender blades as sharp as sabers. A few yards inland, the sawgrass gave way to slightly higher ground where fern husks and weedy gum trees twisted up through piles of humus. His footprints filled with tea-colored water, and the odor of sulfur mixed with the rich smell of peat. He entered the pen only long enough to collect whatever wolf scat he could find and to scoop six cups of dry dog food into each animal's bowl. For the rest of the day he could do as he pleased, as long as he did not leave the lake.

During the first several weeks, Lucash amused himself by exploring in the launch, easing his way among the decaying cypress stumps near shore. Sometimes he ventured north to Albemarle Sound, a broad estuary that flowed eastward and emptied eventually into the sea. The upper reaches of the sound encircled a narrow spur of land; when released, the wolves at South Lake would encounter water in three directions.

The dark, humid woodlands intrigued him, since this was where the wolves would learn to live in the wild. There were no paths except narrow game trails, so traveling over land meant breaking through knots of thorny brier, gnarled myrtle, and slippery gum. Even so, Lucash took to walking every second day, easing the boat into narrow coves and working his way inland until he grew tired.

The houseboat he found comfortable enough, although he hated its dampness. On sunny days fragments of light bounced up from the lake and spread tracings on the walls. When it rained hard, huge drops massaged the surface of the lake until he could not tell where air ended and water began.

He wondered if the wolves were disoriented, or if they realized how far they had come from the breeding center in Washington State. It was frustrating to know so little about them. Except for the few minutes he spent feeding them each day, he was supposed to avoid the pen. The male, 211, weighed more than seventy pounds and was larger than most red wolves, with a heavily brindled coat and dark-black streaks like racing stripes on his front legs. He had already earned a reputation as scrappy; the day of his arrival he had bitten one of his handlers on the leg. The female, 196, seemed more docile, maybe to a fault. One day as Lucash scooped food into the animals' bowls, 196 approached him cautiously, her tail wagging and her head lowered in submission. She liked his smell, apparently; she wanted to make friends. Lucash froze, then snarled at her and threw the food buckets in her direction.

The unbroken solitude began to grow tiresome. One morning he would wake refreshed, then the next morning a restlessness would nag at him as soon as he opened his eyes. He sought out chores that were physically exhausting, such as chopping down small pines and cutting the logs into lengths to shore up the messy trail to the pen. After dark he dozed, listening to the hollow slapping of waves against the houseboat hull. In this way, swaying between moods of contentment and boredom, he passed the first of six long shifts in the woods.

If the public had wholly supported the red wolf reintroduction, there might have been little reason for Lucash and the other caretakers to spend six months living in the bush. But the low rumblings of dissension from local hunters continued, and Warren Parker had decided to take no chances. It seemed well worthwhile to pay the caretakers to act as wolf keepers and security guards.

Opposition to the release seemed to be strongest in Manns Harbor, the largest of the three communities on the Dare County mainland. Manns Harbor was a country town on the rim of the continent. It might have been situated in the tobacco country of the Piedmont or the Mississippi Delta, so typically southern was its character, but in fact it perched rather precariously on the first step of land within the protective arch of the Outer Banks. From the tone of discussion at the public meetings, Parker and his colleagues could tell that the residents of Manns Harbor wanted little more from the government than to be left alone. What Parker could not tell, what no one could perceive without studying the recent social history of the region, was the level of resentment the residents of the peninsula felt toward newcomers.

Due east of Manns Harbor lay the prosperous beach towns of Nags Head and Kill Devil Hills. In the previous ten years the Outer Banks, once known for their quaint communities of shuttered wood cottages, had grown crowded with bars, yacht clubs, and year-round resorts. Until the mid-

1980s, the new coastal residents had stayed mostly on the barrier islands. But urbanization was spreading, and Manns Harbor was one of the first outposts to feel its impact. By 1986 people from out of state were sizing up the town, purchasing waterfront lots—property that in New Jersey or Maryland or New York would have cost considerably more—and building large, fashionable houses. One of these was John Taylor, the manager of the new Alligator River refuge.

Taylor was fair and freckled, with a slim build and a carefully trimmed reddish-blond beard. Within the Fish and Wildlife Service he was known as cheerful, reasonable, and smart. He was also a career administrator and a professional diplomat, the kind of man for whom the independent-minded natives of Manns Harbor had little respect. In his new job he held dominion over a countryside that for decades had belonged to absentee corporations, and that had been open to hunters and trappers virtually without restriction.

Taylor believed the Fish and Wildlife Service had been as lenient as possible in drafting hunting regulations for the refuge. He frequently pointed out that no other federal wildlife refuge allowed deer hunting with teams of dogs. But his arguments failed to sway the opinions of local hunters, who complained that the service had reneged on its promise to keep the refuge open for traditional uses.

About two dozen men from Dare County used the refuge to hunt deer with dogs. Most were natives of the peninsula, and their grievances, if not shared, were at least given tacit approval by their neighbors. They had grown up hunting on what was now refuge property; in turn, they wanted to teach their grandsons the skills they had learned on the land just beyond their back yards. The service had not only gone back on its word, they complained, but it had also deprived them of their birthright. And as development spread from the Outer Banks to the mainland, it seemed to them that their entire way of life—their neighborhoods, their fishing grounds, their backwoods solace—was being dismantled piece by piece by interlopers.

One day shortly after the arrival of the wolves, I stopped by White's grocery, a convenience store and lunch counter in Manns Harbor with a sign over its door advertising "10,099 items to choose from." Wade White, the owner, had been quoted in local papers as being openly opposed to the reintroduction. White predicted that eventually the service would ban hunters from the refuge altogether as an excuse to protect the wolves.

A rotund woman with a waitress pad greeted me at the lunch counter and told me White was not in. She had graying hair swept high off her forehead and the country accent of a mainland native, full of dropped consonants and hollow vowels. When I told her what I wanted, she drew in a breath and shook her head. "Wrong place to ask about wolves," she said.

"Are you against having them here?"

She nodded vigorously, her eyes wide. "We got enough wild animals around here. We got bear that come right in our back yard." Her mouth was pursed, the lips trembling slightly. Her expression startled me. It had not occurred to me that some residents would be so frightened by the prospect of living near wolves. "Other day a bear went through and tore up our garden," she continued, "tore a bean trellis right out of the ground. My husband was back there not ten minutes before it happened. A bear killed a dog out here too, killed it right up the road not long ago. A wild animal gets hungry, it's going to eat whatever it can find."

"You want to hear some complaining about wolves," said an elderly man seated at the counter, "go on down and talk to some of the fishermen."

Benny's Seafood was the largest packing house on the disheveled Manns Harbor waterfront, a row of crooked wood buildings with dooryards of rusting junk. It was owned by Benny Rippons, a thick, blunt-spoken man with a mouth that recalled past hard times in the downward set of its corners. Rippons had moved to Dare County in the early 1970s to escape the crowded fishing waters of the Eastern Shore of Maryland. He considered himself a local, though in terms of social rank he would never be on equal footing with a lifelong resident. In his parking lot were several rusted crab pots, patches of shattered glass, and a pickup with a garish purple sticker in the back window. "Support Wildlife," it said in bright letters. "Throw a Party."

Rippons was at work at his desk. He grunted when I knocked on his door, then laughed cynically when I asked to speak to him about wolves. The low, dingy room was lit by a shadeless fluorescent tube that gave off a throbbing light. The air smelled of fish and rotting rubber. It was like every other fish house I had visited on the coast; scant on comfort, it was the domain of men who cherished their independence and would not shirk from hard physical labor. "Wolves," Rippons muttered. "You know, nobody out here says much about 'em anymore."

"They don't?" I was mildly surprised. From reports in local newspapers, I had assumed the pending release was a topic of frequent discussion.

"Nah." Rippons leaned back in his chair, yawned, and flexed his thick arms above his head. "There's nothing to say. We're gonna have wolves loose in this county. Oh, you got people who're against having them, that's for sure. I'd say," he wrinkled his brow and glanced upward, "I'd say there's maybe ten to twenty percent who're for it and another forty percent who don't much care. The rest are against it.

"I'm not saying, now, that people are going to go out and hunt down wolves on purpose. But you got to understand, some of our boys would kill anything they saw that moved." His accent was so thick I had trouble understanding certain words. In his mouth "kill" became "keel" and "can't" became "cain't."

"It's not the wolves I got the biggest problem with. I can't see what good they're going to do us, but it's not the wolves. It's that John Taylor fellow. He's smooth." Rippons paused, his eyebrows arched. He slapped his open palm on the desk. "Taylor'll tell you he hasn't closed off much to dogs, but look at what they left us. Look how they got it all chopped up. They sectioned it off so much you can't hunt most of it, because you can't keep your dogs in the boundaries. A dog doesn't recognize a wildlife boundary; he'll just keep going after the game."

A blond fisherman had come in wearing rubber knee boots that were filthy with grime and fish scales. Rippons extended his arm to give the fisherman a yellow pay slip, but kept his eyes on me.

"They took all the good parts and left us the junk," the fisherman grumbled, hanging around.

"Look," Rippons said, suddenly belligerent. "You spent most your life in the city, right? In the city you got lots to do for entertainment. We like it out here because we like the woods and the water. Hunting's what we do for fun. Every boy who's old enough to carry a gun goes hunting."

"Do you all hunt with dogs?"

"A number of us do from right around here. Still hunters, they don't have to worry about tromping through swamp. But if you're going to follow the game—and that's how I was raised, that's hunting to me—there ain't a place in this country that's harder to hunt than these swamps, they're so damn thick. You can't do it without dogs."

He paused and shrugged, still emphatic but no longer near anger. "To hell with the wolves," he said. "What do you think? They going to make it out there?"

I conceded there was a reasonable chance the wolves would starve or be shot.

"That's my feeling exactly."

"The food issue is a particularly troubling one," Mike Phillips was saying. "Our overriding concern is how to break whatever association these wolves have made between humans and food. The question is, are we doing enough?"

Phillips, a refuge biologist, spoke crisply and with an enthusiasm I found contagious, if somewhat tiring. He and I were driving in a muddy pickup to Point Peter, a brushy marsh where another pair of wolves was being held. Along the road heaps of myrtle and bay parted long enough for me to catch glimpses into small groves of pond pine. Without warning the jumble of limbs and vines opened, and the forest gave way to a field of shrubs with evergreen foliage that grew to the horizon like hedges. We had entered a pocosin, a unique freshwater bog endemic to the North Carolina coast.

The food issue was troubling indeed, for it encompassed most of the potential problems of returning captive wolves to the wild. During the previous month the wolves had been weaned from their diet of dog food. Now they were fed only meat from prey animals, and only every fourth day, to accustom them to an irregular supply of food. A plywood screen had been wired to one side of each acclimation pen so the caretakers could toss meat over the fence without being seen. Now, in mid-February, the biologists seldom entered the pens.

No one knew whether these precautions would make the wolves less likely to seek out people once they were released. "If you ask the zoo people in Washington, they'll tell you there's no way we're going to break adult wolves of their association between humans and food in six months," Phillips said. "The wildlife biologists on the project think we at least ought to try. At one point we were even talking about rigging up some kind of remote food delivery system with buckets and pulleys. But that would have been difficult, and we weren't sure it would work."

At twenty-nine, Phillips was energetic and ambitious, with a ready wit and an easy, friendly laugh. Deeply devoted to his job, he seemed happier talking about wolves than anything else, with the possible exception of baseball. Blond, square-shouldered, and athletic, he had played ball in high school and college, and had flirted with the notion of turning pro. "I decided that if I wanted to have something to do after the age of thirty," he said, "I'd better go into biology."

For the previous several years Phillips had conducted radio-tracking surveys of deer and gray wolves in Minnesota and had studied grizzly bear habitat in Alaska. In the summer of 1986 he finished his master's degree at the University of Alaska and came to North Carolina to work as a volunteer for the red wolf project. It proved to be a wise move. A month later he was hired to oversee the day-to-day operations of the acclimation period and the release. It was Phillips who carried out Warren Parker's and John Taylor's orders, who supervised the caretakers, who answered questions from the public, who called in a veterinarian when a wolf needed medical attention. If the reintroduction succeeded, much of the credit would belong to him.

We turned east onto a dirt road that was badly rutted and pocked with deep holes. Phillips stopped to open a cable gate and drove on, past signs warning the public to keep out. A mile into the pocosin sat a small travel trailer surrounded by equipment—stacks of box traps, a telemetry antenna, a large wooden spool turned on its side for a workbench. For now the trailer was home to Beth Kennedy, a stocky woman with short, sandy hair.

Phillips called hello and hopped into the bed of the pickup to unload a canister of propane. "This should keep you warm for a little while," he said cheerfully. Kennedy rolled the canister to the side of the trailer and stood it upright. "Any mail?" she asked, dusting her hands together.

"None that came before I left the office."

She looked at me and smiled, disappointed. "Out here you tend to live from mail call to mail call," she said.

Phillips was unloading more supplies from the truck. He set a five-gallon gas can on the roadside and tossed a dark garbage bag next to it. The bag fell with a thud, its contents heavy and soft. "There are two raccoons in there," he said. "As soon as we leave, you need to gut them, dust the cavities with Clovite, and throw them in the pen." Clovite was a powdered vitamin supplement routinely added to the wolf food.

I looked around at the stacks of traps, the gas cans and water jugs, and felt suddenly out of place. Point Peter struck me as a bleak campsite, much farther from civilization than the twenty miles that separated it from Manns Harbor. The stunted vegetation and weak sun reminded me of northern latitudes. Phillips, perhaps noticing my disoriented expression, nodded toward an antenna that had been propped at an angle on the wooden spool. "Show her the telemetry gear," he suggested to Kennedy.

For the past week the caretakers had been conducting a radio telemetry study, monitoring a pattern of beeps given off by the transmitter in each radio collar. By listening to the signals, they could determine whether individual wolves were resting or moving around the pen. "We're trying to see what time of day the wolves are most active, when they'll be most likely to hunt or just be on the move," Kennedy said, escorting me to the workbench and opening the receiver, a small metal box with a scratched meter. "This isn't too fancy, but we don't need anything very strong. The pen's just a hundred yards over there." She motioned north. Her voice was soft and she spoke slowly, a striking contrast to Phillips. He had seated himself on a plank next to the trailer and was drumming his fingers on his thighs.

I asked whether anyone had tried to sneak by her to see the wolves. At a little more than a mile from a major road, the Point Peter pen was the most accessible to the public.

"At first I had a lot of visitors," Kennedy replied. "The week after the wolves got here, about twenty people came down to the gate asking to see them. I just tried to explain that we have to limit their contact with people. Then I had a problem the night before Thanksgiving with some guys firelighting deer." She glanced toward Phillips, who had stopped drumming his thighs and was sitting quietly, as if in thought.

"The biggest thing was, one day I had three hunters walk right up to the trailer. I was sitting by that canal with my binoculars, trying to figure out what a darn little wren was. I heard a voice not three feet from me. It really scared me, because I didn't know there was anyone around for miles. So I stood up and said, hello, or something like that.

"They were pretty big men. They knew they weren't supposed to be back here." She stopped, as if waiting for Phillips to interject something.

"What happened?" I asked.

"They left eventually," Kennedy said, "but they weren't too pleasant about it. They started saying things like, 'This is our favorite hunting spot and we're not going to let you take it away for the wolves.' I thought, oh come on, there are only one or two deer in this whole area.

"They went on for a while, saying stuff like the wolves won't last a week outside the pens, because they're going to put out deer carcasses with strychnine in them. I just kept telling them, sorry, this area's closed. They weren't too happy about it, but they left. I climbed up on the trailer and watched them go." She paused, her square face somber in the slanted morning sun. "It was interesting confronting three large men, each with a couple of guns apiece."

Phillips slapped his thighs. "Look," he said, "I think it's a mistake to dwell on a couple of people with bad attitudes when the public as a whole has been overwhelmingly supportive."

"That's true," Kennedy said. "You can tell by the number of people who bring us road-killed animals . . ."

"Which we absolutely depend on for food," Phillips said. "Also by the number of requests we have for school programs on the wolves. I honestly don't think we're going to have any major problems. We may have one or two animals poached, but we expect some to die." He stood up. "We need to hit the road."

We were miles from Point Peter, bumping along a dirt road on the west side of the refuge, before Phillips mentioned the hunters again. "The refuge isn't very old, and it certainly isn't the playground it once was," he said. "I don't know . . . If this program fails, the biggest problem is going to be assessing cause and effect. If wolves are poached, is it because they were too tolerant of people or because we haven't done a good enough job of educating the public? Everybody's going to want to know what went wrong. And we're not going to have all the answers."

Eight red wolves, ranging in age from three to seven years. Eight sleek, shy, beautiful animals, unwitting subjects in a grand experiment. Eight on whose future depended the fate of their entire species. Even a partial success at Alligator River meant the chance for more releases elsewhere. A resounding failure, or even a problem wolf wandering into a community, might doom the entire project.

They differed in disposition, in the way, even, that they interacted with their mates. They were to be treated equally, but already the caretakers were placing bets on which would fare best after the release. The wolves at South Lake, the feisty male and the female that had tried to befriend Chris Lucash, appeared tough and healthy and were being allowed to breed. The female had already raised two litters in captivity. With her experience as a mother and the male's aggressive temperament, they would likely make good parents.

At Phantom Road, in a pen shaded by spreading hardwoods on the west side of the refuge, a second pair of wolves was expected to produce a litter of pups by early May. Both animals had been raised by hand by Sue Behrns, who cared for the wolves at the breeding center in Washington. The female was high-spirited and dominant; the male seemed to dote on her.

The other two females had been fitted with small hormonal implants, sewn under their skin, to keep them from ovulating in the spring of 1987. Because the stress of raising pups might affect their ability to adjust to freedom, Parker did not want all four pairs released with young.

Which would prosper and which would die? Behind the plywood screens, they ate and slept and walked their pens, their muscles rippling beneath rich, tawny fur, their heads lowered, their ears perked forward to listen, their gait sure, even on dark, sloppy soil. Sometimes they paced like zoo animals in cages, back and forth along the fences, testing their bounds; but at the mere hint of a person they leapt, secretive and silent, into the cover of trees.

III

RESCUE MISSION

The last stronghold of the red wolf in the wild, a sodden, forsaken flatland on the Texas and Louisiana coast, was remarkable only for its proximity to the burgeoning cities of the Sunbelt. It lay on a wedge of coastline surrounded by the belching gray stacks of oil refineries and the outlying streets of suburbia, the tentacles of Beaumont, Houston, and Galveston. Composed mostly of low, poorly drained coastal marsh and prairie, the land was split by earthen cattle walks elevated above the brush through which wolves scouted for prey.

Fields of cultivated rice gave way in gradients to cattails and rushes and finally salt grass, as the land sloped to the Gulf of Mexico. Oil pumps nodded methodically. A mongrel breed of cattle, the only kind tough enough to withstand the constant dampness and bugs, grazed in pastures turned black by their hooves and dung. In the wettest years, when water lingered in stagnant, tawny pools, calves sometimes smothered from the mosquitoes that clung to the insides of their nostrils like bats.

It was in this last miserable corner of its range that *Canis rufus* waited to die. Short-haired and variable in color, from light tan to almost black, it had the tall, oval ears of a German shepherd, rather than the triangular wedges of the gray wolf of the Far North, and a longer snout. Its legs, spindly beyond proportion, were well suited for picking its way through the soft soils of bottomland forests and marshes. To peer over tall grasses, it might stand momentarily on its hind legs. But its survival depended on its ability to slip silently through brush impassable by humans, scooting over downed logs and under thorny brambles.

The last red wolves denned in hollow logs and stream banks, bearing as many as twelve young and raising what few survived the siege of hookworms that set in shortly after birth. Pups often stayed with their parents when grown, integrating into loose groups that traveled through territories of unknown dimensions. They fed on rabbit, muskrat, nutria, deer, carrion, or whatever else they could find, including the young calves that dotted the pastures of local ranches.

It is likely that the majority of red wolves avoided human settlements and seldom preyed on livestock. Especially in the early twentieth century, a portion of the damage blamed on the species was probably the work of feral dogs. Nevertheless, red wolves did kill pigs, sheep, and calves occasionally, and this tendency effectively doomed them as a species. In Arkansas and Oklahoma, a few red wolves, notorious for their attacks on farm animals, earned names like "The Traveler," "Old Guy Jumbo," and "The Black Devil."

"The ranchmen invariably distinguish between [the red wolves] and the coyotes, and with good reason, for the wolves kill young cattle, goats, and colts with as much regularity as the coyotes kill sheep," Vernon Bailey wrote in his biological survey of Texas in 1905. "While paying a bounty of one or two dollars for coyotes, the ranchmen usually pay ten or twenty for red wolves." The red wolf was enough like the gray wolf of Europe to have little chance of survival. Perhaps most damaging of all to the species were the Old World traditions that pervaded eighteenth- and nineteenth-century American culture, and that included a nearly fanatical hatred of wolves. No allowances were made for the animals' role in the natural system; in their eagerness to rid the countryside of wild beasts, the ranchers set out to kill every wolf they could shoot or bait into traps.

Pups were taken from dens and clubbed. Pits were dug, each with a board balanced over it and pieces of meat suspended enticingly over the middle. When a wolf reached for the meat, the board dumped the animal into the pit, where it would struggle to free itself until shot. Steel traps came into wide use in the 1900s, but, from about 1865 to the turn of the century, the most popular method of eradication was poisoning, usually by stuffing carcasses with strychnine. The poison was effective to a fault. In addition to wolves and coyotes, it killed vast numbers of bears, skunks, weasels, hawks, eagles, and other animals that fed on the contaminated meat.

The red wolf might have escaped persecution by fleeing deeper into wild country, if any wild country had remained. But as settlement spread across the southeast quadrant of the United States, the deciduous woodlands the species favored were cut through with roads, first for logging and mineral development and later for houses and farms. The marshes where it hunted were drained for pastures.

Slowly at first, and then with increasing speed, the red wolf was exterminated from the states along the southeastern seaboard, and from Alabama, Mississippi, Tennessee, Kentucky, Missouri, Oklahoma, and Arkansas. The most eastern subspecies, a dark animal named *Canis rufus floridanus*, was last seen in the Everglades in 1903 and in the Okefenokee Swamp of south Georgia in 1908. A second subspecies, *C. r. gregoryi*, existed in isolated pockets along the Mississippi coast until the 1930s, and on the eastern Louisiana coast until the 1950s, when its social structure disintegrated. It managed to survive for another twenty years in east Texas and western Louisiana. The Texas red wolf, which may have been a third subspecies, *C. r. rufus*, ranged from the Trinity River to the rim of the arid plains of the West. It hunted and denned in groves of scrub oak and mesquite until these forests, too, began to be cut by settlers. There it met an eastward invasion by its adaptable cousin *Canis latrans*, the coyote, which has a remarkable talent for prospering in countrysides disturbed by human use.

Scientists believe the hybridization of red wolves and coyotes began in the dead center of Texas, along the line where the two species had probably bumped against each others' range for thousands of years. The coyote, secretive and fond of traveling alone, had evolved in the dry grasslands and deserts of western North America, but it was amenable to living elsewhere. Just as important in terms of its survival, it was more difficult to capture than the red wolf, which blundered into traps with remarkable naïveté. At some point, probably around 1900, the remaining bands of red wolves grew so sparse, and their social structure so tattered, that they began mating with coyotes and dogs, creating what wildlife managers would later refer to as a hybrid swarm.

It would take more than fifty years for scientists to suspect that the red wolf was in the process of breeding itself out of existence. The lag is somewhat understandable, considering that the hybridization of red wolves and coyotes is an anomaly in the animal world. Normally, distinct species will mate only with their own kind, and will go extinct before crossing the reproductive barrier that separates them from their closest relatives. Even as the red wolf was disappearing, trappers paid under a federal eradication program continued to report that the species was plentiful and just as troublesome as ever. As late as 1963, trappers in Arkansas, Oklahoma, Texas, and Louisiana exterminated 2,771 "red wolves," the overwhelming majority of which, taxonomists realized belatedly, were coyotes and coyote-wolf mongrels.

In 1962 Howard McCarley, a biologist from Austin College in Sherman, Texas, published a paper in which he warned that the red wolf was being supplanted by other wild canines of questionable blood. McCarley's doubts about the nature of the "wolves" so common in the South had been brewing for nearly fifteen years. As a boy he remembered few wild canines in the open country around his home in southern Oklahoma, yet by the late 1940s the thickets and hillsides were rife with small, secretive animals that varied greatly in color. "Everybody called them wolves," he said. "But they just didn't look that much like wolves to me."

It was a longstanding custom for ranchers to hang dead wolves from the fences around their pastures. McCarley was in graduate school and interested in wild animals, and he began taking measurements from skulls he found wired to fence posts along country roads. He compared them to the measurements given for *Canis rufus* in *The Wolves of North America,* a definitive work published in 1944 by Stanley Young and Edward Goldman.

In 1950 McCarley began teaching at a small state university in Nacogdoches, Texas, then returned briefly to graduate school to finish his doctorate degree. Occasionally he would visit natural history museums to look over their specimens of southern wolves. "The measurements of the skulls I had taken myself just didn't add up," he said. "I examined skulls at the University of Arkansas, Texas A & M, LSU, and a few other places. By the middle fifties I was pretty suspicious that what we had in most of Texas and Arkansas weren't true wolves. My suspicions really jelled when all the data were analyzed and I could see how much the recent skulls differed from the skulls of red wolves collected by Goldman back in the twenties and thirties."

McCarley's measurements showed a distinct change in the average size of the skulls of wild canines measured by Goldman and those he found in the same locations twenty and thirty years later. "It was a statistical difference. It wasn't anything you could see by looking at individual animals. But the more recent skulls were on the average much smaller—too small, it seemed to me, to be explainable by an evolutionary trend over such a short period of time." The discrepancy could mean only that a substantial number of red wolves had hybridized with coyotes.

McCarley's report appeared in the *Southwestern Naturalist,* a regional publication. It did not go unnoticed by federal officials, who requested all predator control officers to turn in the skulls of animals they killed, especially large, wolflike animals. "There wasn't much else the federal government could do at that time," McCarley said. "There wasn't an Endangered Species Act yet, and there wasn't any vehicle for saving animals like the red wolf." Even so, the Fish and Wildlife Service did little to curtail the trapping and poisoning of predators in the few corners where pure red wolves might still live. And in 1964 the service published an official tally of animal species that needed federal protection to survive. The red wolf was not among them.

By then, however, a group of mammalogists had begun to catalog all the mammals of North America, and to take note of those that were in immediate danger of going extinct. One of these, a surgeon and biologist in Tacoma, Washington, had been conducting simple electrophoretic studies on the genetic make-up of rare mammals since 1957. "Murray Johnson wrote to me several times almost begging me for blood samples from red wolves," McCarley said. "I wrote back that I couldn't send him any samples, because I had never seen a red wolf alive and wasn't even sure the species still existed in the wild."

In 1964 William Elder, a zoologist at the University of Missouri at Columbia, wrote to Secretary of the Interior Stewart Udall complaining that the red wolf had not been seen in Missouri for more than twenty years and that "from all I can learn, the species is in a precarious state in other parts of its range." Elder's concerns were echoed by two Canadian biologists, Douglas Pimlott and Paul Joslin, who conducted a field survey in Arkansas and Louisiana during 1964 and concluded that the red wolf had nearly disappeared.

The following year, after an extensive search along the Texas coast, Joslin heard the howls of what he believed to be three groups of red wolves in a mosquito-infested area east of Galveston Bay. In addition, government trappers turned in seven skulls from canines captured near the Anahuac National Wildlife Refuge, where Joslin conducted his work. "They were very good examples of red wolf skulls, and they were the first clear physical evidence anyone had found in years that the red wolf still existed," McCarley said.

In early 1965 federal officials added the red wolf to the list of endangered species. Nevertheless, the federally financed predator control programs, in force since 1915, continued to be carried out within the species' last range. The perpetuation of the programs outraged members of the American Society of Mammalogists, an organization of respected scientists that had been highly critical of government predator control efforts for forty years. (Its members included Murray Johnson and William Elder.) That summer the society passed a resolution demanding that all trapping be stopped in states where the red wolf might still survive. Members dispatched copies to federal officials and the governors of Arkansas, Texas, Louisiana, and Mississippi.

But as the society drafted its resolution, another group of scientists began a study that would conclude that the red wolf was not a separate species at all. In 1967 Barbara Lawrence and William Bossert of Harvard University published a taxonomic analysis of skulls from eastern gray wolves, red wolves, and coyotes. They noted that the red wolf had always occurred in a very limited range, and that it was the only wolf in the world not considered to be a member of the species *Canis lupus*, the gray wolf.

Where the range of the red wolf overlapped with that of the gray wolf, it tended to be taller and heavier, as if taking on the characteristics of its cousin. But where the range of the red wolf overlapped with that of the coyote, the wolves tended to be smaller and more coyotelike. The correlations struck Lawrence and Bossert as strange. (It is a principle of biology that where two distinct species share part of their range, the differences between them tend to become exaggerated.) Finally, taxonomic measurements taken by the researchers showed the red wolf to have a skull structure very similar to the gray wolf. To Lawrence and Bossert, the findings suggested strongly that *Canis rufus* was merely a form of *Canis lupus* whose evolution had been restricted to the southern United States. The results of the study attracted wide attention. In his 1970 classic work, *The Wolf*, the renowned biologist L.

David Mech asserted that inevitably the red wolf would be found to be nothing more than a fertile cross between gray wolves and coyotes.

Nevertheless, the protests from those alarmed about the fate of the red wolf were vehement enough to convince state and federal officials that something should be done. In 1968 the U.S. Fish and Wildlife Service assigned a biologist to investigate complaints of wolf predation in seven southeast Texas counties. Instead of being killed, the offending wolves were to be trapped alive and removed from the property. Some were to be shipped to Minnesota, where a researcher would conduct an electrophoretic study to determine their genetic make-up. When the first technician left in 1969, a Texan named Glynn Riley was assigned to take his place.

Riley was in his mid-thirties and well seasoned in the field. A native of the mid-Texas community of Wortham, he had trapped coyotes all his adult life, but had never seen a red wolf. Arriving on the Texas coast, he was shocked at the condition of the animals he found. Their coats were ragged from the sharp-bladed sawgrass of the marshes. Many were completely denuded by mange. With no fur to insulate them against the driving rains and tropical storms of the coast, they grew so sick that they would often wander in front of cars as if waiting to be hit or shot. "They were skinny and rat-tailed. If one of them lived five years, he was doing good, the habitat was so bad." But the animals were clearly much larger than any coyotes Riley had ever seen, and they acted differently. "They were more aggressive, for one thing. They'd show their teeth and growl at you and run at you when you'd get 'em in a trap, instead of backing off like a coyote. Their tracks were bigger, and their scat when you'd come across it was bigger."

In his 1976 essay, "Lament the Red Wolf," the writer Edward Hoagland describes Riley as slim and curly-haired, with "that cowpoke look of not putting much weight on the ground when he walks." There is a bit more meat on his frame now, though he is not heavy. Beneath an ever-present white Stetson, his blue eyes have a lazy twinkle and his smile is slow and warm. After spending a day with him I had the impression that Riley is inclined to take life's tribulations more easily now than during his wolfing days.

Before moving to south Texas, Riley worked on a state coyote control project in Lubbock, and he had developed a comfortable way of hobnobbing with ranchers. He soon became adept at convincing landowners not to shoot the wolves they spotted but to let him trap them. The local bounties had been lifted, and few ranchers still poisoned wolves. Most of the animals stayed on land that was being held in large tracts and that had long been ranched or flooded for rice paddies. The paddies provided the wolves with a plentitude of small rodents for food, but a number still haunted the edges of farm fields, picking off calves.

When Riley received a complaint about a wolf, he would visit the landowner and spend some time in the area looking for tracks and eliciting wolf howls by setting off an old hand-cranked siren. The behavior of the wolves intrigued him; they seemed less excitable and more single-minded than coyotes. "I'd put up a scent marker and check the tracks around it, and it always seemed like the coyotes would get real curious. I'd find four or five tracks where a coyote had come up and scratched at the scent. But a wolf, he'd just come up to it, sniff it, and be on his way like he had a purpose and didn't want to be distracted."

"Usually I'd be able to scare up some wolves in here," he told me as we drove past a rice field bordered by scrubby willow and gum. "This is real good wolf country in here." It was two days before Thanksgiving and still in the seventies, with ninety-eight percent humidity. Snow geese honked in the pallid sky. There were, of course, no wolves left in the country Riley showed me, only hybrids and coyotes. Nevertheless, he could not resist stopping to look at several piles of scat and some scratch marks left by wild canines along a white-shell road.

"I caught a couple of nice big wolves in there once," he said as we rolled by a pasture with swaybacked cattle. "One of them was a male that weighed seventy-six pounds. There was a graduate student working down here; we put radio collars on that wolf and his mate. Then one day the wolf was bedded down in some grass and some cowboys rode by. He jumped up and ran, and the cowboys lassoed him just to see if they could get a rope on him. Then of course they had to kill him. Beat his head in with a pair of fence pliers."

I must have looked aghast, because he quickly added, "They didn't mean anything by it. They just figured it was the thing to do. You have to remember, a wolf'll eat a calf worth four or five hundred dollars." He paused. "I sure wish they had called me, though, instead of killing him."

Sometimes Riley found the skulls of wolf-coyote hybrids wired to fence posts. "I've been interested in skulls all my life, and I'd never seen any like that," he said. "Some would have an undershot jaw, with teeth crowded together and pushing up past the top jaw."

Generally he trapped wolves with no trouble. Occasionally he would single an animal out and follow it just to study its behavior. "There was one old male that I followed for maybe a couple of years. He was missing two toes, so his tracks were real easy to follow. I'd go down and track him around, just keeping up with him and seeing where he was going, how far he ranged and all. Somewhere along the line he got to killing calves and I had to trap him."

When he trapped coyotes, he shot them or occasionally shipped them to Minnesota for genetic researchers to compare with red wolves. The wolves that appeared pure he took to a holding facility on the Anahuac National Wildlife Refuge, until he could find a place to house them or set them free. To him the difference between wolves and hybrids was easy to distinguish. "It got to the point, later on, where there was an awful lot of controversy about what a wolf was, and I didn't ever think it was that hard to tell," he said. "If you'd catch two together, they'd always be two big animals—that is, two wolves. I'm not sure they were really associating that much with coyotes."

In addition to mange, the wolves suffered from hookworms and rampant heartworms. One day, some men riding a swamp buggy flushed a wolf that ran a short distance and collapsed. The men caught it and took it to Riley. "When they'd get heartworms, they couldn't stand any kind of stress," Riley said. "I put that animal in a cage, and when I'd go in to feed him, he'd get nervous and fall down. One day he just fell down and died. I cut him open, and his heart was blown up like a balloon. It was full of worms, even the valves; they were just string, like spaghetti."

Riley took sick or injured wolves to Buddy Long, a veterinarian in the nearby town of Winnie who often treated them for free. The medications for mange developed for domestic dogs did not have the potency to kill all the mites that infested the wolves, but Long would shampoo the animals with insecticides and massage them with salve. He pinned leg bones and amputated feet that had been mangled in traps. During the first few years he held the wolves in his office until they healed, but the pungent, musky smell of the animals bothered his customers. One day he came to work to find a wolf loose in the office. It had chewed through the wire screen on its cage. Shortly thereafter he built some pens and dog runs for the wolves, at his own expense, in a pasture near his house.

For several years Long spent most weekends and dozens of hours during the week treating wolves, helping Riley check trap lines, and shooting rabbit and nutria for wolf food. He came to respect Riley as a consummate outdoorsman. "I've worked with a lot of trappers," he said, "and Glynn was head and shoulders above most."

When necessary, Long served as a local guide for the handful of scientists who were interested enough in the species to want to study it. He also talked about the red wolf to the ranchers whose cows he tended. "The hardest part to me of the whole project was getting the ranchers to see that we had an animal that was going extinct," he said. "There were more coyotes and hybrids out there than you could shake a stick at. A lot of people called everything they saw a red wolf; some people still do." Most of the landowners were cooperative, but a few continued to set out poison illegally and shoot wild canines on sight.

As Riley caught more wolves, it became increasingly difficult to relocate them away from ranch lands. None of the zoos in Texas had expressed any interest in preserving the species, and there were few places, besides the small Anahuac wildlife refuge, where Riley could turn them loose. One partial solution was to send some animals to Washington State, where a group of conservationists had begun to talk of breeding the red wolf in captivity. Among the group was Murray Johnson, the mammalogist and surgeon who was then chairman of the board of the Point Defiance Zoo and Aquarium in Tacoma. The board of directors for the zoo had long been interested in saving endangered animals, and in 1970 its members arranged to build three pens where red wolves could be put on display. From late 1970 through 1972 thirteen wolves were sent to Tacoma. Only a few survived.

Sometimes when Riley caught a marauding animal that looked like a pure red wolf, Long would agree to keep it for a while. "There was nothing else to be done," Long said. "It was a shame, because we knew we had our hands on an animal that was very rare. We'd just wait till things cooled down a little bit and then let it out again somewhere away from the ranch where it had been causing trouble." Occasionally Long would simply open the door to the pen in his pasture so the animal could skulk away in the night.

Riley estimates that when he started work on the Texas coast, a hundred red wolves still inhabited the undeveloped pockets east of Galveston Bay. Helpful though his work with ranchers may have been, it was not enough to preserve the species, and Riley knew it. His pleas for more staff and for money to collar some wolves and conduct radio-tracking surveys brought only silence from the federal government. He himself talks little of his frustration with the government bureaucracy. "There was a lot of stuff going on back then," he said, his eyes flashing for a moment. "But, oh well. The red wolf got saved in captivity, eventually, which is what matters, I guess."

Edward Hoagland, in "Lament the Red Wolf," is less polite. He writes of having to make a personal visit to Washington to learn anything about the species, and of lobbying in Riley's favor to unfreeze the funding that had already been appropriated for the red wolf recovery project. He describes the interest of the Texas bureaucracy in the red wolf as flagging beyond reason. Administrators at a state museum refused to display the red wolf skins Riley sent them, and officials at a zoo in Houston elected not to establish even an inexpensive exhibit on red wolves, because, they said, the subject was too controversial.

Perhaps the reason for the reluctance to fund the red wolf project at the federal level was also controversy, in the form of a feared backlash from ranchers and hunters. Or perhaps it was merely inertia and old habit. The agency in charge of endangered species, the Bureau of Sports Fisheries and Wildlife, had once been known as the U.S. Biological Survey and had paid the trappers who exterminated the red wolf through most of its range. (Eventually the bureau would be absorbed by its parent organization, the Fish and Wildlife Service.)

Or perhaps the scientific debate about the status of the red wolf as a species slowed the start of recovery efforts. If so, the subject was soon put to rest. In 1972 taxonomists John Paradiso and Ronald Nowak published the results of a comprehensive analysis of wolf skulls. Paradiso and Nowak compared the measurements of hundreds of gray wolf, red wolf, and coyote skulls—far more than had been included in the study by Lawrence and Bossert—and found no statistical overlap in size between *Canis lupus* and *Canis rufus*. The study eliminated any doubt (at least in the view of federal wildlife officials) about the red wolf's standing as a true species.

The year after publication of the study, Congress passed the Endangered Species Act, making substantial funds available for vanishing wildlife for the first time in the history of the nation. The adoption of the act immediately brought more money and personnel to the recovery efforts on the Texas coast. But neither the funding nor the increase in staff would be dispatched in quite the way Glynn Riley might have hoped for the health of the red wolf.

In the autumn of 1973 a midwesterner named Curtis Carley arrived in Beaumont, Texas, to oversee the operation of the expanded red wolf recovery project. As project leader, Carley was to carry out the objectives of the new federal recovery plan, written by a group of wildlife scientists known as the recovery team. He was to manage the field operations and collect red wolves for a captive breeding program at the Point Defiance Zoo in Tacoma. Above all he was to make sure that a healthy number of wild red wolves survived the invading packs of coyotes and hybrids.

Until that fall Carley had studied coyote behavior and predator ecology out of the Fish and Wildlife Service office in San Antonio, working alongside trappers in the animal damage control program. He knew Riley and had helped him put radio collars on a few wolves. Where Riley's skills lay in his understanding of how wolves moved and where they might be caught, Carley was a research scientist. He had a master's degree in mammalogy, while Riley had no degree at all. Unlike Riley's relaxed, folksy manner of speaking, Carley's voice had a deep, authoritative ring. Their work relationship was not a particularly comfortable one; for the first time Riley's day-to-day activities were closely supervised by someone who knew less about red wolves than he did, and whose personal style was much more formal than his own.

As soon as possible Carley began scouting the countryside, which he found to be bleak, buggy, and not much to his liking. "There's maybe a month out of the year that's pleasant there," he said, "besides the fact that it's a haven for any kind of parasite you don't want a wild wolf to pick up." It appeared that the hundred or so wolves that survived along the coast had not been tainted with coyote blood. But a well-established population of hybrid animals just to the north was likely to overrun the coast in a matter of years. The task, then, was to figure out how to isolate the true wolves. If even a few hybrids infiltrated the remaining bands, the bloodline of the entire species might be compromised.

Before Carley's arrival, federal administrators in San Antonio and Washington had proposed three options for establishing a barrier between the wolves and the hybrids. The first involved building a fence from east to west along Interstate 10, eighty miles long and too high for coyotes and wolves to jump or climb. Carley soon convinced his superiors that this plan was impractical. Not only would the fence be costly but, to be effective, it would have to block dozens of country roads. A second proposal called for trapping and sterilizing all the wild canines in a wide buffer zone, creating a living barrier that would prevent hybrids from expanding their range. This strategy seemed to Carley to be outlandishly expensive, given the cost of performing surgery on dozens of animals, any of which might stray from the area.

A third plan was to create a buffer by removing all the coyotes and hybrids from a fifteen-mile-wide zone, extending from Houston to Beaumont. This option had been considered years before, but funds had never been made available to carry it out. In early 1974 Carley reluctantly agreed to try clearing a strip of countryside of coyotes and hybrids, and to keep it cleared long enough for biologists to rebuild the scraggly bands of red wolves into a functioning population.

In retrospect, it is difficult to understand how administrators within the Fish and Wildlife Service could have given serious consideration to such a plan. For the buffer zone to have succeeded, trappers would have had to empty twelve hundred square miles—much of it thickly vegetated—of animals known for their stealthy habits. Then, as soon as the coyotes and hybrids north of the buffer zone began to invade it, the trappers would have had to repeat the roundup. It seemed at best an unusually heavy-handed management proposal, and at worst an outright declaration of war against a population of wild animals. The project could have worked only if trappers employed the most radical means, such as extensive poisoning and the use of the "coyote getter," a makeshift gun that shoots cyanide gas into an animal's mouth when the animal pulls on a knob baited with scent. But in 1972 the Nixon administration, bowing to pressure from environmentalists, banned the device.

Curtis Carley, like Glynn Riley, is a quiet hero of the red wolf recovery project. Without his dedication the species might have gone extinct while service officials argued how to save it. Former colleagues describe him as being so committed to his job that its difficulties gnawed at him ceaselessly, fraying his nerves. And the difficulties were many in those early days of endangered species biology, when no one knew how to go about saving an animal that was under siege in its last range.

Carley's memories of his early months on the project are not pleasant. "The one thing I've never been able to express is the sense of absolute frustration at every turn in the road," he said. "We literally did not know how to proceed, or what we would find if we took a certain path.... We had an administration in Washington that was eager for the buffer zone to be established, but then that same administration outlawed the tools we needed. A cyanide death is not a pretty thing. But if you're going to do something and succeed, it should be done with no holds barred. You set an objective and you do it."

Without coyote getters and poison, the only alternatives were for field technicians to shoot whatever animals they could spot and to set leg-hold traps for the rest. All the land was privately owned, much of it in small tracts. Before work could begin, the technicians had to contact the owners for permission. "Some people didn't want the government on their property, some didn't like the idea of using traps, some we couldn't find. . . ." Carley chuckled without mirth.

But the course of action had been chosen and would be followed to its conclusion. In the spring of 1974 Riley and two other technicians spent day after day traveling through field and brush to check their trap lines. In two-and-a-half months they killed only thirteen animals—a small fraction of the number believed to inhabit the buffer zone.

By then, however, Carley had come to suspect, to his horror, that wolf-coyote hybrids had already reached the Texas coast. Each week he spent some of his time trying to locate pure wolves that could be used to expand the captive breeding program. In his ramblings he frequently saw tracks that were smaller than those left by red wolves—that looked, in fact, very much like coyote tracks. "We were trapping animals right in the supposed 'pure' wolf territory, and we were not seeing what we thought we should see. We were getting big animals, and little animals, and animals in between.

"The whole recovery plan was based on the belief that only wolves lived along the coast and that we could easily identify wolves from hybrid animals. I hadn't been there three months before I started getting a bad feeling in my gut that something was not as we had believed. For one thing, we couldn't find a single area that had a good population of wolves and no hybrids or coyotes. It took six months for us to really document this. But by mid-seventy-four the front-line troops were convinced that the buffer zone concept would not work, that the only way to save the red wolf might be to capture all the animals to protect them from hybridization."

The realization threw the recovery program into a state of crisis. Carley knew it was only a matter of time—and probably a short time—before the red wolf would be extinct in its pure form. But he also knew that the Fish and Wildlife Service was not likely to change tack quickly. The bureaucracy set up to manage the red wolf program was unwieldy; for most decisions Carley had to consult members of the recovery team, as well as administrators in San Antonio and Washington. "We had gone into Texas ready to do good things for God and country, and we immediately found out that it wasn't going to be that easy," he said. "We were slightly panicked, because we felt like our backs were to the wall. But I think the panic caused a rush of adrenalin, because then we were able to roll up our sleeves and get to work."

The most pressing problem was figuring out which of the animals on the coast were true wolves, and which contained degrees of coyote blood. Riley maintained that the largest animals did not associate with coyotes, and so could be considered pure. Carley was not so sure. "Appearances of hybrid animals can be deceiving," he said. "We could not afford to take a chance that a hybrid would be let into the captive breeding program. We had a number of people who were coming up with tests we could use that said, 'Coyotes do this and wolves do that.' But no one really knew how a hybrid animal would behave, and that's what we needed to determine."

One of the best ways to test the bloodlines of the dozens of animals that were being trapped, Carley believed, would be to take careful measurements of their skulls and compare them with red wolf skulls that had been preserved from the 1940s. But obtaining the needed measurements meant killing the animals in question. Or so Carley thought. One evening he mentioned the problem at a presentation before the Houston chapter of the Sierra Club. A few weeks later he received a call from Donald Shaefer, a public relations agent at a local hospital who had attended the meeting. Shaefer and his wife, Donna, a radiologist, knew of a study in which biologists had used x-rays to examine the skulls and brain volumes of a species of Pacific seal. Had Carley thought about using x-rays to measure wolf skulls?

"I went to the hospital and talked to a number of the doctors who had been involved in the seal study," Carley said. "Then we flew Don and Donna Shaefer up to Washington to the National Museum to look at wolf skulls there. We had to go through the collection very carefully, because even at this prestigious museum there were some 'red wolf' skulls that were obviously hybrids."

Working with the taxonomist Ronald Nowak, Carley and the Shaefers selected more than two hundred skulls collected from areas they believed to be free of coyote infiltration at the time of the animals' death. By comparing x-rays taken from the top and the sides, they calculated a set of minimum measurements for the brain-volume/skull-size ratio of true red wolves. Any animal with a smaller skull, they agreed, was probably not a pure red wolf.

The skull parameters became one of several yardsticks by which Carley and his colleagues judged each animal's suitability for the captive breeding project. If an animal had the skeletal proportions typical of a red wolf, if it howled like a wolf, weighed the right amount, and had certain skull dimensions, it was probably a purebred. Unfortunately, the use of such strict standards meant that any unusual wolves would be killed as hybrids.

The implications were disturbing to Riley, who insisted that wolves—both the gray and red species—varied greatly in size. In Tacoma, Murray Johnson and others involved in the captive breeding program worried that, by euthanizing the most diminutive wolves, biologists might bias the gene pool of the entire species.

Carley acknowledged the potential problems, but saw no alternative. "The only sure way to draw a line between animals with coyote blood and pure wolves was by breeding them," he said. "If there's coyote blood in a parent, you can usually see evidence of it in the pups. But we didn't have the time or the room to breed every animal we thought might be a wolf. We had to rely on the x-rays and other criteria to make decisions about what to do with certain animals in the short run.

"We all worried that we would make mistakes and kill some wolves. And we knew that by taking this approach, we might be creating a new species of red wolf that was different than the species had been originally. But under the circumstances, we felt we really had no choice."

From mid-1974 on, skull x-rays and other measurements were taken of every animal trapped along the east Texas and west Louisiana coasts. Those judged to be hybrids were removed so that more habitat might be opened for wolves. Although the biologists and trappers were able to work quickly, they found it difficult to locate any animals that even resembled wolves. With the failure to establish a buffer zone, there was no way to know whether the red wolf still existed in the wild. And more trouble lurked ahead.

In early 1974, as Carley began going over the data sheets for the animals being held at the Point Defiance Zoo, he noticed that several had features that resembled coyotes. The previous fall the Fish and Wildlife Service had designated the zoo as the official government breeding center for red wolves. Additional pens had been built on property owned by a mink rancher near the town of Graham. Six wolves taken to the zoo in 1970 and 1971 had given birth to eight pups, and four more wolves had since been obtained from the wild and from a zoo in Oklahoma City. These eighteen animals were to be the founders of a captive population, the one safeguard against the red wolf's extinction in the wild. Yet, Carley realized with dismay as he thumbed through the photos, some of the captive animals were of questionable blood.

In December 1974 Carley flew to Washington State with Howard McCarley and Buddy Long to examine the animals in the breeding program. What they found confirmed Carley's worst fears. In examining the "wolves," the three men agreed that only six had the proper physical characteristics to be included in the breeding program. The other eleven—including the ones that had bred, and all their pups—would have to be destroyed. "It was a very emotional, very disturbing finding," Carley said.

"But we had no choice. We recommended to members of the recovery team that we clean out our own program. They concurred."

The six surviving animals produced no young the following spring. And with coyotes still spreading along the Texas coast, it seemed likely that the red wolf was doomed in the wild. The only way to save the species, Carley believed, was to concentrate on trapping as many wolves as possible. By 1975 he had begun to lobby for allowing *Canis rufus* to go extinct in the wild. To try to maintain it, he argued, would be to waste time and resources that might be better spent on finding a site for the reintroduction of wolves bred in captivity. Although hesitant at first, by the middle of that year service administrators and the recovery team agreed.

"We were not even sure, when the decision was made to take every wolf out of the wild, that we would be able to breed them well in captivity," Carley said. "You know that, to a wild wolf, the cages at a captive breeding center must seem like a concentration camp. We anticipated that we would have a percentage of animals refuse to breed in cages; we just did not know how high the percentage would be, or if it would significantly affect the gene pool.

"We were concerned that we were going against the very intent of the Endangered Species Act, which was to save animals in the *wild*. We knew also that if the political climate of the country changed, we might not ever be able to reintroduce the species, no matter how many animals we had in captivity. It was a real gamble. We felt we might eventually be blamed for the extermination of the red wolf."

By the close of 1976, howling surveys could locate no red wolves in Chambers County, where three years before biologists had believed the population to be stable. The few animals left in Jefferson County resided in a swampy section along the Intracoastal Waterway. In Louisiana, groups of three or four wolves were thought to exist in Cameron and Calcasieu parishes.

It would take another three-and-a-half years for the field staff to capture the last wild red wolves. When finally the project was abandoned in the summer of 1980, four hundred wild canines had been trapped. Only forty were admitted to the breeding program. And the appearances of those animals were deceiving, as Carley had feared. In the end, the survival of the red wolf would rest with only seventeen animals that had escaped the plague of coyote blood.

The wolves chosen for the breeding program in Washington State were transplanted to a countryside that, but for its temperate winters and frequent rains, could not have been more different from the Texas coast. Where they had denned among short, wind-battered shrubs and tallow trees on a coastal plain with scant changes in elevation, they now lived on verdant hillsides beneath the shadows of fifty- and sixty-foot Douglas fir. Where the harsh Texas sun had bleached the landscape of all but subtle hues, the slanting Northwest sun threw a light that was at once bright and filled with color, as if cast through a pale-blue filter.

Dale Pedersen's mink ranch lay in rolling pastureland crossed by narrow country roads and bordered by split-rail fencing. From certain peaks and dips along the roadsides drivers could glimpse the stark triangular mass of Mount Rainier, but most vistas were blocked by the bowed, blackish-green branches of hemlock and fir. Pedersen had worked on a mink ranch as a teenager and had spent some time in New York learning to grade pelts. Later he set up his own operation outside the small community of Graham, building long complexes of pens from which the inquisitive faces of mink peeked whenever he passed. Year after year he paired the animals with mates and gassed them at the end of the season for their pelts; somewhere along the line he became a conservationist. "I decided that as long as I was going to breed animals, I may as well breed some for a good cause," he said. In the late 1960s he accepted an appointment to the mammal committee of the Point Defiance Zoo and Aquarium, which is where he first heard of the red wolf.

Pedersen is slim and dark-haired, with a quiet, unassuming visage. Where Curtis Carley is expansive on the subject of red wolves, Pedersen is taciturn, answering most questions in short phrases. In a program where many people have given freely of their time and money, Pedersen's generosity ranks among the highest.

By 1973, when the federal government signed a contract with the Point Defiance Zoo to run a breeding project for the red wolf, Pedersen had already been housing the animals at his own expense for more than a year. The zoo had constructed only three pens for red wolves, and Pedersen, concerned about overcrowding, had put up two more pens on a game farm he owned near the mink ranch. "The red wolf needed help badly, and no one really seemed interested," said Norman Winnick, who was then the director of the zoo. "We all knew that someone had to do something. Dale was willing to put a lot into the project without expecting any recognition."

To this day Pedersen downplays his own role and points to others— Winnick, Carley, Buddy Long, Sue Behrns, Murray Johnson, and Mike Jones, a local veterinarian—as the reasons for the success of the project. "You wouldn't have a red wolf today if it wasn't for those guys. Nobody was after any professional glory; we all just wanted to see the species saved. And we couldn't have done what we did if there had been a lot of politics involved, like there are with a lot of other wildlife projects." Pedersen paused. "But we did it on peanuts too. We didn't have anywhere near the budget some of these other programs have."

The pens, each a hundred feet square, were isolated from all human activity. Bits of vegetation were left inside to give the wolves cover and ease the stress of living in captivity. Small concrete bunkers made from septic tanks served as dens. The Point Defiance Zoo gave Pedersen equipment for handling the animals, including a number of large-meshed, long-handled dip nets with bright-blue cord, the kind commonly used in the Northwest salmon fishing industry. "We learned pretty quick how to make one kind of tool do for something else," he said.

The first animals to arrive did not look much like wolves. "Their coats were shaggy and thin, like they didn't need a lot of heavy fur down there because of the heat," Pedersen remembered. "Some were old and didn't have good teeth, some died pretty quick from stress and heartworms. We worked with them the best we could. We knew they were different from coyotes, but that's about all we knew. I didn't have any idea what a red wolf was supposed to be."

Most of the time the wolves were left alone, except for daily feedings of dry dog food. When one needed to be handled, Pedersen and the zoo staff would go into the pen and corner the animal as quickly as possible, dropping a net over it to subdue it. "If it started running a lot and getting overheated, we'd leave it alone and come back the next day," he said. "We didn't stress them anymore than we had to."

Slowly the animals responded to the improved habitat and the scrupulous care of Pedersen's staff and Mike Jones. There were fewer problems with mange in Washington State, and no heartworms. The pelage of the wolves thickened and grew darker; it was "almost like the sun down in Texas had been bleaching it out," Pedersen said. The first pair of wolves that came to live at the mink ranch bred in the spring of 1974. "As soon as those pups were born, it started raining and rained steady in sheets for a week," he recalled, grinning a little. "I'd go back and try to get a look at them, thinking they were probably dead. Finally the rain let up a little, and I went in to look at the mom. She had just dug a little hole for her puppies, and she was lying over it keeping them dry."

The instinctive care the mother had shown for her pups invigorated Pedersen's hopes for saving the red wolf. But a major setback was soon to follow. That fall Carley arrived in Washington with Long and McCarley and discovered that eleven of the animals were probably hybrids and needed to be removed from the program. The decision was upsetting to everyone involved. "We knew it had to be done, and we all supported Curt," Pedersen said. "But it bothered me some that they might have been killing pure wolves, when all we had was this one little group in captivity."

In retrospect Pedersen does not believe any pure wolves were killed as hybrids. Murray Johnson is not so sure. Retired now from his practice as a surgeon, Johnson holds a curator's appointment in mammalogy at the University of Washington in Seattle. White-haired, ruddy-complected, slow-spoken, and eminently gracious, he lives in a rambling stone house tucked among the Victorian mansions of Tacoma.

One rainy spring afternoon Johnson and I sat in an upstairs study crowded with books and journals, cabinets and files. Letters from the early days of the red wolf breeding project were spread on a table before us. Through a window I could see storm clouds and the forboding waters of Puget Sound; but in the small study an overhead lamp emitted a circle of warm, cozy light.

Johnson, thumbing through a four-inch-thick folder of papers, was having difficulty reconstructing the story of how the wolves had received sanctuary at the Point Defiance Zoo nearly twenty years before. The folder held memos from Howard McCarley, William Elder, Douglas Pimlott, and other mammalogists from across the country. Through his own work with genetics, Johnson had contacted Ulysses Seal, the Minnesota researcher who ran electrophoretic tests on the red wolf in the 1960s. "He had completed his studies and had some red wolves he couldn't use anymore," Johnson said. "A group of people—some in the scientific community, some from the zoo—hit on the idea of bringing them to Point Defiance."

Other details of the story eluded him. Nevertheless, Johnson was openly proud that the efforts of independent conservationists had contributed so much to the rescue of the species. "You can see that there were a lot of people who wanted to save the red wolf, even before the federal government got deeply involved," he said with a small smile. But when I mentioned Carley's attempts to rid the breeding program of hybrids, his face clouded. "Whether or not all the animals removed from the program were hybrids is a matter of opinion," he said. "Some of the scientists involved disagreed with the viability of the x-ray technique.

"I know as a medical doctor that we must frequently require more refined techniques than x-ray views, and there is always a question of biological variability. I don't think you can take a set of measurements and say, definitively, this is a certain species. But bureaucracies like to use clear, simple comparative data to make decisions. Curt Carley was a good biopolitician, and we supported him. . . . We decided the important thing was not to quibble over technique. The important thing was to save the genes, the animal itself, and we had enough wolves to do that."

It is doubtful that scientists will ever know what number, if any, of the animals destroyed from 1974 to 1980 were in truth red wolves. But, as they suspected, federal biologists learned that just because an animal had the physical characteristics of a red wolf did not guarantee it was a purebred. After 1974 all pups born into the program were carefully measured and examined every three months. Frequently the members of a litter would be radically different in size and color, a clear sign that they were hybrids. Carley worried that the recovery efforts may have come too late, that already the integrity of the red wolf had been destroyed by interbreeding.

And answers to his questions and fears would be excruciatingly slow in coming. In 1975 and 1976 the wolves failed to raise a single healthy pup.

Meant to Be Wild

Discouraged, Carley and his associates pored over records to see how frequently the animals had been handled and what drugs they had been given. "We were looking for anything at all that might have thrown their breeding cycles off," Carley said. Finally in 1977 a pair of wolves dubbed Mr. and Mrs. Sabine, after the Sabine Pass in Texas, where they were caught, produced four healthy pups. All appeared to be pure wolves. Three other litters were born that year as well. Although two of the litters proved to be hybrids, the breeding program was finally building steam. "When the Sabines bred, it was as if the dam had broken," Carley said. "After that we had litter after litter. The only explanation I could ever come up with was that the mated pairs were all fairly new to captivity, and it simply took some time for them to get comfortable in their surroundings."

In the spring of 1978 Pedersen hired Sue Behrns to care for the wolves and the other wild animals he had taken under his care. Twelve pens had been constructed on the mink ranch with funds from the Fish and Wildlife Service, and he received an annual fee for the animals' upkeep. Pedersen had come to enjoy keeping orphaned and injured animals, and had built another complex to house his menagerie of deer, foxes, panthers, wolverines, and badgers.

Behrns, a strong redhead with a cherubic face, was then in her mid-twenties. She had been educated as a horticulturist, but after college decided she had chosen the wrong field. For as long as she could remember, she had been drawn to animals. "I was one of those kids who would see an animal, any animal, and want to walk up to it and put my arms around it, regardless of what it was or who it belonged to," she said.

Dressed every day in the garb of an animal handler—black rubber knee boots, muddy jeans, and a pea-green coat with a dozen pockets—Behrns comes home to a trailer around which animals congregate in knots. At a stranger's approach, cats scatter in all directions from the dooryard. Most of the space inside, where not occupied by great orbs of plants, is taken up by cages and animals. At the time of my visit Behrns claimed a Steller's jay, a gerbel, ten cats, and five dogs, including a lion-sized Rottweiler named Alec. She chatters to animals in snatches of baby talk as she goes about her chores, showing an undisguised favoritism, referring to a cougar fondly as "a little twirp" and a badger disparagingly as "a nasty old thing."

When Pedersen hired Behrns in 1978 the condition of the red wolf in captivity was still precarious. The animals were breeding, but most of the pups were not living as long as a month. That year the wolves bore seven litters, each with from three to seven pups. It was the policy of the breeding program to handle the animals as little as possible, and Carley instructed Behrns to peek into each den just long enough to ascertain that the mother was nursing. "There were all these wives' tales like, if a pup gets human scent on it, the mother will kill it," she said. "We didn't know what was true and what wasn't."

Over several weeks most of the pups died, until only eleven were left. "Hookworms were just wreaking havoc with them, for one thing," Behrns said. "There was also a problem with puppies getting infected sores on their feet from pushing against the floor of the concrete dens. At least that's what we figured out later. The pups would just disappear because the parents would eat them. There was no way for us to tell what had killed them."

The following year, after much discussion, it was decided that Behrns should examine the pups shortly after their birth and check their sex. When they were ten days old she wormed them and gave them other needed medications. Most of the five litters that year seemed healthy. But when the pups reached the age of about three months they began to die. "I'd go into the pen one day and there would be these puppies all shriveled up and dehydrated, some of them already dead," Behrns said. "I'd think, geez, I must have missed something yesterday. They were good-sized, frisky animals by then, not little helpless pups, but they would just change completely overnight. There'd be a bad odor and a bloody stool, or just parts of them lying around from where the parents had eaten them. If they were alive I'd rush them to the vet."

By late summer laboratory tests showed that the pups had contracted Parvo, an animal virus largely unknown in the Pacific Northwest. "I'd worked in a vet's office, and I'd never heard of it," Behrns said. "But the next year there was a big scare about it; a lot of dogs caught it. By then we had started vaccinating the wolves."

In spite of the outbreak, twelve pups survived. Several of those were raised by Behrns in her trailer. "We decided that as long as there were still three puppies in a litter they could stay with the mother, but if the litter got down to two, I'd pull them both and take care of them myself," she said. "We were still trying to figure out which of the parents were hybrids, so it was important to have some young alive that we could look at."

It was the first opportunity that Behrns, or anyone, would have to become well acquainted with the nature of young red wolves. They were like little wet weasels, slick and black, with ears that flopped over instead of standing up. Once, when she was hand-raising a very young pup, she howled quietly in its ear to see what it would do. The pup's eyes had not even opened, but it threw back its head and tried to howl, giving short squeaks comically high in pitch.

The abundant energy of the strongest pups surprised her. Two small females, the first she raised by hand, managed to crawl out of their incubator box, which she had placed inside another box. The walls of the large container were too high for the pups to climb, but they kept walking, their heads butted against it as if trying to push it down. "They were unstoppable.

The bottom was slick, and I guess they thought they were going somewhere." She named them Tanky and Bulldozer.

Although the wolves were all vaccinated for Parvo beginning in 1980, some continued to fall ill. In 1981 Behrns discovered a young pup hiding in a corner of a pen, dehydrated and near death. "I had hand-raised her, and she was real tame. Whenever I'd go near the pen she'd come running. And then one day she just wasn't anywhere around. I finally found her under a bush, all shriveled up and looking just awful. I grabbed her and put her in my truck and took her right to the vet."

Behrns had named the pup Chewy, "because when she was real young she reminded me of a little Chihuahua, for some reason." She was reluctant to leave Chewy at the veterinarian's office for fear the pup would become too depressed to get well. "Often I think a wild animal loses all hope at a vet's," she said. "They can be just traumatized by all the people and other animals in such close quarters, and the barking dogs.

"There was a nurse working for the vet who was real nice. I talked her into going over to Chewy's cage every day, petting her a little and letting her lay her head in her lap, just to give her a little reason to live."

Chewy recovered. Six months later it was determined that she was a hybrid. As was the custom, Behrns administered the lethal injection. The strain of destroying pups she had raised affected her deeply. "I just had to grit my teeth and do it, figuring it was for the good of the species. For a while we joked that whenever a puppy lived any length of time, it was a hybrid."

The pups were x-rayed to determine their cranial measurements at the age of three months, and again at six months, nine months, and a year. "At first we used the old x-ray machine that had been brought here from the field program in Texas," Behrns said. "Curt would take an x-ray, then I'd dash down into town to a chiropractor's office to get it developed and see if the wolf was lined up right. Then I'd dash back and he'd take another. It was hilarious."

To Behrns's great relief, Tanky and Bulldozer met the criteria to remain in the breeding program. So did two females she delivered herself by Caesarean section in 1981. On the morning of the delivery, Behrns peeked into the den where the mother was lying and noticed that she looked strange. She was not due to give birth for more than a week. "I can't really tell you what was wrong; she just didn't look right," Behrns recalled. "So I went down and called the vet. When I checked on her a little while later, she was dead.

"She was still warm. I picked her up and carried her down to the ranch kitchen, and one of the mink workers gave me an old knife. It wasn't the most sanitary of conditions, but I figured what the heck.

"I cut her open and there were nine puppies inside. I worked on all of 'em, pumping their chests and blowing through their noses as fast as I could.

The guy who had given me the knife helped. If they didn't respond immediately we just put them down and went on. Two of them hiccupped. We must have worked on those puppies for three hours." Behrns's face grew thoughtful at the memory; the gaze from her blue eyes wandered downward and anchored on the surface of the table where we sat in her kitchen. "I've always thought that I could have saved more, maybe, if I'd had another couple of people to help.

"So I brought those two here and started feeding them, thinking maybe this isn't worth it, maybe they have brain damage. Or maybe they'll be okay. They had a sucking response, but who knows if a brain damaged puppy would have that?"

The pups thrived under her care. Behrns, growing increasingly attached to them, was reluctant to name them for fear they would still succumb to disease or be culled out of the program. "I called them Blackie and Brownie, interchangeably. Finally, at three months, I put them in with the other pups, and they did fine." One was assigned the studbook number 194. At the age of five she was shipped to the Alligator River National Wildlife Refuge in North Carolina, where she was placed with her mate in a pen surrounded by sweet gum and red maple, just off Phantom Road.

Tanky and Bulldozer grew to be fine, frisky wolves. In 1982 Tanky gave birth to a litter that included wolf number 211, the feisty male that would be taken to the pen at South Lake on the Alligator River refuge. A few years later she died of unknown causes. Bulldozer lived longer, becoming so large and dominant that she would challenge males to see if she could make them cower. Slowly the last of the hybrids were eliminated from the breeding program, although the process required years longer than expected. "We had one female we kept breeding and breeding because we were so sure she was pure," Behrns said. "She'd have a litter of pups that were all different and we'd say, oh, it must be the male."

In 1979, with the wolves starting to breed well in captivity, Carley and Warren Parker began talking of reintroducing the species to the southeastern United States. Their efforts to locate a site would not come to fruition until 1986, with the plan to release wolves in North Carolina.

Behrns, though heartened that the animals would be given a chance to live in the wild, found herself growing nervous about the impending release. It did not particularly bother her that other people would soon be responsible for the animals she had raised. But she was distressed by the possibility that residents around the Alligator River refuge would try to kill them. "Apprehensive is the best word to describe it, I think," she said. "I didn't want anyone hurting them—no, wait," she paused for a long time, thinking deeply, her cheeks still burnished from her hours outside that day, her short strawberry-blond hair sliding across the side of her face. It was late; we had sat in her trailer, talking intently, for hours. "This just now

Meant to Be Wild

occurs to me. I think I was worried that someone would hurt the animals and—I don't know—get some sort of perverse pleasure out of it, like a sadist. That someone would kill one of the wolves and go off bragging about it to his friends. I still worry about that some."

"So you took them to Alligator River and came back here," I said.

"I left them there, not feeling bad about it, really, but not knowing how well they were going to do. I knew—we all knew—some of them would probably die, but that that was what the species needed to live in the wild. And that was okay."

It was nearly midnight, well past Behrns's normal bedtime. We said goodnight and I spread my sleeping bag across her couch. I closed my eyes, relaxing into the blackness of a rainy Northwest night, conjuring up images of wolves nestling with their young, of Behrns gently tugging mouth from nipple just long enough to cradle the warm roll of flesh in her hand, and of pups with their heads thrown back to howl, though they lacked the strength to open their eyes.

I awoke sometime later to a sound I could not place, a sound that did not belong to the humid peace of a spring night. Somewhere nearby a group was giving war cheers and cries, whooping it up, letting their voices rise and break, warble, and rise again, sirenlike in pitch and tremolo. I glanced groggily at a clock as I rose to go to the front door. There was no stadium nearby and, besides, what crowd would attend a sporting event at two o'clock in the morning?

The answer lay off Behrns's front stoop. Up a small rise, on the far side of the office from which Pedersen ran his ranch, forty wolves were crying in the night, their howls rising together like Indians steeling themselves for battle. The sound ebbed except for a few lingering yaps, but within seconds it had begun to build again, a feverish crescendo, lilting and rising, with short barks surrounding a single high, unbroken note. It was a chorus so different from what I had expected of wolves that I was startled completely awake; sleep would evade me most of that night. The wolves, having howled themselves out, began to fall silent, their mingled cries dropping off one by one like the lights of a house blinking out, each leaving a tidy emptiness in the drizzling night.

IV

A NEW FREEDOM

Seven summers and seven winters after the last mange-ridden red wolf was hauled from the Texas marshlands, the species' new incarnation was made ready for release in the North Carolina swamps. Larger, redder, and more robust than its ancestors, it had been bred to survive disease and harsh conditions. The red wolf had not been pampered in captivity. It behaved furtively, especially around people. But how much remained of its wild character could not be determined as long as it was caged.

There would be no flags or fanfare the day the red wolf returned to freedom. Nothing would seem out of the ordinary to the animals themselves. Early that morning a person would walk up the trail to each pen. Pieces of meat would be hurled over the plywood screen, as they had been every fourth day for months. Then, as the wolves grabbed for the bits of rabbit and deer, hands would quickly unlock the door. It would be swung open for good, and the wolves would be free to leave.

But even as Warren Parker and Mike Phillips worked out the strategy for tracking the wolves, even as the anticipation of the caretakers increased, events were unfolding that would push the release further into the future. It had been scheduled for early May, when the yearly population of young rabbits, rodents, and deer would be at a peak. With so many inexperienced prey animals in the woods, the biologists hoped the wolves would have an easy time learning to hunt.

The first hint of trouble came in mid-April, when Warren Parker received a call from officials at 3M, the company designing the radio capture collars. Tests on the collars had been proceeding more slowly than anticipated, Parker was told. Would it be a problem if the release was postponed a few weeks?

Every detail, from the timing of the arrival of the wolves in North Carolina to the budget for the caretakers' salaries, had been pegged to the release of the wolves in May. Parker was not pleased. But the public had been assured the wolves would not be freed without the capture collars. Until 3M could test the mechanics of the dart-firing device to the satisfaction of its engineers, the animals would remain in their pens.

The spring of 1987 was unusually wet and windy, even for the blustery Outer Banks. For most of April the sky remained cloudy, and the constant dampness turned the air heavy and dank. A ceaseless southwest wind drove the waters of the coastal sounds hard against the shoreline of the Alligator River refuge and into its swamps, saturating the soil in the pens where the wolves waited.

One morning in late April a line of pickups and jeeps threaded its way into the western portion of the refuge, passed the caretaker camp at Pole Road, and turned south onto Phantom Road, toward a wolf pen. At a trail-head the trucks stopped to disgorge ten people and an assortment of equipment—scales, kennels, catch poles, and toolboxes filled with medical supplies. That day the pregnant female at Phantom Road was to be examined in the pen, and her mate was to be taken to a veterinarian's office for a more thorough examination. If all went as planned, it would be the last time the wolves were handled extensively before their release.

Eighty feet above the road, warblers in bright spring plumage chittered and hopped through a canopy of maple and gum. New buds brushed the woods with rose, orchid, and pale green, a collage as colorful as the foliage of autumn, but more subtle. The Phantom Road pen, situated in an open hardwood grove, was one of the prettiest sites on the refuge. But it was also buggy, and the prolonged rains had left water pooled on the surface of the dense, peaty soil. As he stepped out of a truck, Parker groaned. Except for a few high spots and some exposed roots, the trail into the woods was submerged. "Don't believe I've ever seen it this wet out here," said John Taylor, a note of concern in his voice.

At the back of the line of trucks, Phillips was taking an inventory of gear with Roland Smith, the curator of wolves at the breeding center in Washington. Smith served as a member of the red wolf recovery team, the group of scientists that was overseeing the project. He and Phillips gathered an armload of equipment and began making their way to the pen. The rest of us picked up the gear that was left and followed. The male wolf was simply to be caught and put in a kennel for transport. The female, believed to be two weeks from giving birth, was to be checked by Larry Cooper, the veterinarian for the project. Two button-sized radioactive chips would also be sewn beneath the skin on her back to release traces of radiation into her muscles and digestive system. By collecting scat samples and checking them for radiation, the biologists hoped to construct maps of the territories used by each animal.

The footing on the trail was slippery. I carried a metal toolbox and a large, bright-blue salmon net that kept snagging in vegetation. In front of me two biologists struggled with the cumbersome kennel. Parker, Phillips, and Smith had already entered the pen by the time the rest of the group made its way into the woods. I could see the female darting along the far side, but Phillips and

An adult red wolf. Photo courtesy of U.S. Fish and Wildlife
Service.

Curtis J. Carley with an anesthetized male red wolf on Bulls Island, South Carolina, in 1976. Photo by Kenneth Stansell.

Coyote-wolf hybrids, such as these year-old siblings, can vary greatly in appearance. Photo by Curtis J. Carley.

The wolf pups were like little wet weasels. Photo courtesy of U.S. Fish and Wildlife Service.

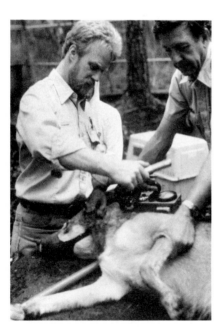

John Taylor, left, and Warren Parker examine a captive red wolf. Photo courtesy of U.S. Fish and Wildlife Service.

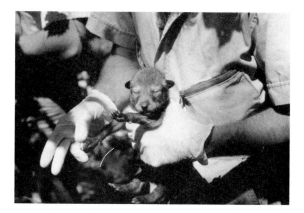

A week-old red wolf pup. Photo courtesy of U.S. Fish and Wildlife Service.

Mike Phillips administers a shot of heartworm medicine. Photo courtesy of U.S. Fish and Wildlife Service.

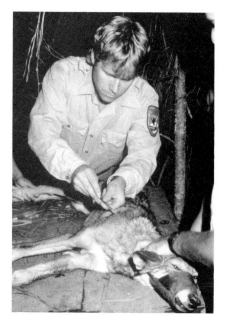

These two male wolves were born on Bulls Island, South Carolina, and later released at the Alligator River National Wildlife Refuge. Photo courtesy of U.S. Fish and Wildlife Service.

The Alligator River refuge is a maze of marshes and thick woodlands. Photo courtesy of U.S. Fish and Wildlife Service.

A red wolf being released into a pen. Photo courtesy of U.S. Fish and Wildlife Service.

Chris Lucash found this female wolf
dead from an infection on the bank of
Pamlico Sound. Photo courtesy of U.S.
Fish and Wildlife Service.

Smith were ignoring her. They stood together in a corner, talking excitedly. At our approach Phillips turned and yelled, "We've got pups in here."

A single pup, slick, brown, and only a few inches long, had been struggling to swim in a puddle just outside a small, wooden wolf house. Smith had heard a sound like the cheeping of a bird and had gone to the corner to investigate. Standing with the pup cradled against his broad chest, he looked befuddled and a little worried. "Let's wrap it up, quick," he said.

It was a male, about three days old. Apparently the mother had been nursing inside the house when she heard the approach of people and bolted, throwing the pup into the water. "He's soaked," Smith said. "We've got to keep him warm." As he spoke Taylor handed him a rough, black blanket.

Several men fanned out around the pen, searching for other pups. But if more had been born, there was nothing left of them. "Let's get her before we do anything else," Smith said, nodding at the female. Frantic, she was leaping high against the fence as if trying to clear it. Smith handed the pup to Cooper and grabbed a salmon net, brandishing it as he walked slowly toward her. At the sight of the net she retreated to a corner and stood hunched in a foot of water. Smith slipped the noose from a catch pole over her neck, muzzled her, and helped Phillips carry her over to another small house, where Cooper had set up a makeshift operating table. They laid her carefully on the flat roof and stretched her legs in front of her as if she were a ham ready to be dressed. Her shoulders quivered as Cooper shaved a patch of fur from her back where the tracer chips were to be sewn.

Using broad, casual strokes Cooper painted the patch of naked skin with a saffron solution, then covered it with a surgical cloth and cut an inch-long incision. He was not accustomed to working outside, and next to the tanned, woods-wise biologists he seemed pale and soft. But his hands were sure, and he worked quickly. Beside him, Bob Crabtree, a research biologist from Idaho, gave instructions on how to implant the two brittle, translucent chips.

The surgery lasted only a few minutes. Cooper stitched the incision closed and turned the wolf on her side to draw a sample of blood from her front leg. That done, Phillips and Smith looped some rope around her feet, hogtying her so she could be hung upside down on a scale suspended from a long stick. They lifted the trussed wolf gingerly to the scale and let her swing free. The metal hook pulled down to fifty-eight pounds.

It was time to leave. Those of us who had been watching began gathering equipment and ferrying it out to the road. Smith had taken the pup and was placing it gently in one of the houses. A caretaker stood nearby, holding the mother from behind with the catch pole. "Should we leave the blanket to keep the pup warm?" Parker called.

Smith shook his head grimly. "It's too unfamiliar. The mother would just tear it up," he said. We could hear the pup whine as the mother, released from the catch pole, bolted to the back of the pen. We stood outside uncertainly. "Let's go," Smith said. "She'll find him."

I climbed into a jeep with Smith and Phillips. The two were good friends, and normally they kept up a ceaseless, jovial banter. But Phillips was uncharacteristically glum, and Smith, a burly man, sat slumped in his seat. "What do you think, Roland?" Phillips asked at length.

"He may make it. He may even have a good chance, since he's the only pup left and he's been nursing. You can assume he's bonded with the mother. There's just so much water . . ." Smith's voice trailed off. The grove of maple and gum, so appealing in drier seasons, had become a hellish quagmire. The mother had probably given birth to five or six pups, and all but one had died from disease, parasites, exposure, or some unknown cause. In any other spring it might have mattered less. But this first year the biologists wanted to release two pairs of wolves with young, as an experiment. It was thought that pups might adjust easily to freedom, since they had never known life in a pen. Instead of a thriving litter, though, the Phantom Road wolves would raise only one pup—if that one survived.

He did not. That night, when the biologists returned the male to Phantom Road, they found the pup unconscious in the back of the pen. Although Smith revived him with artificial respiration, he died a few hours later.

The following day, at Smith's insistence, the biologists moved the pregnant female 196 and her scrappy mate 211 from the pen at South Lake, which was also badly flooded, to the pen on the dry ridge at Pole Road. Smith was convinced that pups could not survive at South Lake; the entire enclosure, except for a few square feet, was underwater. The wolves being kept at Pole Road were moved to South Lake. (That female was not pregnant.) Parker worried that the move, coming so close to the release, would disorient the wolves. But he trusted Smith, and the death of the young pup had rattled him.

That morning Phillips and Parker had also planned to have Cooper examine the wolves being kept in the windy Point Peter bog. Unlike the other wolves, the Point Peter female had been raised at a private breeding center in St. Louis. She was nervous and extremely submissive; Parker worried that she was not a good choice for release. Her mate was timid and sluggish and had gained so much weight that Phillips had ordered his rations cut. Otherwise he appeared to be in good condition. But in examining him, Cooper noticed a major problem. Three of his canine teeth had been broken off, and the fourth was worn down to a nub.

"I'll tell you," Mike Phillips said over a glass of beer at the end of the day, "you could have knocked me over with a feather. There's no reason on earth why a healthy male wolf—a fat, sluggish wolf—should have those kinds of problems. His teeth make it look like he's twenty years old. If he were high strung, the kind of animal that chews on chain link a lot, maybe I could see it. But him? C'mon."

I asked Phillips if he thought the male at Point Peter would have difficulty hunting. "He can kill a raccoon in the pen," Phillips replied, "but it takes him a while. He's going to have a hard enough time feeding himself in the wild, much less trying to support pups, if he lives long enough to sire any."

The reintroduction was not getting off to an auspicious start. And there was still the problem of the capture collars, scheduled to be delivered in early June. As Memorial Day approached, engineers at 3M announced there would be a further delay. The radio transmitters in the collars did not have as great a range as they had anticipated. Technicians hoped to be able to fix the problem within a few weeks.

If Parker had known what lay ahead that summer, he might have considered dropping the capture collars from the project. But he and Taylor had presented the collars as an integral part of the release, and both were reluctant to rewrite their own rules. In addition, the plan for reintroducing the wolves had been published in the Federal Register, which carried the weight of law. Without posting a change of plans in the register—a process that could take months—Parker felt legally bound to release the wolves with capture collars. As summer arrived, the caretakers grew restless. Two decided they could no longer live at the isolated camps in the woods and quit, leaving Phillips to scramble for fill-in help.

The days had grown long and humid; the maple and gum were in full leaf. After rains the swamp filled with the thick, exotic cackles of frogs. At dusk dragonflies and damsel flies peppered the air, their bodies bright metallic streaks of blue, green, and bronze. In the summer heat the Alligator River refuge had taken on the feel of a jungle, with its dripping air and draping vines. The warm sun and sporadic rain brought clouds of mosquitoes and deer flies from which neither the wolves nor their keepers could escape.

Phillips began to wonder how well the animals would withstand the parasites of the swamp. Ticks were rampant, and that spring the male 211 and his mate, now at Pole Road, had developed a slight case of hookworms. In mid-May the female gave birth to her pups, but all of them disappeared. With no evidence, it was impossible to say what had gone wrong.

June passed quietly. In the first week of July, engineers at 3M reported that they were making progress on the collars but that their tests were still incomplete. On the fifteenth, Phillips used the last of the prey meat he had stockpiled over the winter. The caretakers had no choice but to revert to feeding the wolves dog food. It was as if the project was regressing, and the mood of the staff deteriorated day by day. The caretakers, sick of biding their time, grumbled among themselves about the heat and the bugs. Parker, too, was frustrated almost to the point of anger, an emotion he rarely displayed.

August arrived, and with it a promise from the engineers at 3M to hold a workshop by the middle of the month. They would bring at least two capture collars that could be fitted on wolves the following day. The news imbued Phillips with fresh energy. He and Chris Lucash, who had become his main assistant, could scarcely believe that the release might finally be at hand.

Early on a Thursday morning the entire cadre of scientists responsible for running the red wolf release gathered in a conference room at the headquarters of the Alligator River refuge. Parker was there, haggard from weeks of stress. Roland Smith had arrived from Washington two days before. John Taylor took a seat at the back of the room, along with several members of the refuge staff. Phillips came in last, his face grim, and dropped into a chair. For the past hour he had been closeted in his office with two engineers from 3M. From what he had seen and heard, he was skeptical about the engineers' abilities to make the collars work.

The workshop was to be conducted by Hugh Sontag, a slim, pale man with ebony hair. Both he and Greg Juneamann, a vice president for the company, were dressed in red polo shirts with the 3M logo embroidered over one breast.

In front of the robust outdoorsmen of the service, Sontag, with his slight build and boyish face, seemed uncomfortably out-of-place. His audience was not friendly, and he knew it. Yet the concept of the capture collar was truly innovative. If the computer mounted on it could do all he claimed, the biologists would be able to turn the radio signal on and off by remote control to save battery power. At the punch of a button they would also be able to retrieve data collected by the computer on how active the wolves had been during the previous twenty-four hours. Sontag quietly explained the construction of the collar, including the circuitry of the computer, which would ride in a small box below the animal's throat. The two tranquilizer darts would protrude from the back of the neck like an extra pair of ears. Each consisted of a drug vial attached to a twelve-gauge needle and fitted inside a sturdy plastic cover. "There's a pyrotechnical device that enables you to fire the darts remotely," Sontag said. "For that reason we treat the darts like a loaded round."

Fitting tiny explosives on the neck of a wolf did not strike me as an appealing prospect. But as Sontag continued, describing the capacity of the computer and the triggering transmitter—the control box from which commands were sent to the computer—Taylor, Smith, and Phillips began to look increasingly impressed. Parker sat at the front of the room, his earnest face growing happier as he dialed commands into the triggering transmitter and sent them to the computer, which responded with encouraging beeps.

But a desk-top demonstration was not enough to satisfy Phillips that the collars were ready to be put on the wolves. He had confided earlier that he was afraid to trust the radio transmitter designed by 3M, the most important component in the whole package. "It's been only marginally tested," he had told me. "Sure, it'll be nice to have the tranquilizer darts and all the stuff that comes with them. But face it, if the beeper goes out, we're stuck. We'll have a wolf out in the bush with a fancy piece of equipment around its neck that we can't even track to."

Sontag claimed the radio signal emitted by the collar could be heard from two miles. But the range had never been checked when the collar was on an animal. "What if it turns out the signal can be blocked by the wolf's body?" Phillips now asked. "Every time he lies down we're going to lose him."

It seemed some field tests were in order. As the workshop ended, Phillips suggested a plan. The biologists would fit one of the collars on a dog and test the range of the radio signal both from a helicopter and from the ground. Taylor agreed to use his dog, a yellow lab named Kuichak, as long as the darts were not fired.

It was already late afternoon. Phillips and Smith tossed a telemetry receiver and a metal clipboard into a truck outfitted with a large, metal antenna. "I have a feeling this may be a complete disaster," Phillips muttered. Sontag would check the signal range from the helicopter, and Parker would follow in a second truck. Juneamann and I agreed to ride with him.

A northeast breeze had gathered, steady and cool beneath a brilliant, blue sky. The sun had just begun its southward slant; its yellow light, bouncing off bean stalks in a farm field, signaled the end of summer soon to come. Parker drove slowly. Phillips and Smith had left only a few minutes before him, and there would be nothing to see until they had contacted Taylor by radio and chosen a point to begin the tests. Riding beside Parker, Greg Juneamann was quiet and polite. Parker kept his voice calm, although his mouth had become a jagged line. "I sure hope this thing works," he said.

"Oh, so do we, so do we," Juneamann said. "We wouldn't have come down today if we didn't think we had everything ready to go. We certainly wouldn't want to do anything to hold this project up." He paused for a long time. "Of course, we do want to make sure the product works."

Parker nodded once and turned onto a dirt road. Ahead, Smith leaned against the telemetry truck, looking at the ground. I could see Phillips inside with a set of headphones, his face screwed into a scowl. The antenna, shaped like the letter H, protruded from the roof of the truck. It was attached to the floor of the cab and could be rotated from inside. As we watched, it spun north, then south. Parker got out of his truck and motioned to Smith to come close. "Where are John and the dog?" he whispered.

"Over there a little less than a mile." Smith pointed toward an agricultural field, one of several that local farmers leased from the refuge. Phillips had chosen to try the easiest test first. Ground tracking in the Alligator River was tricky; the thick forests, interrupted by roads and clear cuts, could cause radio waves to bounce. Across the low vegetation of the farm field the signal would have no interference.

Phillips called to Parker from the window of the truck. "Which way is my antenna pointed?"

"That way, northeast. You picking up anything?"

"Nothing." He ducked back into the truck and continued to turn the metal shaft. On the horizon the helicopter appeared, its bulbous shadow bouncing across trees. Phillips reached for the radio mike to call Sontag. "Hugh, what are those frequencies again? I just want to double check."

Sontag answered over the crackle of static, and added that he had picked up the signal from three miles. Phillips slid to the driver's seat and started the truck. "I'm moving closer to John," he said. "I can't get anything from here."

We followed him a third of a mile up a dirt road, where he parked and kept manipulating the antenna. Kuichak and the collar were straight across the field, a little more than a half mile away. The helicopter landed and dispatched Sontag, who jogged over to the telemetry truck, opened the tailgate, and began assembling a hand antenna. No one spoke as he pulled on a set of headphones and walked up the road.

Phillips started the truck abruptly and drove another quarter mile, parking at an intersection. Taylor and his dog were now visible in the distance. Parker pulled behind Phillips's truck, got out, and began kicking at a knot of grass in the road. His boot undercut the green clump and sent it flying, but he continued to kick, forming an oval pit in the dust. Smith walked over and put his hand on Parker's back. At his touch the older man's shoulders sagged. "I think I'll walk on down toward John," he said.

Phillips glanced up as Parker walked by. He pulled the headphones off and looked after him. "Goddammit," he said.

"Are you picking up anything?" Smith asked.

"A real weak signal, but it keeps fading in and out. Maybe we should just buy a couple of dartboards and start practicing."

"Maybe we can teach the wolves to come when we whistle," Smith said sardonically.

"Maybe we can just borrow a couple of bombs from the military and nuke the wolves if we need to recapture them."

"Or nuke Three M."

"We can't do that—they make a helluva roll of Scotch tape," Phillips said.

Down the road Parker and Taylor were conferring, their heads bent close together. The rest of us joined them. Kuichak bounded happily after tossed sticks, oblivious to the group's black mood. Parker had picked up a couple of metal hooks from somewhere and was banging them together nervously. Of everyone, Sontag seemed coolest. "Earlier this week we decreased the signal range slightly to save on battery drain," he said. "Maybe we overcompensated."

"I guess that's possible, but, gee, I didn't think it would make that much difference," Juneamann said.

Parker clanked the metal pieces together and said nothing.

"Could the signal strength be adjusted up?" Phillips asked.

"Possibly," Sontag said, "but I'm not sure we could do it here. It might mean going back to Minnesota."

There was nothing left to discuss. We piled into the trucks for the tedious ride back over the rutted roads. As Parker's truck reached the highway, I asked Juneamann again about the results of the ground tests in Minnesota. "We could hear the collar two miles away, at least," he said. "I really can't understand what the problem is here. It's not like we've used inexpensive components or anything."

What about the range of the triggering transmitter, I asked. Had Sontag been able to give commands to the computer in the collar from two miles?

"Oh, no," he said, "the ground range on that was much less. I think it was only about a quarter of a mile."

If a wolf holed up in a roadless section of the swamp, it would be impossible, with a range of only a quarter mile, for the biologists to use any of the special features of the collar, or to fire the darts. And Sontag's results could not be duplicated. The following day Phillips tested the triggering transmitter Sontag had brought and found it to have a range of only a tenth of a mile. "From that distance," he told me, "you could lasso a wolf."

There had been, it seemed, a very basic misunderstanding. Phillips had always planned to track the wolves from the ground, because of the expense of hiring a plane. The 3M collars had been intended for use mainly from the air. Most radio collars could be used either way. But the power needed to run the computer in the capture collar was so great that the signal had to be turned down or the batteries would go dead in a matter of weeks.

The Fish and Wildlife Service had gambled and lost. Parker and the administrators in the southeast regional office had assumed the capture collars would be tested and working well by May 1987, as the engineers at 3M had promised. From the standpoint of public relations, the delays had the potential to be disastrous. That summer Taylor had been quoted several times in local papers as saying the release was imminent. But the public perception of the project was hazy; people who asked about the wolves often expressed surprise when told that the animals had yet to be freed. (Frequently people also asked Phillips how the "red foxes" were doing, a misimpression that showed, perhaps, a general lack of understanding about the significance of the project.)

The local hunters were not so charitable. Their animosity toward the Fish and Wildlife Service had not diminished, and since spring they had been lobbying heavily to force a change in the rules about hunting with dogs. In past years they had trained their teams during the summer, but now dogs were banned from the refuge except during fall hunting. Recently a group of men from Manns Harbor had asked Congressman Walter Jones and Senator Jesse Helms to force the service to drop the new rules.

Jones and Helms were two of the most powerful men in Congress, and they had contacted the service regional office in Atlanta. Administrators there had refused to relax the regulations, much to Taylor's relief. But until the wolves were freed, there was still a chance, however small, that the hunting group could throw up some roadblock that would keep the animals from being released at all. The course of events was out of Taylor's and Parker's control, and frighteningly so. Everything was riding on the increasingly doubtful ability of engineers at 3M to make the capture collars work well enough to be tracked from the ground.

On Wednesday, September 9, Parker appeared in Phillips's office with a look of determination on his face. "We're going to let some wolves go this weekend," he said.

"It would be nice," Phillips replied.

"Oh, there's no question this time," Parker retorted. "There are going to be some wolves out of their pens this weekend."

The previous three weeks had brought added frustrations. Engineers from 3M had returned to the refuge to test the capture collars, which once again proved to be unreliable. Deer season was due to open in mid-October; the wolves needed to be released as soon as possible so they would have time to learn to hide when hunters approached.

Although he was not saying so publicly, Parker had decided to free the wolves at the South Lake pen even if it meant fitting them with both the 3M collar and a conventional radio collar. The two collars would be cumbersome, but Parker had been assured by gray wolf biologists that they would not constrict the movements of the animals or affect their ability to hunt.

Sontag, Juneamann, and a design engineer were due to fly in from Minnesota with two revamped collars that week. They arrived Friday morning, only to discover a problem with the mechanisms designed to fire the darts. But the range of the radio signal had increased—it could now be picked up by ground receivers a mile away—and the engineers corrected the malfunction with the darts overnight. By Saturday afternoon a brief round of tests had showed the collars to be working properly. No one on the refuge staff placed much faith in the tests. Nevertheless, on Sunday, September 13, the capture collars would be put on the South Lake wolves. The pen door would be opened the following day.

Early Sunday morning the biologists and engineers met at a boat ramp just off South Lake and began ferrying equipment to the pen. Parker had slept fitfully, waking every twenty minutes or so to check the clock. Both he and Phillips were more tense than the rest of the staff had ever seen them.

A videotape of that morning shows the South Lake female, thin and big-eared, lying on the roof of a wolf house, her topaz eyes narrowed and blinking in distress, her jaws hidden by a blue cloth muzzle. Lucash crouches above her, holding down her quivering shoulder. Larry Cooper monitors a rectal thermometer. At one point the wolf's temperature elevates to 105 degrees, and Roland Smith appears and begins sponging cold water onto her belly. Although no one says so, the thought of losing a wolf to stress-induced heat stroke is frightening.

The last shot of the wolf shows her pinned to the house with two wide, shiny strips around her neck: the 3M collar and, above it, a standard radio collar. The two look bulky and uncomfortable, especially at the front of the animal's throat, where the box containing the 3M computer rides.

Neither Parker nor Phillips relished the notion of leaving both collars on the wolves. After the pen had been cleared and the equipment loaded into boats for the trip back, the two men sneaked up to the pen for a final glimpse of the wolves. At their approach the animals dashed to the rear fence, where they sprinted back and forth as if looking for a way out. The biologists left, satisfied that the two collars would cause the wolves no hardships.

"Today was absolutely the most exciting time," Parker crowed. "My goodness, I'll tell you, there were days I thought it would never happen."

A victory breakfast in Manteo. Outside, rain spilled from gutters in long streams. Phillips and Lucash had stayed on the lake to move the houseboat to a cove south of the pen and were undoubtedly getting drenched. The rest of the group had returned to town. I found Parker and Juneamann eating together, their faces aglow, all tension between them gone.

"I'm just so relieved to have that done, and without a hitch," Parker said. "It couldn't have gone better, though I did get a little nervous when that female started to overheat. If she had dropped dead on us, you would have had to bury me too."

"It was something to see the collars actually on wolves," Juneamann interjected.

"That's a wonderful package, very impressive," Parker said. "The capture collars are going to have so much benefit that I think it'll easily outweigh any problems caused by the delay. My word, yes."

He looked straight at me, his brow wrinkling happily. He seemed to have forgotten the anxiousness that had plagued him all summer. "So what happens now?" I asked.

"We'll let the South Lake pair go tomorrow, of course, and then I'm told there's a seventy percent chance that we'll get the other collars by October first."

"And if you don't?" It was only Parker and me in the conversation now. Juneamann concentrated on eating.

"We're optimistic. We've made a lot of progress and things have worked out fine so far."

"But if you don't get them?"

"We'll hold the wolves in pens until January. October first is absolutely the last date they can be released." Parker paused. "We just all feel—Roland and Mike and I—we strongly think they shouldn't be let out without at least two weeks left before hunting season."

"Will that be long enough to make them wary of hunters?"

Parker shrugged and half smiled. "No one has any idea," he said.

Monday, September 14, was a bright fall day. Shortly before 9:00 A.M. the principals of the red wolf project clambered into a launch for the culmination of ten months of work. Parker was mostly silent, but Taylor, Phillips, and Smith joked together as usual. "Warren, this is a big step," Phillips said jovially. "Are you absolutely, positively sure you want to let them out?"

The release was to be carried out with few people around. Phillips and Smith unloaded their gear at the houseboat, where they were to begin monitoring the wolves' movements immediately. Lucash piloted Taylor and Parker to the trail that led to the pen. He would remember later that Parker seemed distracted, as if he could not believe what he was doing.

The trail, underwater from a recent rain, was slippery and difficult to negotiate. Moving with great care, Parker and Taylor lifted a seventy-pound deer carcass from the deck of the boat and carried it into the woods. They set it near a patch of shrubs ten yards from the pen, where the wolves could easily find it. Then they returned to the boat.

Taylor shouldered a video camera, while Parker picked up several more chunks of deer meat and made his way back up the trail. The animals were to be fed just as if it were a normal day, so they would be under no pressure to leave the pen. Parker stumbled, caught a branch, and righted himself. Black holes gaped in the ground, showing where others had fallen to their hips in muck. "All I need now is to fall and break a leg," he said under his breath. Halfway to the pen he heard the wolves splashing through puddles, as if they were running hard along the back edge. From the sound of it they were moving well, despite the stress of the previous day. He positioned himself so he could toss the chunks of meat over the plywood screen. Although the wolves obviously knew he was there, he would keep up the feeding ritual to the end.

The meat dropped heavily to the ground on the far side. He reached into his pocket and extracted a key, then glanced back toward Taylor before unlocking the heavy chain that held the pen door. He had to nudge the door slightly to swing it open. Flashes of tawny brown were visible through the brush as the wolves bolted along the rear fence. He took a breath and walked away as quickly as he could. Forty yards down the trail he paused with Taylor to look back at the gaping door.

Lucash had already assembled a small telemetry antenna when Parker and Taylor came stumbling back to the boat. It was nine-forty-five.

"I did it," Parker croaked nervously. "I did it."

And nothing happened. For days the wolves did not venture from the pen. Phillips had scheduled tracking crews to work around the clock the first week, and since the wolves did not move far enough for their compass bearings to change, no one had much to do. But the weather stayed fair, with light wind. At night the trackers watched for comets among thousands of wavering stars.

It became a sort of game to listen through headphones and try to sketch an electronic portrait of the wolves. The standard telemetry collar contained a special activity switch; the pattern of beeps would change when the wolves were moving and stay the same when they were still. If a wolf was eating, as I sometimes imagined, the beeps would come unevenly for several seconds when it bent to pick up food, then remain fairly steady as it chewed. The thin electronic filament left me dissatisfied; I wanted to know more, not just about the wolves' movements but about what they were thinking and feeling.

Tuesday evening after dark Lucash and I anchored a skiff on a point of land northeast of the pen to take a different compass bearing on the wolves. In front of the skiff spread the darkened surface of the lake, like a thin, undulating membrane stretched over the earth. The stars became more brilliant with each passing hour. Nothing stirred. "Sure wish I knew what the wolves were doing," Lucash said after a long period of silence.

"Sure wish I knew if they've even come out of the pen. What do you think, Chris?"

"About what? The wolves leaving the pen? Yeah, they've probably at least stuck their noses out the door. Maybe they've even found that deer Warren and John tossed back up in there. But I bet they're real skittish. It's like Warren said—all the bad things they've ever known have come through that door. Meaning us."

By Wednesday afternoon Parker had begun to fret. Twice he had dreamed that the door to the pen had somehow swung closed, although he had locked it open with a heavy chain. Now, with no discernible movement for more than two days, he was downright rattled. "I just wish I knew what they were doing," he said repeatedly. "I think tomorrow, if they still don't go anywhere, I'll take the canoe up in that cove, real quiet, and see if I see anything, maybe take a quick look around."

Thursday morning around six o'clock, Phillips and his tracking partner noticed that the signal for the female had drifted slightly to the northwest. The movement was enough to set Parker's mind at ease. By Friday noon the female had wandered about two miles away from the pen, far enough to be out of range of the houseboat. To map her location, the trackers had to motor west around two points of land, easing their way past crab pots and stumps. She remained in the same area for the next three days. The male did not follow her.

What could be going through the female's mind? Was she confused and depressed? Lost? Or had she caught a deer? I knew it was anthropomorphic to think the wolf would react to her new-found freedom with the same emotions I might experience. Still, sitting alone in a boat near midnight, listening to a faint, fluctuating beep, it was easy to imagine what she might feel as she wandered through unfamiliar thickets. This was the wolf I had seen in the videotape quaking beneath Lucash's hands. Over the past eight months she had been outfitted with radio tracking equipment, implants to make her sterile, tracer chips to mark her scat. None of those devices, whatever their value, could tell me anything about what mattered most, the strength of will and spirit she would need to make a life in the wild. I could only pray that her sacrifices, physical and mental, would ultimately be worthwhile.

The nights grew darker and starless as the week ended. The spired trees along shore became invisible in a patchy haze. On the eighth day after his release, the male finally left the vicinity of the pen and moved west, then north. As he meandered through the swamp, the female drifted south. She arrived back at the pen the following day.

Phillips and Lucash had worked sixteen straight days. Both badly needed a break, but Parker was worried that the wolves would not be able to find each other, and Phillips had promised to watch them closely.

Wednesday afternoon, ten days after the release, Lucash motored around the narrow spur of land to check the location of the male. The all-night tracking sessions had ended a few days before, but he was still cranky from exhaustion. The boat had not been easy to start, probably because of the high humidity, and a telemetry receiver was acting up. By the time he anchored the boat and set up the telemetry antenna, he was in a foul mood.

The male had been staying close to the water, and Lucash thought he knew where to find him. As he dialed the receiver to the frequency in the wolf's collar, the radio signal came booming out, stronger than Lucash had ever heard it. He was looking at the shoreline when suddenly he noticed something. "It was right in some trees, just a small movement. A wolf nose. He stuck his head out of the brush and touched his nose to the water. Just in and out, like that, for a couple of seconds.

"I don't think I would have seen him if I hadn't been looking right there. I honestly don't think I would have known it was a wolf and not a deer. He almost moved like a deer."

It was the first time in a century, maybe longer, that anyone had seen a wolf roaming free on the North Carolina coast.

Meant to Be Wild

V

A Chorus of Wolves

Shortly after four o'clock on the morning of October 2, Chris Lucash and George Paleudis rumbled past the caretaker trailer at Pole Road in a telemetry truck. It was the day after the day they had awaited all summer; not just two but now eight wolves were free to wander the swamps of the Alligator River. Despite misgivings, Phillips and his staff had fitted the wolves at Point Peter, Pole Road, and Phantom Road with capture collars two days before. Just in case, they had also put conventional radio collars on two of the males. Then, on October 1, they had opened the door to all three pens.

The preceding weeks had been full of uncertainty as the engineers at 3M rushed to build six more functioning capture collars by Parker's deadline, the end of September. No one was sure how long the collars would continue to work, since their batteries were operating at peak capacity. But the final barrier to the wolves' probational freedom had been removed. From this point on, every milestone in the project would be measured from autumn 1987.

Lucash and Paleudis were too numbed by exhaustion to feel the elation they had expected. Paleudis was a wiry man, small but strong, energetic, and a bit brash. He had graduated from college in June and immediately gone to work as a wolf caretaker. On this morning both he and Lucash were exceptionally cranky. They had worked until eight o'clock the previous night, and then they had been too keyed up to sleep. Easing their way around ruts and potholes in a truck with a broken muffler had done nothing to sooth their weariness or improve their moods. As Paleudis pulled up to the intersection of Pole and Phantom roads and turned off the engine, Lucash motioned to the telemetry receiver. "Help yourself," he said. Paleudis made a face, slipped on the headphones, and began rotating the large antenna on top of the truck.

At 1:00 A.M., when Phillips had last checked the wolves' location, all the animals had been in their pens. The signals from the wolves in the Phantom Road pen, a half mile to the west, should have been easily audible. Paleudis could hear nothing. He checked the frequency settings and antenna connections. Nothing. "Something's weird," he said. "They're not there."

He pulled off the headphones and handed them to Lucash, who was more experienced as a tracker. Lucash tuned the frequency dial to the setting for the Phantom Road male and spun the antenna slowly in a circle. To the northeast he heard a faint high-pitched pulse. "They've moved," he said, surprised.

If the Phantom Road wolves had begun to explore, perhaps the wolves in the Pole Road pen had, too. Intrigued, the men turned the truck east, retracing their way. The signal for the Phantom Road male grew stronger; although Lucash could not tell for sure, it seemed to be coming straight from the north. Paleudis eased toward the caretaker trailer until Lucash motioned him to stop. The Pole Road pen was directly north, but the signal for 211, the Pole Road male, was strong from the east, from the road in front of the truck. "I don't believe it," Lucash said. The Phantom Road pair had run the Pole Road wolves off their home turf.

Both men sat for a minute in surprised silence. In captivity the frisky Pole Road male had seemed much more dominant than the Phantom Road male. But with the opening of the pen doors all rules of behavior had been redrawn. "Look down the road," Lucash said. "I bet the Pole Road wolves are sitting in the middle of it. I bet we see 'em." In reply Paleudis started the truck and eased slowly ahead. The vehicle lurched and bumped along, its headlights flashing into the trees. Paleudis had driven only a few yards when a pair of topaz eyes appeared. It was the Pole Road male, running right at them.

"Chris," Paleudis said, "I don't feel very good."

The wolf was clearly not frightened of the headlights or the unmuffled roar of the truck. Behind him the trackers could see the fainter glint of the female's eyes. She seemed to want to follow her mate, but she was hanging back, unsure. If Lucash and Paleudis had been hunters gunning for wolves, they would have had two easy marks. Fifteen feet from them, the male turned and trotted off toward a broken-down Ram Charger, parked fifty feet away. The female followed.

Paleudis turned off the engine and sat, silent, incredulous, and very, very tired. If the wolves had retained a shred of the species' natural wildness, he thought, they would have bolted into the underbrush at the first sign of humans. As the animals neared the Ram Charger they reversed direction and trotted back toward Lucash and Paleudis. Twenty yards from them the male suddenly slowed, then turned and ran back up the middle of the road.

It was the last thing the young biologists had expected; they had not the faintest idea what to do. The animals continued to trot between the two vehicles, cautious but more curious than afraid. Lucash muttered cynically that they looked like pet dogs. Finally the male, wandering toward them, moved into some grass at the side of the road. The female refused to turn aside. She walked up to the left bumper of the truck, sniffed it, crossed the road, and came alongside the door on the passenger side.

Meant to Be Wild

Lucash was ready. He flung open the door and leapt out, screaming and clapping his hands. The female, caught off guard, skittered sideways into the cane grass on the shoulder of the road. It was too dark to see her, but Lucash could tell by the rustling stems that she was crouched within ten feet of him. He started toward her, kicking hard into the grass, and heard the soft sound of stalks being crushed as the female lay down. She reminded him of a dog that had been beaten until its spirit had shattered beyond repair.

Over the next two weeks, reports of wolf sightings poured into the refuge office:

On October 3 Phillips and some trackers saw the male from Phantom Road running down the middle of a road. He ran clumsily and refused to duck into the bushes, although the trackers followed him for a hundred yards with the horn blaring. "It was pathetic," Phillips said. "We could have run him over."

A man walking along a dirt road saw a wolf briefly and reported that the animal disappeared into the brush very much like a deer.

On a road deep in the refuge, a man in a truck noticed a wolf running forty yards in front of him. He followed it for a mile, until the wolf collapsed on the roadside, exhausted. It lay there panting while he pulled up next to it.

A couple driving on the highway south of Manns Harbor saw a wolf in the middle of the pavement "acting like a lost German shepherd." They stopped, but it refused to move until they honked their horn.

A hunter was walking down a path when a wolf swam across a canal, climbed the bank, and strolled up to him. The regular hunting season did not open until the middle of October, but the refuge held a brief season for muzzle loaders earlier in the month. The man assumed the animal coming toward him was a pet German shepherd, since it had a large collar and seemed so friendly. When he noticed the darts in the collar, he realized he was eye to eye with a wolf. He shouldered his gun and began to yell. The wolf cringed and walked away sideways, its tail between its legs.

As far as Phillips could tell, the sightings included every one of the six wolves just released. The animals did not seem to understand that trucks and cars could harm them. More discouraging, a few of them were not at all frightened of direct contact with people. Maybe they could be trained to be warier. Lucash, Paleudis, and Phillips started carrying rifles loaded with cracker shells so they could haze any wolves that refused to move off the roads. The male at Point Peter had taken to lingering along the shoulders of the main highway. Unless he could be taught to keep off the pavement, it was only a matter of time until he was hit.

And it was only a matter of time until hunters from Manns Harbor began to encounter wolves. When the regular season opened October 13, Taylor had the roads through the preserve posted with small pink signs.

"RED WOLVES roam free on Alligator River National Wildlife Refuge," they read. "Please report sightings." Below was a warning that anyone caught harming a wolf could be fined $100,000 and sentenced to a year in jail.

Phillips grew noticeably tense. It seemed to him that the wolves had started moving off the roads more quickly when they were approached by cars. The frequent sightings continued, and it was conceivable that the hunters from Manns Harbor would make good their promise to put out traps or deer carcasses laced with strychnine. Still, he was gratified by the number of calls he received from local residents who had seen the wolves. Most of the callers sounded concerned and excited to have spotted such rare animals. Toward the end of the first week of hunting he had begun to relax a little. For the time being, the delicate balance of tolerance between the wolf and the human population of the Dare County peninsula appeared to be holding.

But the largest problems still lay ahead. In the excitement of the latest release, the trackers had not had much time to monitor the activities of the wolves at South Lake. On October 15 the male was spotted by a motorist just north of Highway 64, the major road west from Manns Harbor. To reach the highway he had walked seven miles around the lake. When he encountered a canal a few yards north of the road, he stopped. For several days he lounged on the canal bank, picking at a dead fish and ignoring the trackers, who fired at him with cracker shells to try to drive him back into the woods. On October 18 he began moving slowly east along the canal toward Manns Harbor. Four miles separated him from the houses on the outskirts of town.

Phillips, though concerned, had another, more immediate crisis at hand. The female from the Phantom Road pen had left her doting male behind and taken to wandering. By October 19 she had meandered south for fifteen miles. Early that morning she crossed the refuge boundary and ventured onto private farmland.

The terrain was no different, except a bit drier. No livestock ranged there that she could kill. But she was off federal property. Phillips and John Taylor decided she would have to be captured and returned to Phantom Road. Parker, who happened to be in town, concurred.

Early that afternoon, Parker, Taylor, Phillips, and two refuge staff members set out with nets, catch poles, telemetry sets, and, in case the capture collar did not work, two air rifles loaded with tranquilizer darts. Lucash and Paleudis had taken a rare day off to go fishing. Before driving south, Taylor posted another staff member, a Manns Harbor resident named Jim Beasley, in a truck along Highway 64. No one really expected any trouble from the South Lake male. But if he began to move toward town, Beasley was to radio for help.

Shortly after seven o'clock Lucash and Paleudis stopped by the refuge office to make a phone call. The two-way radio was on, and they could hear Beasley calling Taylor and Phillips, who were apparently out of range. Lucash grabbed the mike. "What's going on, Jim?"

"We got, uh, a situation here. He's at the post office."

In the previous three hours, the South Lake male had walked to the fringe of Manns Harbor along a ridge north of the highway. Beasley could do nothing but call for help and follow. By seven-thirty the wolf was walking between houses, sniffing at flower gardens.

It was just before dark. Lucash keyed the mike and called loudly for Phillips. The radio in the office was more powerful than the one in Beasley's truck, and there was a chance the capture party would hear it. No reply. Lucash reached behind Phillips's desk and grabbed one of the triggering transmitters used to fire the darts in the 3M collars. He called into the mike a last time. To his surprise Phillips answered, his voice urgent and broken by static. The group had given up trying to catch the female and had started back. They were a little less than an hour from Manns Harbor.

Jim Beasley was a friendly, talkative man who, like most Manns Harbor residents, was well known to his neighbors. As he sat in the parking lot of the post office holding a small telemetry antenna, he could not help but attract attention. By the time Lucash and Paleudis arrived, three local men had stopped to see what Beasley was up to. One claimed to have seen a wolf on private property. The men leaned against their pickups, muttering among themselves. Lucash started to set up the triggering transmitter, but changed his mind. Neither he nor Paleudis believed the capture collar would work, and there was no sense trying to fire the darts in front of an audience.

By then the wolf was circling through a pecan orchard across the road. Beasley drove slowly out of the parking lot, turned into the driveway of a house next to the orchard, and let Paleudis out. Paleudis carried a hand telemetry set—a small receiver, a set of headphones through which he could hear the beeping radio signal, and a two-foot-high antenna that looked like a chopped-off television antenna. He inched his way around the side of the house, holding the antenna in front of himself like a divining rod. The signal grew stronger. In the weak light of dusk he could see the wolf running back and forth along a fence on the edge of the orchard. He wished he had carried the triggering transmitter; the wolf was only fifty yards away. Suddenly the animal ducked through a gate and disappeared into a trailer park. Paleudis groaned.

He walked back to the post office to find that the capture party had arrived. Phillips was talking tersely to Beasley. Taylor, though calm, was clearly worried. The Fish and Wildlife Service had no legal right to set foot on private property without permission, even to catch a wolf, and Taylor did not want anyone to get hurt. Parker wandered nervously from man to man, asking each what he thought. No one had any clear plan. Ideas tumbled out to be picked up, turned over, dropped, and picked up again. All agreed that they needed to stay poised and quiet; they needed to catch the wolf with little ado.

The easiest thing would be to try firing the darts in the 3M collar. Parker and Lucash crossed the road on foot and made their way to the fence on the edge of the trailer park. Lucash propped the triggering transmitter on a post and dialed the code to open communication with the computer in the collar. "Oh my Lord, I hope this works," Parker prayed. "I just wish it would work, if only this one time." The computer was supposed to respond with a long beep, signaling its readiness to fire the darts. But the transmitter remained silent. They would have to try darting the wolf with air rifles.

Paleudis and Beasley, meanwhile, had followed the wolf into the trailer park. Walking along with the headphones on, Paleudis suddenly picked up a strong, pulsing signal from an open pasture. If they could corner the wolf in the pasture they might get several clear shots at him.

Now the men split into two teams, each with a radio tracker, a marksman, and two other people to run interference. Parker and Taylor carried the guns. Both were skilled marksmen, but both were also nervous. If they hit the wolf in the head or the belly, the tranquilizer in the dart would probably kill him.

The teams moved quickly through the pasture, toward a small pond where the wolf had stopped to rest. The wolf, sensing trouble, began to pace. The radio signal that betrayed his location glided back and forth, a restless ghost. Parker, sneaking through some tall grass, thought he saw movement in front of him. Squinting hard, he made out a dark profile within fifty feet, a reasonable shot. He brought the gun to his shoulder and fired. The wolf, untouched, skittered to the west.

Taylor, Parker, and most of the other men ran back to the trucks, hoping to catch up with the wolf and drive him farther west, away from town. Beasley and Paleudis, left behind, looked at each other, bewildered. With both dart guns now hundreds of yards away, there was nothing for them to do. "I just hope the wolf doesn't come back this way, because we're not going to be able to do anything but sit and watch him come," Paleudis grumbled. He paused to listen to the signal. "I just hope . . . Hey, I think he is; I think he's moving back here. Hey, you guys, he's coming this way, fast."

In a sudden burst the wolf dashed for cover, running far around Paleudis, darting into the pecan grove. Beasley had called for help on a two-way radio he carried, and Lucash and Phillips pulled up just as the wolf bounded into some brush next to a house. They were forty feet from him, with no gun. Phillips reached for the triggering transmitter and dialed several commands to the computer in the 3M collar. No response. He got out of the truck cautiously and skirted the house with a large salmon net, thinking the wolf might see the net and freeze. As Parker and Taylor drove up, the wolf exploded from the brush and turned toward an open sand quarry. Paleudis sprinted after him, with Parker and Beasley at his heels.

The quarry was east of the pecan orchard and farther off the main road. Slowing down, Paleudis could see lights from several trailers on its edge. Parker and Beasley caught up, breathing heavily. The signal moved toward the trailers, and half a dozen dogs began to bark.

The night had taken on a surreal quality. What had been a small knot in Paleudis's stomach solidified into a dull ache of dread. The three men walked quietly toward the trailers, catching glimpses of the wolf as it paused to sniff at a bush or a tree. As Parker neared a trailer, the back door flung open with a loud bang. "Who's out there?" a man's voice called gruffly.

"It's just some of us from the Fish and Wildlife Service," Parker drawled in his most friendly tone. "We heard there might be a wolf roaming around back here, and we just wanted to check."

The man peered out into the darkness, grunted once, and said he had a brother who worked for Fish and Wildlife. He admonished Parker to be careful—didn't he know he could get shot, sneaking around like that?—before ducking back inside. Paleudis sighed in relief.

A dirt road ran from the quarry to the main highway. As the men moved away from the trailers, Parker saw the wolf standing in the middle of the road, looking back at them. He drew up his gun and fired. The wolf bolted, crossed the highway, and disappeared into a thickly settled section of Manns Harbor.

Harold Butler was settling down to watch television about ten o'clock when his teenage son glanced out the window and saw a strange profile in the front yard. "Daddy," the boy said, "there's a red wolf outside."

"No there's not," Butler said, even as he got up to look. But there it was, standing in the driveway sixty feet from his doublewide trailer. "Huh," he said, opening the front door and stepping out. The wolf looked at him and walked slowly away, toward the road. With a flashlight, Butler could see the wide collar with the two darts sticking up. It was a red wolf, all right. His three hunting dogs were barking loudly, and over the racket Butler could hear voices and a truck. Hesitant, he started down the drive. A gunshot sent him scrambling behind a bush for cover.

The capture party, in disarray, had begun cruising the roads through town, trying to stay on public right-of-ways instead of running through yards. Occasionally, Parker ventured down driveways or back alleys, getting off another several shots. Once he thought he had hit the wolf in the base of the tail; the animal seemed to slow down a little, but still eluded them. All the group could do was ride through town and wait for a stroke of luck. And luck had appeared, momentarily, when the wolf paused in Butler's yard.

As Butler recalled the incident later, the men from the Fish and Wildlife Service tried to downplay the seriousness of having a wolf loose in the neighborhood. "They kept telling me things like, this animal's almost like a pet and we know where it is all the time. They gave me the impression it had been hit with a tranquilizer and it'd be down any minute. Then they started coming through my yard with their . . . " he stopped to grope for a word, "their detectors and stuff."

For the moment, though, Butler had no choice but to go back inside, grumbling about what he might do if the wolf threatened one of his children. The capture party checked his yard and left. But an hour later Butler glanced out his window and saw the wolf sniffing his daughter's bicycle, which was parked ten feet from the trailer. Angrily he grabbed a revolver and burst out the front door, firing in the air as he stepped off the porch. As the wolf fled he went back inside and called the sheriff.

Phillips, Lucash, and Paleudis were riding together when they heard the shot. "That's it," Phillips groaned, "that's it. We're looking for a body."

Taylor was the first back to Butler's trailer. He and his wife were on the front steps, waiting for the sheriff. "What happened?" Taylor asked as calmly as possible. "Did you hit the wolf?"

"No I didn't hit the wolf," Butler retorted. "I didn't try. But I damn sure might if he gets near my property again."

A few minutes later two deputies arrived. Parker and Taylor, embarrassed and worried, apologized for the commotion. The deputies' presence, they knew, would draw even more attention. But with the officers standing by, they could stalk the wolf aggressively, even on private property. They regrouped and hatched another plan. The wolf was holed up in a yard next to Butler's. Parker and Taylor, each accompanied by a tracker, would try to box the animal in, until he could be darted. Everyone else was to stay on the road. Tired and discouraged, the group had little faith that the plan would work, but they had no other ideas.

Parker and Taylor moved toward the wolf carefully, shining flashlights on bushes, fences, outbuildings, cars. If the animal had been hit by a dart, he showed little sign of it now. The signal drifted through a side yard as if he were sizing up the advancing men, looking for a way out. As the biologists closed in, he burst forward, slipping neatly across the road, running toward Pamlico Sound.

Paleudis, standing near a patrol car, saw the wolf move into a grassy yard. "He's here," Paleudis hollered, "you guys, he's over here." The shouts brought Parker running.

Near a grove of trees the wolf paused to look back at his pursuers. As he turned, Parker was on him. The dart gun fired a final time; the wolf flinched, fled through the trees, and collapsed near the water.

"I'm amazed someone didn't get hurt."

"My goodness, yes. There was a lot of potential for trouble."

"But it was our worst-case scenario, and we got him, we got him."

No one had much gumption for work the next day. The South Lake male had regained consciousness and was curled up in a kennel with a bowl of dog food. The day before his release he had weighed seventy-two pounds, but in four weeks he had dropped to sixty-one pounds. Parker and Phillips came to work prepared for a flood of phone calls, but received only a few, mostly from reporters. No one pressed them about why the capture collar had not worked; with the wolf in custody it did not seem to matter.

The mood of the wolf crew was grim, and there was still the problem of the Phantom Road female, loose on private property to the south. Phillips had hoped she might wander back north, but she had settled into thick forest and not moved even a quarter of a mile. There was nothing to do but try to corral her too.

Around noon the following day another capture party assembled near a dry woodland twenty-five miles south of Manns Harbor. The group included the people who had worked to catch the South Lake male, plus a couple of extra refuge staff members and me. The wolf was holed up in an old cypress grove that had been logged forty years before. In the decades since, gum, oak, and bay had pushed up past the cypress knees and the stumps that had been left to rot. Looking into the woods, I could see skeins of cat brier and stands of a thorny plant known as devil's walking stick. I wondered how an animal could make its way through such vegetation without getting badly cut.

Farm roads edged the grove on the west and south. From the strength of the radio signal, it seemed likely that the female was resting only a hundred yards from where the roads crossed. "This is good; this could be real easy," Phillips said when all the members of the party had arrived. The rest of us looked dubiously at the wall of undergrowth that awaited our attack.

"Here's what I thought we'd do," Phillips continued crisply, dropping to a crouch and picking up a stick to draw in the dust of the roadbed. He made an X to mark the intersection. "We'll send several people in this way," he pulled the stick along one road, "and some more from this other road. They'll form a semicircle behind the wolf and try to drive her out so we can get a shot at her." It sounded like a reasonable plan, although any number of things could go wrong.

I pulled on a jacket to protect against briers and picked up a hand radio. Lucash and I were to work our way in from the west. Before we set out, Phillips issued a final warning over the radio. "Be careful to watch for holes where the cedar stumps have rotted away. They can be deep and treacherous." He did not have to remind us to watch also for poisonous snakes.

Bay and oak swayed high above us, and the air smelled richly of humus. We thrashed our way into the woods, jogging in short bursts to break through bushes and vines. Fragments of sunlight brightened the orange pine straw that was scattered thickly across the uneven ground. I jumped onto a log and followed it through brush as far as I could, then hopped off and chose another, feeling clumsy and slow. Fifty yards in I was directionless; if not for the distant breaking light that showed where the road had been cleared, I would not have known west from south.

I pushed through the undergrowth for what seemed a long time. Lucash, wearing a telemetry set, was only fifteen yards behind me, but I kept losing track of him through the branches and vines. A bush crackled in front of me, and I saw Beasley waving his arms to get my attention. With two other people beyond him, the semicircle was in place.

"She's fifty yards that way," Lucash called, slipping off the headphones and waving one arm. "Go to your right, slow."

If the wolf decided to run between us, there would be no way to stop her. We turned and began working our way toward the road, drawing the circle closed. I moved in front of Lucash slightly, stepped off a log, and fell through humus almost to my waist. By the time I had extracted myself neither Lucash nor Beasley was in sight. I started forward hesitantly, but stopped when Lucash appeared through some vines. He waved to me to backtrack around him.

"She's right . . . Dammit, the signal keeps fading. She's right in here." His voice trailed off as he stepped down from a small rise toward a tight ball of myrtle. I ventured closer, brushing away branches. The slivers of sunlight had begun to play tricks on my eyes. In that one small patch of roots and limbs were dozens of places where a cinnamon-colored wolf could hide. I combed carefully through some bushes, wondering what I should do if my hand fell on wolf hide.

"Here she is. I see her." Lucash was half diving under branches and downed logs. "Move around slow—real slow—and help me block her off."

The Phantom Road female had flattened herself into a hollow formed by the exposed roots of a large oak. Her body was pressed into a bank of dirt directly beneath the trunk. I moved slowly into position beside Lucash and peeled off my jacket, holding it in front of me, hoping to block her only avenue of escape. She had tasted freedom, and she could still bolt. Instead she buried her face in the roots and turned her back to us. "Steady girl," Lucash said edgily.

But she had no intention of moving. As we waited for the nets and catch poles and, finally, the kennel—the trappings of a life she had supposedly left behind—the wolf began to shake, and a deep sadness descended over me. She was beautiful, with her short, tawny fur and her alert, elegant ears. Her instincts had served her well; without the trickery of telemetry we never would have found her.

The men arrived with the kennel. Lucash and Phillips pulled the wolf from beneath the tree, staying clear of her jaws. They dumped her into the upturned kennel, snapped the door closed, and hauled her unceremoniously to a waiting truck.

That fall and winter of 1987–88 was one of the hardest periods the men and women who worked on the red wolf reintroduction would ever face. The recapture of the Phantom Road female stripped us of any naïveté. We had known all along that the wolves might wander, and that under law their movements would have to be restricted. What we did not anticipate was how quickly one would leave the refuge, and how cunningly wild she would behave when we arrived to bring her back.

I was not part of the full-time staff for the project, just an occasional helper. Within weeks after all the wolves were released, however, I had begun to wonder how much the reintroduction could accomplish. The cypress grove where the Phantom Road female had chosen to hide was dry, full of good cover, and right next to farm fields that attracted rabbits, rodents, and deer. In contrast, the land within the Alligator River refuge was some of the swampiest, most inhospitable in the region. It was the kind of land the wolves themselves would probably avoid if given their pick of places to roam. Was it fair to restrict them to such marginal habitat and expect them to thrive?

Yet, Phillips reminded me, the wolves were being granted a chance to live in a facsimile of freedom, which was more than they had been granted in the East for a hundred years. By December they were seldom spotted from the roads. And slowly they began to show evidence of being able to kill prey larger than rabbits and raccoons. The first clear sign came in mid-November when Phillips, sorting through samples of wolf scat, found the remains of a hoof from a young deer. The scat was from the Pole Road male.

Shortly after their release, the Pole Road wolves had reasserted themselves and driven off the Phantom Road male and his wandering mate. They took to traveling south of their old pen and swinging back north in a regular circuit, sticking to a well-defined home range. The change in their behavior was nothing short of remarkable. They were acting like wild wolves.

The toothless male at Point Peter and his mate seldom strayed far from their pen. The wolves at South Lake were not so sedentary. After his capture in Manns Harbor, the male was taken back to the South Lake pen, locked in for ten days, and let go. Three weeks later, trackers found him within a mile of Manns Harbor, looking thin and worn. They set out traps and caught him easily. A few days later they also trappped the female at South Lake, although she had never shown any inclination to seek out human settlement. Phillips replaced the two collars on each of the wolves with the radio telemetry collars he had favored using all along.

The South Lake wolves were taken to a marsh twenty miles south of Manns Harbor—far enough, Phillips believed, to keep the male out of trouble—and let go. In mid-December the female was found dead on a nearby beach. An autopsy concluded that she had died from an infection of unknown cause.

The Phantom Road male, abandoned by his wandering mate, stayed in the vicinity of Pole Road for most of October. Trackers frequently found scat and long scrape marks on the roads where he and the Pole Road wolves were vying for territory. On November 8, after the Phantom Road male had not moved for several days, Lucash and Phillips trapped him. He had been bitten in the neck several times during a fight. Without medical treatment he likely would have died.

After her capture in the thick forest south of the refuge, the Phantom Road female was held for several days in her old pen and released. Again she wandered far south, but this time she proved even more wily. When a group of people tried to surround her in the woods, she ran neatly around them and disappeared. Although somewhat chagrined, Phillips was pleased by her wild behavior. He and Lucash set traps for her and caught her two days later.

The Phantom Road wolves were reunited in captivity in mid-December. It is impossible to say how they may have reacted to each other, but one thing seems clear: neither was happy to be caged while the Pole Road wolves were running free. Over the next two weeks, the Pole Road pair visited the Phantom Road pen frequently enough to wear a path around the outside.

The day after Christmas, Phillips, Lucash, a volunteer named Marcia Lyons, and I drove to Phantom Road to check on the doting male and his wandering mate. The female's foot had been cut when she was trapped, and the wound had not shown signs of healing. Phillips, concerned that it might need stitches, had arranged for her to be taken to be examined by Larry Cooper, the veterinarian. It was cool and rainy—pneumonia weather, someone called it—but the biologists had celebrated Christmas together, and we were all in good moods. We reached the pen about midmorning.

The wolves were in one of the small wooden houses, lying tightly together. Lucash popped the roof off, shoved the male to one side with the end of a catch pole, and slipped the noose around the neck of the female. He remarked that she seemed unusually listless. As we grabbed her shoulders and began to pull her out we all gasped. Her left front foot was gone, and the leg had been stripped of flesh. All that remained were blood-streaked bones and tendons, swinging free.

"Bring her up, get her in the kennel," Phillips barked. We had frozen, but his words brought us back; we lifted together and swung her as gently as possible into the cage. Phillips latched the door, turned the kennel on its side, and leaned on it, breathing as heavily as if his lungs were about to explode. My own chest felt as if it had been rammed by a timber: I wanted to cry but couldn't find my way past the shock of what I had seen.

Later, Lucash would discover a bowed section of fence where the female had leapt against it, probably as a gesture of threat against the Pole Road wolves. One link of the fence was lined with bits of fur and flesh, dried by the cool winter winds. The female's foot must have gotten caught in the link, and the Pole Road wolves had pulled the leg through and attacked it.

Just then, though, there was no ready explanation. Lucash and Phillips picked up the kennel, carried it to the truck, and slid it into the bed. As I stood by the tailgate the Phantom Road female looked out at me, her almond eyes glazed and without spark. It occurred to me that I had never before looked a wolf directly in the face.

"Honey," Phillips said, his voice raspy with emotion, "you're outta here."

The Phantom Road female was euthanized by Larry Cooper that afternoon. Her value to the red wolf project had been in her wildness, and while she might have survived with three legs the biologists agreed that only the healthiest wolves should be part of the first reintroduction.

In the spring of 1988 the biologists released two more female wolves, a yearling and a two-year-old, to replace the South Lake and the Phantom Road females. They were chosen for release with the thought that younger animals might be less tolerant of people and cars. In mid-April, the Point Peter wolves began spending large amounts of time around the farm fields within the refuge, which were several miles north of their pen. For a week they marked the dirt roads in the area with large scat and deep, obvious scrapes. After that, the female kept to one small thicket, moving little if at all. It was classic breeding behavior. At exactly the same time, the Pole Road wolves stopped traveling through their territory and confined their movements to a dry hardwood grove, as if they were digging a den. No one had expected the wolves to breed the first season after their release, but Phillips, Lucash, and Paleudis grew cautiously excited.

In late April, the Point Peter female abruptly moved back to her old pen, as if she had lost her litter. Through mid-May the biologists watched the dirt roads carefully for the small round tracks of pups, but found none.

One morning Phillips was flying low over the refuge when he spotted the Pole Road wolves. The budget for the project had been increased, and he had started radio tracking from a plane several times a week. The wolves trotted briskly along a dirt road, paying little attention to a small clumsy animal that followed. From the air the pup looked young—too young, Phillips thought, to be away from the den—but it moved quickly, and the parents seemed to trust it would keep up. If any others had survived, they were out of sight.

Near the end of May, the toothless male at Point Peter was hit by a car and killed. Phillips decided to put out meat supplements for the Point Peter female, just in case she was caring for young. And indeed, one Saturday the female was seen moving slowly along the highway with a pup too small to travel through the brushy pocosin. Phillips blocked traffic in one lane, until the female could escort her pup safely across the road and into cover.

With the two births and the two new releases, the number of free-ranging wolves stood at nine. In mid-June, however, the male that had wandered into Manns Harbor the previous fall was hit by a car. A short time later, the Pole Road female returned alone to her old pen. She crawled inside, stuck her head halfway into one of the wooden houses, and died. When the biologists found her body, she weighed less than thirty pounds. An autopsy showed that she had contracted a uterine infection, probably after she had given birth.

She and her mate had been by far the most self-sufficient of the wolves in the wild, and the favorites of the wolf crew. No one had suspected she was sick. And, now the biologists could only assume that her single pup had starved.

"You've got to wonder," Chris Lucash said, "whether this project is ever going to go anywhere."

We were sitting in a tavern on a late June night lamenting the fate of the wolves. One of the young females, released at the South Lake site, had just been retrapped after she began making nightly trips into the small community of East Lake. She was being held in a new complex of pens. Her intended mate, the old Phantom Road male, had wandered from South Lake into the outskirts of Manns Harbor and had been recaptured without incident. Phillips leaned far back into a booth with a preoccupied look on his face. Paleudis stared dejectedly at the table.

"We talk to the locals," Lucash continued, "and they tell us we're crazy to put wolves out there because of the chiggers and ticks and deer flies. Then hookworms start showing up in the scat. They've probably got heartworms too. There are hardly any foxes out there; the locals say they can't survive. How the hell can we expect wolves to make it?"

"It isn't all that bad," Phillips interjected, "but I don't think this place is ever going to be anything more than a showcase."

I asked what he meant.

"There's no way we're going to have a viable population of wolves out there. We don't have the room. South Lake obviously isn't a good area for wolves, probably because there isn't enough prey. We've put four wolves there, and three of them have left. Based on what we've seen so far, these wolves depend on roads for travel. And on the south end there are more roadless areas.

"Now, maybe a pup that's born and raised in the wild will be able to move through the woods better. But for now we can only conclude that not as many wolves can be put on the refuge as we originally thought. We're going to have a heavily managed population, with animals being moved in and out for genetic diversity. We're probably going to have to supplement them, too, not just with food but with heartworm medicine, for example."

We were all silent for a few minutes. From the beginning we had known that the red wolf project was being watched by wildlife scientists nationally, and that a failure would not bode well for the release of predators elsewhere. What constituted failure, I wondered.

"If nothing else," Phillips said, "at least the public has learned a lot about the problems involved with saving endangered species. If this project goes belly up in five years, and people are asking why the federal government spent so much money on it, we'll at least be able to point to the educational value. I would argue that increasing public awareness is one of the most important things you can do to save endangered species."

As the biologists continued to talk, I thought about the original objectives of the project—the plans to use the capture collars, and to confine the animals strictly to the refuge. By spring the capture collars had all been replaced with conventional telemetry collars. I had never been comfortable with the concept of the capture collars; it was based, it seemed to me, on two contradictory promises. The devices were not supposed to change the animals, or inhibit their wildness, in any significant way; yet they were also supposed to keep the wolves under a constant thumb. One could not expect to have both.

When the wolves arrived in North Carolina in 1986, biologists had predicted that the refuge would hold thirty to thirty-five healthy, free-ranging adults and pups. The number had since been scaled down to between ten and fifteen. It occurred to me that Phillips was right; with so few wolves, the Alligator River project would be little more than the skeleton of a wild population. I had not expected the project to turn out like this. I had not anticipated that the species would be faced with so many demands to behave in a manner acceptable to human society. Neither had it occurred to me that wolves would be rotated in and out of the refuge like interchangeable parts.

During the time I had spent getting to know Phillips, I had heard him talk several times about conservation as a slow, incremental process. He was optimistic; he predicted that wolves would eventually live in North Carolina without an edict restricting them to public land, maybe even without radio collars. But I was growing impatient. In the bleakness of that night, I tended to doubt that the red wolf project would ever accomplish much more than it already had.

For the rest of the summer and into the fall the biologists concentrated on analyzing the telemetry data they had gathered since the release of the wolves. Much of their time was also spent caring for the animals in the new pens, built in a dry, piney grove known as Sandy Ridge. There was talk of making Alligator River a second major breeding center for red wolves. A number of animals had been brought to the refuge from Washington State on the understanding that they would not be released that year, if at all.

Parker had reduced the time he spent at Alligator River and was looking for other areas where the red wolf could be reintroduced. In the autumn of 1987, a pair of wolves had been flown from the Washington breeding center to South Carolina, where Parker hoped to start a new program on Bulls Island. He had decided that the best way to cultivate wild behavior in red wolves might be to free them at a very young age. The recovery team had agreed to let him experiment with releasing two-month-old pups on the island.

The wolves bred in their pen on Bulls Island that spring, and in July they were released with two pups. A few weeks later, the mother was killed by an alligator; but the father and pups adjusted easily to freedom and seemed to have little trouble learning to hunt. If all went well over the winter of 1988–89, Parker planned to move the Bulls Island pups to the refuge in North Carolina. The father would be trapped, paired with a new mate, and freed to raise a second litter in the wild. In this way, Bulls Island would become a training ground for young wild wolves. Parker also hoped to release wolf families on barrier islands off the coasts of Florida and Mississippi.

One cool October afternoon I drove alone to a remote site near a series of clear cuts on the south side of the Alligator River refuge. Phillips and Lucash had driven down earlier in the day with some volunteers and John Windley, a biological technician on the wolf crew.

That evening we were to hold a vigil for the Pole Road pup. Although we all had assumed the pup had starved after the death of its mother, in midsummer Phillips spotted it from the air, tagging along after its father. For weeks the biologists would find no sign of it, then someone would come across its prints or glimpse it from a plane. Recently Phillips and Parker had decided the pup was large enough to wear a radio collar, and plans had been made to trap it.

The thought of handling the pup intrigued everyone. Phillips needed only two helpers to set the trap lines and check them during the night, but more than a truckload of us asked to go along. Just before dark we assembled at the intersection of two dirt roads, a little less than a half mile from where Lucash and Phillips had positioned the first trap.

A half moon lit the hazy sky and illuminated the furrowed bark of two great pines beside the road. Their trunks, thick black shafts, rose straight and branchless for fifty feet. I took a seat in a truck with Phillips, Lucash, and Windley and settled in for the night. Each trap had been equipped with a radio transmitter that was designed to start beeping when the jaws sprang closed. Every fifteen minutes Phillips pulled on a set of headphones to listen for signals.

We were on edge, but pleasantly so, like children waiting in the dark for unknown treasure. We traded jokes, shared bag lunches, and slapped lazily at stray mosquitoes. Eight o'clock came, then nine. Windley remarked that the prime time for catching the pup had passed; the wolves were usually most active at dusk and dawn. "You going to stay out all night, Mike?" I asked.

"As long as it takes," he replied with a yawn. The Pole Road pup was not likely to be wary of traps, and he expected to catch it by morning.

A military jet passed low overhead, its tail lights flashing red and green. Its vibrations startled me. Normally the jet activity ended at dark. In the distance we could hear the low rasp of lazers being fired. "That pilot just spent more on one bombing run than the wolf project will see all year," Phillips said, disgusted.

We waited, mostly silent. It occurred to me that our quarry, being raised utterly wild, was one of only two such red wolves in the world. No human had ever touched it. The thought of it wearing a radio collar saddened me. But hunting season had just opened, and Phillips believed a collar might protect the pup somewhat from poachers. The refuge was not open wilderness, I reminded myself.

Without speaking, Lucash eased open the door of the truck and stepped outside. I caught the door before it closed and followed him. The night air, though damp, was cool and refreshing on my face. Across a canal I could see nappy spires of cedar rising into the silvered sky. Somewhere in the silent woods two wolves rested near each other, no doubt hoping to avoid whatever human mischief was afoot. I trotted a few steps to catch up with Lucash, who was strolling down the softly lit road. "Chris," I whispered, "does it ever bother you to think this refuge is nothing but a huge zoo?"

He turned to me and frowned, then smirked. "Why do you ask, you nosy writer?"

"Because it bothers me."

He resumed his ambling pace, kicking at a rock. "Yeah, it does. But not as much as it used to. I've gotten more philosophical since so many wolves died this summer. I figure it this way: At least some wolves are out, and they're learning to be wild. They're not wasting away in pens. They're building up tolerance to parasites and learning to catch what they eat. Maybe it is a big zoo, but at least it's preparing them for a wild existence, if that ever becomes possible."

Not long before, there had been hope of removing the red wolf from the endangered species list by 1995. I knew that no one, not even Warren Parker in his most optimistic mood, talked in those terms anymore. Besides the lack of good habitat, a recent study on the red wolf gene pool had drastically changed the thrust of the program. Until 1987, administrators in the Fish and Wildlife Service had assumed that the red wolf could be safely restored by building a captive population and three separate wild populations from

250 animals. Now researchers feared that, with so few wolves, the diversity of the species would slowly melt away. According to the new analysis, the population would not be stable with fewer than 550 animals—330 in captivity and 220 in the wild.

The total population of red wolves was still less than 125. In the past year everyone—the biologists, the breeders, the field technicians, the highest service officials—had come to realize that the future of the species depended on its genetic recovery. And genetic recovery could not be accomplished without decades of careful breeding.

We had perhaps as long as half a century before the red wolf could be restored to a healthy condition. Meanwhile the species would have a chance to rebuild its wild skills, and its social and cultural heritage, at Alligator River. Just as important, it would once again become part of the natural and cultural landscape of North Carolina. Lucash's words had struck me and shaken the pessimism I had felt for months.

In fifty years human values could be reshaped; it had happened before. Enough space might still be found for the species to live almost as freely as it did before settlers waged their crusade to kill it. Who could say? The resurrection of the red wolf would not unfold as quickly as I might like, or in exactly the way I might hope. But unfold it would.

The Pole Road pup did not step into a trap that night, nor the next, nor the next. By the end of the week, Phillips conceded temporary defeat. All winter the wolf crew tried, off and on, to catch the pup. More than once the biologists found tracks where the pup had walked up to a trap, sniffed it, and fled. Its wily behavior pleased them, although it cost them many long nights of waiting.

That fall the Pole Road male began keeping loose company with the second of the young female wolves released in the spring of 1988. In late December, the male was discovered dead in the woods from a freak accident; he had been devouring a raccoon and had somehow strangled on the animal's kidney. To Phillips, his death was as disheartening as the loss of a friend.

Across the peninsula, near the farm fields on the east side of the refuge, the Point Peter pup grew up healthy and wild, but not as sly about traps as the pup from Pole Road. The wolf crew captured her without trouble, weighed her, fitted her with a radio collar, and released her immediately. Two months later they trapped the pup's mother, the Point Peter female. They planned to pair her with the old Phantom Road male and keep her in captivity until any pups she bore were old enough to be wormed and vaccinated against disease.

It was a mild January day when the biologists trapped the Point Peter female and drove her to Sandy Ridge. As soon as they released her from the kennel into a pen she ran to the fence and began to pace back and forth, gazing at the woods beyond. "She clearly wanted out. She absolutely refused to look in toward the center of the pen," Phillips said. "It was real sad. It made me feel like we were going backwards."

And this animal, I remembered, was the highest strung of the original eight wolves, the one believed to have the slimmest chance of adapting to freedom, along with her toothless mate. What would happen to her daughter, I asked Phillips. "Who knows?" he replied. "She may have a hard time for a while. I don't know for sure that wolves get lonely, but I can imagine they do, they're such social animals."

The Point Peter female was freed again in August 1989, along with her new mate and their four young pups. There were, suddenly, quite a number of animals for the biologists to track. Twelve wolves roamed the Alligator River—the family group, five new animals that had just been released, and the Pole Road pup, which finally stepped into a trap in late spring. She was petite but feisty, with the brindled black markings of her father. In the last days of summer, the biologists released another four wolves, all of them two years old and paired with potential mates.

Phillips, his mood buoyant, kept copious records on where the animals moved in relation to each other. Parker, too, seemed to have a renewed optimism. He talked of using the Alligator River refuge as a halfway station where captive-born wolves could begin to adapt to freedom. Eventually, he predicted, a much larger population of red wolves would be established on a large tract of land, possibly in the Great Smoky Mountains National Park. He also hinted that the day might soon come when the owners of the farms around the Alligator River refuge would agree to let wolves wander onto their property.

That fall of 1989, a sound that had been too long absent from North Carolina reverberated through the forests of the Alligator River, piercing the humid nights and sending chills through those of us lucky enough to hear. It was the sound of wolves, a quavering chorus that, with grand irony, reached its most frenetic pitch whenever a military jet roared through the sky.

On cool nights we rode around in trucks just after dark, tracking the faint, pulsing signals from the collars of sixteen wolves. We still had our favorites, but there were too many, at this point, to keep track of easily. We rode in a pickup, Mike Phillips, John Windley, and I, to see what we could hear. A half moon rose, hidden off and on by scaly clouds. Phillips stepped outside the truck, cupped his hands to his mouth, threw back his head, and began to howl. A minute later we heard the response. The wails, coming from two directions, began low and ominous but quickly rose in pitch. They blended with a beautiful dissonance and built to an eerie crescendo. They filled the night, but carved out a hollow in my chest. I turned from them, aching.

There are only two things left to tell, for now. One night during that same period, I went out with Phillips to track the movements of the Point Peter female and her pups. Occasionally the biologists monitored some of the wolves all night to see where they moved and when. It was an uneventful evening, and we spent most of our time chatting and reading. At daybreak, though, Phillips suggested that I go with Mike Morse, one of the biologists on the project, to look for scrape marks the wolves had made during the night.

To reach Morse I had to pole a small skiff across a canal and walk through some grasses to the edge of a farm field. Morse had worked full-time as a volunteer for the project until a few months before, when Phillips had found the money to hire him. He was good-natured and gregarious, one of those people born with a natural effervescence. We were both groggy from lack of sleep, but he seemed better able to cope with the coming of daylight than I.

We climbed into an army range jeep, circa 1965, and bounced down a dirt road to an intersection near where the wolves had bedded down. Morse pulled up to a large pile of scat and stopped. "Oh yeah," he said with gusto, "oh yeah!" There was scat on both sides of the intersection, accompanied by long dark slashes in the dirt. Judging from the surrounding tracks, one side of the intersection had been claimed by the males from Bulls Island, the other side by the Point Peter female.

Morse crouched to examine a set of fresh prints in the damp dirt. "Look here," he said. "This is neat as hell." Beside the tracks were those of a smaller animal, one of the new Point Peter pups. They led to a pile of scat and two scrape marks, one very short, one two feet long. One made by a novice, one by a pro. I caught Morse's eye and smiled widely.

Hope is born of small things. The promise of life for red wolves grows both from their long evolution as a species and the wild behaviors that have emerged from captivity intact, if not unscarred. To me, the scrape mark made by that five-month-old pup symbolized the beauty and tenacity of the natural world. We have not yet crippled it, not completely. With luck, we never will.

One afternoon I wandered over to the North Carolina Aquarium on Roanoke Island, which is next door to the Dare County Airport. I had been to the aquarium many times before. Once, in November 1986, I had listened as Warren Parker explained to a gathering of reporters there how red wolves would be prepared for release. On this particular day I entered the aquarium with some trepidation, for I was going to see a new exhibit on red wolves.

I found the exhibit in a free-standing octagonal glass case. Inside were the bodies of two red wolves I had known, stuffed and lifelike. One, the female from South Lake that had died of an unknown infection, was standing on an arched log with a rabbit in her mouth. The other, the

Phantom Road female, was crouched on her stomach below the log, her head raised, her severed left front leg hidden by a pile of leaves. Both wolves were looking into the distance as if they had been alerted by some noise.

I walked to the front of the case, where I could study the Phantom Road female's bright glass eyes. The last time I had seen her, her gaze had been dulled by pain and shock. She had been granted life when Sue Behrns plucked her from her dead mother's womb. She had been brought to her death by the wild impulses that Warren Parker and Mike Phillips and all the rest of us wanted so badly to save.

It was a weekend, and the aquarium was crowded with parents and young children. A couple with a baby in a stroller walked up to the red wolf exhibit and paused. "They look smaller than I thought," the man said.

"Yeah," said the woman, "but they're pretty. I wonder why they call them red wolves. They're more tan." The baby whimpered in sleep.

I looked back at the Phantom Road female's topaz eyes. The taxidermist had done a good job; it almost seemed that there was still a spark of life in them, and in the body itself, poised to whirl and run. The couple moved on. "Thank you," I whispered to the wolves, turning to leave. "Thank you."

BIRDS IN THE HAND

VI

A RUINOUS LEGACY

When settlers from Europe arrived in North America in the late sixteenth and seventeenth centuries, they discovered a wilderness of lavish abundance and beauty, with great numbers of game, flowers and berries in fields of riotous color, forests that stretched inland as far as they could travel. It is interesting to look back at their writings, in light of what has transpired during the past four hundred years. For some time now I have tried to envision the continent as it existed in 1587, the year the first ragtag band of English colonists made landfall in the New World.

The one hundred and fifty people who would become known to history as the Lost Colony had left behind a country of formal gardens and poplar-lined streets, of stone fences and open, rolling fields. On the shores of what would become North Carolina, they discovered a bewildering land of dark forests and mysterious Indians who kept "dogs of the woods" for pets. "They go in great Droves in the Night, to hunt Deer, which they do as well as the best Pack of Hounds," the explorer John Lawson wrote of these animals. ". . . When they hunt in the Night, that there is a great many together, they make the most hideous and frightful Noise, that was ever heard." They were not dogs but wolves.

The wolves of Alligator River were freed just west of where the first English colonists are believed to have made their failed stand against the wilderness. The reintroduction of the species, occurring in the first part of the continent to come under English axe and plough, inspired in me a deep curiosity about aboriginal America. The wolves were released in forests that were swampy and dense and threatening, perhaps something like the forests of the sixteenth century. But there were no longer extensive stands of white cedar and bald cypress on the land adjoining the Alligator River. There were no "dogs of the woods," no cougars, no plentiful schools of herring and shad and rockfish in the sounds. There were no eagles, no parakeets, no Indians, and only a remnant population of bears.

The early colonists, and those who followed, came from a Europe stripped of biological diversity. In America they had before them a world that was truly new, virginal in beauty and bountiful in possibilities. If they had chosen, they could have forged a new, more humble relationship with the natural world. Instead, they plundered the countryside with blind avarice, granting no refuge to the animals and people they displaced in their mad scramble for territory. Within a hundred years of settlement, the great forests of the East Coast were reduced to ragged patches. The wildlife had withdrawn to the west or, in many cases, had simply disappeared.

Until settlers timbered and tilled their way across the continent, the skies over North America were filled with clouds of shorebirds and pigeons. Bison and elk grazed east of the Appalachian Mountains. Sea sturgeon ran in Atlantic coastal rivers, and lake sturgeon, as long as eight feet and weighing more than three hundred pounds, were abundant in the Great Lakes, the Ohio River, the Mississippi, and other large waterways. To the west, jaguars ranged as far north as Arkansas, and condors foraged from Mexico to the Columbia River. Otters swam in the streams in such numbers that at times they hindered the passage of explorers' canoes. The plains teemed with bison, bears, antelope, elk, ferrets, foxes, badgers, wolverines, and wolves—one of the greatest congregation of mammals to have developed in the history of the world.

The breaking of the wilderness was quickly fatal to animal species that were not widely distributed and that lacked the ability to adjust to sudden change. In his remarkable book, *Wildlife in America*, Peter Matthiessen describes the decimation of these animals, region by region, as settlement advanced across the continent. He writes of the Labrador duck, a pretty, pied species with snowy cheeks and chest and a black necklace that widened to a diamond on its back. It may have been slaughtered beyond recovery by parties of New England colonists that raided the Labrador coast in the 1750s to procure feathers for their bedding. Whatever the cause, the species grew rare by the mid-1800s and was extinct by 1875.

Among the sea rocks of Maine, Nova Scotia, and New Brunswick lived a mysterious coastal mammal that apparently fed and played in the swirling eddies of tidal rivers. Known as the sea mink, it was much admired for its shining black fur, sleek body, and long, tapered tail. Early settlers described it as larger than a fox, with the streamlined shape of a greyhound. It may well have been one of the most beautiful of the mammals encountered in the New World, but it disappeared before naturalists could observe it in any detail.

The ivory-billed woodpecker, a large, conspicuous bird with striking black-and-white wings and a red topnotch, had a smooth white bill used by certain Indian tribes to make the coronets worn by great warriors. It was, according to the late eighteenth-century naturalist Alexander Wilson, "a majestic and formidable species." But it fed almost exclusively on grubs found beneath the bark of dead pines, and it could not adapt to the extensive logging of the forests where it lived. By the 1950s only a few nesting pairs remained in the Southeast. Ornithologists had given the species up for dead, when a few pairs of breeding birds were discovered in Cuba in 1986. Its condition is still precarious.

When one considers the pace at which the settlers cleared forest and fenced the land, it is not surprising that animals as uncommon as the sea mink disappeared. Yet especially now, from the comfortable distance of a century, it is difficult to comprehend the thoughtless killing of such species as the Eskimo curlew, the bison, the golden plover, the passenger pigeon, and the Carolina parakeet. These animals once gathered by the thousands, and perhaps hundreds of thousands, but within a few decades they were hunted out. Only the plover, the bison, and possibly the curlew survive today, in greatly diminished numbers.

The passenger pigeon, which congregated in flocks so large that the noise of their flight drowned the roar of guns, and the bison, slaughtered by the millions for their meat and hides, are among the most famous of North America's decimated animals. The Carolina parakeet is somewhat less well known, though it must have been one of the most beautiful creatures on the continent. Its feathers were bright green tinged with blue, and it had a yellow-gold head with a startling orange mask. When a flock alighted, Alexander Wilson reported, the ground "appeared at a distance as if covered by a carpet of richest green, orange, and yellow...." Unfortunately, the parakeet was fond of grain, which earned it the reputation of an agricultural pest. It was hunted by pet traders, sportsmen, feather collectors, and farmers, who would bait the birds into fields with grain and shoot as many as possible. Once a flock was fired on, the survivors took flight and circled back over the field, as if trying to rouse their slain companions—an endearing but fatal characteristic the parakeet shared with the Eskimo curlew and several other birds that are now extinct.

In 1800 the parakeet was reported to be abundant throughout the eastern United States, from Michigan and New York to Florida. In 1918 the last one died in captivity.

It is immensely disturbing to study the natural history of America, to read about the decimation of the barren-ground caribou, the right whale, the gray wolf, the Colorado River squawfish, the manatee, the ocelot. It is just as disturbing to look at the modern landscape and try to see what once was, to imagine that each bird in flight, each butterfly and rodent and toad, might be one of the species lost to civilization. But through all the stories of abundance and senseless waste, nothing strikes me more powerfully than Wilson's description of Carolina parakeets hopping over the ground: green and gold, with slivers of blue and flashes of orange, a living kaleidoscope of color, a startling tropical anomaly in our drab temperate clime. They must have seemed like animals from a fairy tale, bedecked with jewels, rejoicing in the sheer fact of their existence.

We do not have much left from those early days. The fantastic wilderness has been defaced by our own hand, the scars of economic greed left deep in its soils.

In light of what we have already lost, the wild animals we have preserved are precious beyond measure. Yet the destruction of their habitat continues, and the money spent on their behalf is not enough to save more than a fraction of those in danger. Although the United States has enacted one of the most important wildlife conservation laws in the world, the cavalier attitude of our ancestors remains deeply ingrained in the national spirit. We still do not believe, or do not care, that we are capable of destroying whole species of animals and whole natural systems. And so we merely watch while the killing continues.

If Americans sincerely hope to save our remaining native flora and fauna, we must make a vast commitment of money, scientific resources, and political will. Just as important, we must have a clear vision of what we hope to accomplish, a guiding philosophy that can unite biologists, conservationists, and government officials and carry them through times of ecological crisis or political change. So far the national vision has been clouded, and progress toward saving our rarest animals has been piecemeal and desultory. With the restoration of the red wolf, and with the recent reintroduction of several other rare species, it is time to take stock of where we are and gauge how we should proceed.

In 1973 Congress passed the Endangered Species Act, a landmark bill, and the cornerstone of federal policy toward wildlife and environmental conservation. Its fruits are most apparent in the thriving population of brown pelicans in the eastern United States. The brown pelican was once critically endangered; in Louisiana, Texas, and California it is endangered still. The species was nearly destroyed by a single environmental poison, the pesticide DDT. The presence of DDT in the food chain produces freakish physiological changes in birds of prey; among other things, it causes them to lay eggs with unusually thin shells. After the pesticide was outlawed, populations of pelicans, eagles, ospreys, falcons, and other predatory species slowly began to recover.

Most vanishing animals suffer from a range of problems that are far more difficult to correct. Their migration routes have been blocked by cities, power lines, and other trappings of human settlement. The rivers and lakes where they breed have been polluted by agricultural runoff, acid rain, or the residue from sewers. They no longer have an adequate choice of nesting sites. In some cases, they have been isolated from others of their kind and are becoming vulnerable to the complex processes of extinction.

During the first fifteen years after the passage of the act, only two of the four hundred ninety-three species listed as endangered or threatened within the United States—the brown pelican and the American alligator—

were declared to have recovered. (The alligator is still listed as threatened to protect the American crocodile, which resembles it.) Three species of birds from the Pacific island of Palau were also removed from the list after biologists discovered previously unknown populations of each one. Eighteen species are believed to have gone extinct, including the dusky seaside sparrow, the Palos Verdes blue butterfly, and possibly the Bachman's warbler. Ten species were reclassified from endangered to threatened.

This is hardly a stellar record of accomplishment. It suggests that federal management of endangered species programs is scattered and inefficient—a suspicion born out by several recent Congressional studies. In 1988 the General Accounting Office issued a report on federal efforts to preserve rare animals and plants. The General Accounting Office is an investigative agency of Congress. It examined the programs run by the U.S. Fish and Wildlife Service, which administers the Endangered Species Act, and the National Marine Fisheries Service, which is responsible for preserving rare marine species. Although the investigators took care not to be overly critical in tone, their findings were damning. Neither service had maintained detailed information on the status of different animals and plants. As far as the investigators could determine, the condition of only sixteen percent of the species was improving. Thirty-seven percent were believed to be in stable condition, while thirty-four percent were growing worse. Two percent were thought to be already extinct, although not declared as such. The status of the other species was unknown.

The recovery plans for many species were not being carried out, investigators found, largely because of inadequate funding. (According to a federal audit conducted in 1990, the service received about $8.4 million annually to conduct endangered species recovery programs. Yet the agency might need as much as $4.6 *billion* to run effective programs for every species listed as threatened or endangered.) And the little money available had not been distributed on an equitable basis. From 1982 through 1985, nearly half the recovery program budget was dedicated to twelve species with "high public appeal," such as the bald eagle, the gray wolf, the grizzly bear, and the California condor. Some of the programs for plants and less popular animals barely received enough funds to pay the salaries of their biologists.

Service administrators acknowledge that they routinely place more emphasis on recovery programs for appealing animal species than for plants and invertebrates. By highlighting certain animals, officials believe they will increase public awareness of endangered species and generate enthusiasm for their programs. Animals like the California condor will become topics of household discussion, school children will learn about radio tracking and captive breeding, and Congress will find it politically beneficial to allocate more funds for recovery projects. At the same time, political pressure will increase to save the habitat needed by not only the favored animals but by rare toads, spiders, herbs, and other neglected species.

The usefulness of the Endangered Species Act in preserving wild habitat can be seen clearly in the controversy over the northern spotted owl, a rare subspecies endemic to the Pacific Northwest. Nearly ninety percent of the ancient forests where the spotted owl nests have been destroyed, and about half the remaining stands are in danger of being logged. Conservationists hope to win prohibitions against logging in millions of acres of old-growth forest, on the grounds that deep, undisturbed woodlands are critical to the survival of the spotted owl. (Ironically, in this case the act has been invoked over the initial objections of the Fish and Wildlife Service. In 1987, when service administrators were first asked to include the northern spotted owl on the list of endangered and threatened species, they refused. Their decision was challenged in federal court by environmental groups, and in 1990 the owl was officially listed as threatened. It remains to be seen whether conservationists can use the listing to win permanent injunctions against further logging.)

Administered effectively, the Endangered Species Act could be a powerful tool for preserving what remains of the American wilderness, since the law requires the federal government to protect critical habitat for all species listed as threatened or endangered. However, the Wilderness Society, the Defenders of Wildlife, and other environmental groups complain that federal agencies such as the U.S. Forest Service and the Bureau of Land Management have traditionally ignored this provision in favor of timber and mining companies, ranchers, hunters, and other special interests. This has occurred not only on private land and within national forests but on federal wildlife refuges. In 1989 the General Accounting Office found that sixty percent of the nation's wildlife refuges allowed "secondary" activities—mining, grazing, boating, hunting, off-road vehicle use—that were harmful to wildlife. And in 1990 a federal audit concluded that the Fish and Wildlife Service was so hindered by mismanagement and a lack of resources that hundreds of animals and plants will almost certainly disappear before any effort is made to save them.

As of 1993 the federal list of endangered and threatened species included seven hundred seventy-five American species and five hundred thirty foreign species. (The United States cites listings from other countries to promote global conservation and discourage trade in products made from rare animals and plants.) Scientists believe the list encompasses only a fraction of the species in danger of extinction. Between three thousand and four thousand animals and plants in the United States have been identified as candidates for the list. Little or nothing is being done to preserve them. Instead, federal administrators have chosen to concentrate on expensive recovery programs for a dozen species, several of which are nearly extinct.

Many environmentalists hope the Clinton Administration will devise a saner process for managing endangered species, perhaps by establishing a national biological survey. Nevertheless, it will be difficult to reverse what wildlife biologists describe as a policy of ecological brinkmanship. Imperiled species receive the full attention of the Fish and Wildlife Service only when they are within a few years of extinction. By then they may have suffered irreparable genetic harm, and their habitat may be gone. Meanwhile, plants and animals that are still healthy but declining in number are left to fend for themselves. Few attempts are made to improve their well-being or preserve their habitat until their condition, too, becomes critical. This strategy is flawed on two counts; it is both short-sighted and too piecemeal to be effective. It attempts to reconstruct animal populations one by one, usually in spoiled natural areas. And it relies on management techniques that require extensive manipulation of animals, as in the costly, chancy art of captive breeding.

In the past decade conservation scientists have begun to experiment with rebuilding entire ecological systems, in which native animal and plant groups are reintroduced and exotic, or transplanted, species are rooted out. This philosophy of restoration ecology, as it is known, has grown partly from increased knowledge about the complexity of nature. It stems also from concerns that many animals and plants need extensive habitat—much more than scientists suspected even ten years ago—to survive.

In an article published by the British journal *Nature* in 1987, wildlife ecologist William Newmark reported that an alarming number of mammals have disappeared from national parks in the western United States, apparently because they did not have enough land to survive. In a comprehensive review of reported mammal sightings, Newmark found that even relatively large reserves, such as Yosemite and Sequoia-Kings Canyon national parks, had lost a quarter of their mammal species since being established in the early 1900s. Isolated by the destruction of surrounding habitat, the mammal populations apparently grew too insular to survive. Smaller parks, such as Zion and Bryce Canyon, lost thirty-six percent of their mammal species as the lands bordering them were timbered, cut through with roads, or otherwise developed. The largest national park system, Yellowstone and the adjacent Grand Tetons park, lost only four percent of its mammals. In the Canadian Rockies, where four interconnected national parks spread across an area twice the size of the Yellowstone-Grand Tetons preserve, all the indigenous mammal species were still present.

A later study cast some doubt on the legitimacy of Newmark's findings. Nevertheless, scientists have come to accept that the rate of decline in wild mammal populations is higher than previously believed. It appears that most of our national parks are too small, and that the United States probably does not have enough wild land set aside to preserve its remaining native flora and fauna. To stem the loss of animal species, scientists believe that national parkland and the properties surrounding it need to be managed under integrated plans that allow wildlife to move freely through broad territories.

Other recent studies have shown fish and bird populations to be in equal peril. A report issued in 1989 by the American Fisheries Society estimated that a third of the freshwater fish species found in the United States, Canada, and Mexico have been seriously affected by the degradation of lakes, rivers, and streams. Of a thousand fish species, the society listed three hundred sixty-four as either endangered, threatened, of questionable health, or living in such small populations that minor ecological changes could place them in danger. The society found that freshwater fish were being adversely affected by a range of problems, including pollution from sewage and chemicals, acid rain, development, the diversion of water for drinking and irrigation, and the introduction of exotic fish that have disrupted breeding and food chains. The worst problems existed in the southwest United States and in Mexico, where virtually every species was believed to be living in a precarious state. The authors concluded that until wildlife managers make a commitment to preserving entire ecosystems, instead of continuing "inconsistent recovery efforts for individual species," very little is likely to improve.

While fish suffer from poor water quality, songbird populations may be faltering because of wide deforestation, especially in the East. In *Wildlife in America*, Matthiessen cites an old saying that before the settlement of the continent by Europeans, a squirrel could run from the Atlantic to the Mississippi on treetops. The unbroken woods provided hiding places and shelter for millions of songbirds and other species. Since the 1700s, however, so much forest has been cut that many scientists believe predators are destroying the nests of certain songbirds with alarming frequency.

In 1983 David Wilcove, now a senior scientist for the Environmental Defense Fund, conducted an ingenious experiment to see whether birds that nested in large tracts of forest were more likely to raise their young successfully than those that nested in small, scattered groves. Wilcove fashioned artificial nests, filled them with quail eggs, and placed them in several distinctly different stands of forest. In the Great Smoky Mountains National Park, he found after a week that only two percent of the nests had been disturbed by predators. At a half million acres, the forest within the park was the largest tract he studied. When nests were placed in woodlands of between ten and thirty acres, the predation rate rose dramatically. In small wooded tracts in rural Maryland, nearly half the nests, on the average, were disturbed. And in wooded patches near suburban development, predators damaged eggs in an average of seventy percent of the nests.

Wilcove's results do not bode well for the songbirds that construct open, bowl-shaped nests—among them, ovenbirds, redstarts, and certain warblers, vireos, and thrushes. Such nests are much harder to defend than the tree cavities used by, for instance, chickadees and titmice. And songbirds that winter in tropical rain forests and nest in the eastern United States face pressures from extensive deforestation on both ends of their range.

While certain bird species may be faring poorly in the new American landscape, parasitic species such as the brown-headed cowbird are faring quite well. Cowbirds do not make nests of their own; instead, they lay their eggs in existing nests and leave them to be hatched and raised by other species. Many host birds do not recognize the cowbird eggs as different, and the large, aggressive cowbird chicks monopolize the food brought back by the unsuspecting parents. The songbird young suffer doubly; if they survive predation, they may be crowded out by nest mates that are not even of their own kind.

Similar stories can be told about great black-backed and herring gulls, which prey on the young of terns, skimmers, and puffins. Similar stories can be told, in fact, about almost every kind of animal in almost every kind of habitat. Adaptable coyotes flourish in altered country and interbreed with red wolves. Common black racers replace the scarlet king snakes and diamondback rattlesnakes of the East, which prefer to live in large forests. Bullfrog tadpoles crowd out the tadpoles of rarer species, such as the red-legged frog of southern California.

The victors in each of these contests are known as generalist species. They are the hobos of the animal world; they can live virtually anywhere and find food and shelter in the heart of human civilization. Some, such as great black-backed gulls, tend to be aggressive interlopers that drive other species from their homes. Others merely invade disturbed habitats, pushing out more specialized species.

This is the legacy we have created, a world of gulls and cowbirds and rats and roaches that grow healthier with each passing day. This is what we have to look forward to, barring radical changes in the way we leave our mark on the natural landscape. Fish will die out, except for hardy varieties of hatchery-raised trout and catfish. Warblers will fall silent. The countryside will be filled with house sparrows and starlings, gulls and grackles, mice and men. And little else.

How have we arrived at such a miserable state of affairs? We are caught, it appears, in a classic cycle of poverty: We have squandered our riches, and now we are too confused and demoralized to break our spendthrift habits.

It has become popular for conservationists to talk of the need for a major shift in the way Western society views itself in relation to the natural world—as separate from nature, and somehow superior to it. The shared European and American value system is so strongly entrenched in industrialized countries, and has been embraced so readily by developing nations, that it has gained control of the world's economies and usurped local cultures. Our disregard for nature has been exported and proliferated, until the earth now suffers from what may be an unprecedented wave of extinctions. "Curiously," writes biologist Paul Ehrlich of Stanford University, "scientific analysis points toward the need for a quasi-religious transformation of contemporary cultures. Whether such a transformation can be achieved in time is problematic, to say the least."

Pleas for change in our callous attitude toward nature are nothing new. John Muir made them in the nineteenth century; so did Ralph Waldo Emerson and Henry David Thoreau. In the 1948 introduction to *A Sand County Almanac*, Aldo Leopold bluntly described what he saw as the central problem with our prevailing belief system. "Conservation is getting us nowhere," he complained, "because it is incompatible with our Abrahamic concept of land. We abuse land because we regard it as a commodity belonging to us." One might extend that reasoning to include plants and animals.

In *Silent Spring*, published in 1962, Rachel Carson chastised the "practitioners of chemical control who have brought to their task . . . no humility before the vast forces with which they tamper." Carson was referring to the Western propensity for trying to engineer solutions to the environmental problems we create. This is a curious and comparatively new phenomenon, viewed against the backdrop of human history. Unlike the Christian settlers, North American Indians defined their place in the world through mythology; they found their gods and holy spirits in the hills, plants, and animals around them. The thought that humans could control nature, and improve it, never infiltrated their way of life.

In contrast, the contentious relationship that modern Americans have developed with nature is a logical outcome of the Judeo-Christian religious tradition so dominant in our society, despite attempts to divorce church from state. The historian Lynn White, Jr., has described Christianity as "the most anthropocentric religion the world has ever seen." Anthropocentrism is the tendency to view oneself as the center of the world. In an address before the American Association for the Advancement of Science in 1966, White blamed the Judeo-Christian tradition for creating a belief system that encourages people to exploit their natural surroundings.

Before the rise of Judaism, White noted, most cultures relied heavily on mythology to order their worlds. It was believed that every rock and river and creature had its own *genius loci*, or guiding spirit. If a tree was cut or an animal killed, its *genius loci* had to be appeased. But the Jewish religion provided a version of the creation story in which everything in the world was fashioned by God for the benefit of humankind. Christianity dismantled what little belief remained in the concept of *genius loci* by condemning nature worship as idolatry. "By destroying pagan animism," White said,

"Christianity made it possible to exploit nature in a mood of indifference to the feelings of natural objects."

Through religious crusades and economic imperialism, Christianity gradually became the dominant religion in Europe, and then in the world. When Christian settlers reached North America, they saw themselves as undertaking God's work as they laid waste to the wilderness and purged the countryside of wild beasts. This interpretation of history has been supported by several other scholars, notably Frederick Turner in his book *Beyond Geography: The Western Spirit Against the Wilderness*. Yet Christ's teachings do not necessarily encourage an anthropocentric view of the world. Professor White, who described himself as a troubled man of the church, pointed to Saint Francis of Assisi as a model of how Christians could live in harmony with their surroundings by humbly viewing themselves as an integral part of the natural world. Unfortunately, Francis's visionary beliefs were never absorbed by the church.

It is a distinctly Christian characteristic to believe in perpetual progress and the continual accrual of knowledge. Day by day the world is becoming a better place, largely through the mystical powers of science and technology. These values have come to be accepted even by Westerners who abhor the Christian religion. We do not have to turn to nature for solutions to problems; we can create solutions ourselves. "Despite Copernicus," White noted, "all the cosmos rotates around our little globe. Despite Darwin, we are *not*, in our hearts, part of the natural process. We are superior to nature, contemptuous of it, willing to use it for our slightest whim.... More science and more technology are not going to get us out of the present ecological crisis until we find a new religion, or rethink our old one."

Such calls for a religious transformation have provided theoretical meat for several diverse conservation groups, including a philosophical school known as deep ecology, founded in 1972 by Arne Naess, a Norwegian philosopher and environmental activist. Naess rejected what he viewed as shallow attempts to correct ecological problems through short-sighted, self-serving, technological solutions. Shallow ecologists generally work within the existing social structure to convince people that nature is inherently valuable and should be preserved. They pass laws limiting pollution; they create wilderness areas for the enjoyment of hikers and campers. A deep ecologist, in contrast, assumes that natural systems have a *right* to evolve without serious disruptions from people. All life has value, and all living things should be allowed to exist in their natural state. The difference between Naess's philosophy and traditional conservation philosophy is profound: under the tenets of deep ecology, true conservation cannot occur without broad social change.

Critics of deep ecology dismiss the school as utopian, naïve, and too inclined to consider the good of animals and plants over the good of humankind. Nevertheless, Naess's teachings have strongly influenced the direction of the Greens political party in western Europe and that of several conservation organizations in the United States, including the radical group Earth First! The doctrines of deep ecology echo beliefs held by aboriginal cultures for millenia—that the earth and its animals are sacred, and that the human race will cease to exist unless it lives within the balance of nature.

I mention all this because I believe the way our society perceives nature will be one of the most critical factors in determining whether we can restore the rare species we have nearly destroyed. If we think of an animal as a predictable organism that can be easily manipulated, if we believe it will in no way be harmed by radio collaring, constant surveillance, and frequent disruption of its breeding, we may seriously misjudge what it needs to survive. If, on the other hand, we view it as an intelligent, complex, and somewhat mysterious creature, if we accept that it has a unique way of moving through its world, we are likely to perceive it with greater sympathy, respect, and insight.

It is interesting to consider the rationale behind captive breeding in light of these thoughts. Since the 1970s, dozens of animal species have been rescued from extinction by government breeding programs and by zoos that devote thousands of dollars to raising rare animals, many of which will never be put on display. Captive breeding has often been likened to a modern Noah's ark; it is a way to purchase time for species that are only a few years away from extirpation in the wild. Time, it seems, has become the holy grail of conservation. With more time we might manage to educate people, change social values, save vital habitat, learn more about the way animals live in the wild. And, to stop the passage of time, we must breed animals under artificial conditions.

If we are to accept this line of reasoning, we must look closely at its implications. By preserving endangered species in captivity, we continue to operate from within an anthropocentric view of nature. We assume the right to manipulate individual animals as we deem necessary for the good of the species and (a hidden agenda, perhaps, but a very real one) as we deem politically expedient. We make unavoidable choices about which will live and which will die.

There is a curious similarity in the way American settlers went about homesteading the wilderness and the way scientists are trying to rescue the country's disappearing animals. Early Americans feared the unpredictable twists that living in wild country could bring. Their solution was to fell the forests, drive out the wolves, mountain lions, and bears, and erect the institutions needed for a thriving economy. Controlling wilderness meant keeping it fenced and at a distance. Now scientists have turned to taking the rarest animals into custody against the dangers—many of them natural—

that might drive them to extinction. Once again, the uncontrollable wilderness is being shut out.

How might we be changing the animals? The question demands continual reexamination. Under deep ecology, and the religions of many aboriginal cultures as well, a wild creature cannot exist outside its natural context. Take it from the mountainside or the forest where it lives, and it becomes something else. If our self-centered view of the universe is ever to change, we must begin to understand the world as a fragile filigree of earth, water, plants, and creatures. Wrenching animals out of their ecological niches does little to help us toward this goal.

We have backed ourselves, unwittingly, into a philosophical corner. By accepting captive breeding as a necessary evil, we cast our lot unequivocably with the technocrats. We can only hope that careful science will shape the animals as nature would have shaped them, without subtly altering their physical and psychological make-up. We must gamble that we will not drastically disrupt the process of their evolution. And we assume, perhaps wrongfully, that we are not destroying important social and cultural traditions that may affect the ability of reintroduced animals to survive.

If the techniques of captive breeding and reintroduction are to be used as tools of conservation, we must recognize them for what they are—desperate attempts to patch things together until, through some miracle, the world situation improves. The danger is that we will become so absorbed in our jury-rigging that we will forget to search for real solutions.

This, then, is the task before us: to recognize the uses and limitations of breeding animals in cages, and to devise a careful, comprehensive national policy that preserves whole natural systems instead of individual species. To begin, we must look closely at the failures and successes of the captive breeding and reintroduction programs conducted in the past.

VII

THE NEED FOR COMPASSION

The Whooping Crane and the Peregrine Falcon

In examining the endangered species recovery programs that depend, to one extent or another, on captive breeding, one senses that the biologists involved spend much of their time working close to the edge of panic. When the population of a species drops below a certain number, it becomes inbred and especially vulnerable to disease outbreaks and environmental disruptions. The problems caused by inbreeding tend to reinforce each other, pulling the species into a downward spiral, until the noose of extinction slowly tightens. Most endangered species programs are run as crisis operations—which is, of course, the reason scientists turn to captive breeding.

But with many animal species, as with humans, courtship and breeding can be a delicate process, and infertility can be caused as much by stress as by some physical malfunction. When wild animals are penned and given a limited choice of mates (or no choice at all), a constant nervousness may overpower their natural urge to procreate. The challenge for scientists is to give the animals ample privacy, while at the same time enticing them to breed quickly and frequently.

This is a difficult balance to reach and sustain. It requires a deep, thorough understanding of the species and a natural talent for handling animals. It also requires a stable political environment, which is rare in programs that tend to be as politically and emotionally charged as endangered species work. A scientist may hone a breeding regimen for years, making miniscule adjustments to help the animals feel more comfortable, only to have the program demolished by a sudden shift in personnel or a change in policy on how the animals must be managed.

Of the earliest breeding and release programs, four are of particular interest because of their various successes and failures. Two, the recovery programs for the whooping crane and the peregrine falcon, were carried out entirely within the United States and Canada. The others involved mammals reintroduced abroad—the Arabian oryx in Oman and the golden lion tamarin in Brazil—but drew from the expertise of American wildlife scientists. Each of the programs faced similar challenges. The animals needed to be put at ease in captivity, but also handled in a way that would preserve their wild character. Once they had begun to breed, their offspring needed to be conditioned for their eventual release.

In the end the programs that proved most successful were the simplest. They combined pure biology with a new, artful brand of husbandry. They were not hampered by overzealous scrutiny. Instead, they were conducted quietly, out of the glare of public controversy, and under the guidance of a few key scientists with clear, well-conceived goals.

In the spring of 1954 a biologist at Wood Buffalo National Park in northern Alberta was flying over a swampy plain when he happened to spot four whooping cranes and a rusty-colored chick. The cranes were feeding in a burned-out lowland of marsh and muskeg, and had apparently nested there. The discovery was startling, for the whooping crane was nearly extinct. American and Canadian biologists had been searching for the breeding grounds of the last wild flock for nearly a decade.

At that time the world population of whooping cranes consisted of about twenty wild birds and a few others kept in captivity. The wild flock spent winters at the Aransas National Wildlife Refuge on the Texas coast, sixty miles northeast of Corpus Christi. The marshes of the refuge were bordered by the Intracoastal Waterway, a major route for pleasure boats and commercial barges. The surrounding lands and waters were becoming sullied by oil and gas development, and just to the south was Matagorda Island, where the U.S. Air Force had established a bombing range. It was hardly a pristine habitat for rare cranes, but it was all that remained within the United States.

Two hundred years before, the heart of the whooping crane's breeding range had lain in the prairies and pothole wetlands of the upper Midwest, from Illinois and Iowa north to Minnesota, North Dakota, and Canada. Despite reports from the 1700s of flocks of whooping cranes that filled the sky, it is believed that the species was always much rarer than the smaller, grayer sandhill crane to which it is closely related. But its appearance and behavior were conspicuous enough to guarantee it notice among early American naturalists. Standing five feet tall, its white satin plumage, scarlet crown, and long, slate-colored legs made it unmistakable among wading birds. It became well known for its throbbing bugle calls and leap-filled mating dances.

By the 1890s, however, so much land had been timbered and planted that the species had all but disappeared from the northern United States. The southern marshes where the birds spent the winter were also being quickly developed, and during migrations the cranes were easy targets for hunters. By the late 1930s only two flocks still survived, the Aransas flock and a flock of about thirteen birds that lived year round in a remote swamp along the southwest coast of Louisiana. In 1940 a fierce windstorm flushed the Louisiana cranes out of their secluded wetlands and blew them north, toward the roads and canals and insulated communities of the bayou. Only six returned. A seventh, crippled by a gunshot wound, was eventually turned over to the New Orleans Zoo. The others were presumed to have been killed and eaten.

In 1945 the National Audubon Society and the Fish and Wildlife Service agreed to fund a joint research venture known as the Cooperative Whooping Crane Project. The next year a National Audubon Society biologist named Robert Porter Allen was hired to undertake a study on the behavior and food needs of the cranes. By then all but two of the birds in the Louisiana flock had died. It appeared that the flock on the Aransas refuge, which numbered eighteen, was the only one that remained.

Allen, aghast that the population had fallen so low, spent every possible hour studying the birds in the buggy Texas wetlands where they foraged for small marine animals. He found them to be much less gregarious than sandhill cranes; they stayed in pairs or small family groups, growing ill-tempered and aggressive when approached by unfamiliar cranes. From what he could tell, the birds seemed to mate for life. In spring, usually during the strong southerly breezes of March and early April, the adult cranes began to leave the refuge in pairs, taking their offspring from the previous year. They flew due north, passed over the Dakotas, and vanished into Canada. It concerned Allen greatly that no one had been able to locate the flock's nesting grounds so they could be protected. For several consecutive summers he and other researchers conducted aerial surveys of Alberta and Saskatchewan, searching for nesting cranes. And then in 1954 the flightless chick was spotted at Wood Buffalo Park.

The cranes had chosen to nest twenty-five hundred miles from the Aransas refuge. The flock's migration took it due north over the Great Plains and the wheat fields of Saskatchewan, where farmers sometimes shot at the birds for feeding on their grain. Surveying the sphagnum bogs and muskeg flats, Allen was surprised to find several pairs of cranes raising two chicks. The parents had always returned to Texas with a single chick, if any. Yet the females apparently laid two eggs each time they bred. Half or more of the chicks seemed to be disappearing during the long migration.

Allen believed the flock would slowly increase in the wild, if only its wintering ground could be kept unpolluted, and if hunters and farmers could be dissuaded from shooting migrating birds. But to his dismay, the discovery of the breeding grounds acted as the catalyst for a full-scale captive breeding program, the first ever to be run by the federal government. For years a number of influential scientists had maintained that the whooping crane was destined to disappear unless biologists could keep it from becoming genetically inbred. Among them was S. Dillon Ripley, who in the mid-1950s served as the curator of zoology at the Peabody Museum at Yale University. (He was later appointed secretary of the Smithsonian Institution.) Ripley bred waterfowl domestically, and he was convinced that under the right conditions whooping cranes could produce substantial numbers of chicks in captivity. He believed a breeding program could offset the annual losses from migration and help preserve whatever remained of the species' genetic diversity.

Ripley's position was supported by scientists in Canada and by John Lynch, a Fish and Wildlife Service biologist from Lafayette, Louisiana, who also raised waterfowl privately. In 1955 Lynch prepared a formal proposal on breeding whooping cranes, in which he noted that the annual production of chicks might be doubled, or even tripled, in a controlled setting.

Allen was not opposed to a limited breeding program. But he feared that under Lynch's proposal all the cranes eventually would be captured, and he questioned whether the offspring of caged birds would ever be returned to the wild. He objected that the proposal would only increase the chances that the species would perish completely in the wild. Trapping such large birds was tricky; no one could guarantee that it would be done safely. The cranes were extremely territorial and had a rigid social structure that Allen doubted could be duplicated in captivity. He argued that whooping cranes had already been shown to be high strung and difficult to breed in confinement.

Allen's views were shared by the officers of the National Audubon Society, which had considerable sway in Congress. But Ripley was an important and persuasive man. In 1956 the Bureau of Sports Fisheries and Wildlife organized a meeting of wildlife scientists from across the United States and Canada. (At that time the bureau, a division of the federal Fish and Wildlife Service, was responsible for managing endangered species.) After seven hours of vitriolic debate over the merits of captive breeding, the participants admitted they had reached an impasse.

In frustration, the director of the bureau, Daniel Janzen, appointed a dozen scientists to a new committee, the Whooping Crane Advisory Group, and asked them to write a list of recommendations for preserving the species. The creation of that committee brought an end to the dominant role held by the Audubon Society, and Allen, in the recovery of the whooping crane.

In 1955 an outspoken midwesterner moved to the Washington offices of the Fish and Wildlife Service to take a job in the research section of Sports Fisheries and Wildlife. Then in his early thirties, Ray Erickson had come east from the field research station at the Malheur National Wildlife Refuge in southern Oregon.

Erickson's invitation to work in Washington had grown out of an incident at Malheur the previous year. J. Clark Salyer, the chief of the wildlife refuge program, was touring the refuge and was being roundly critical of its management. When he began denigrating the refuge research program, Erickson took offense. "I told him there were very good reasons why things were being done in a certain way, and I listed several examples," Erickson recalled. "Afterwards I was sure I had ruined my career with the Fish and Wildlife Service." But Salyer, rather than being offended, was impressed by Erickson's spunk.

Slim and graying now, Erickson has been retired from the service since 1980. He lives on a hillside on the outskirts of Salem, Oregon, overlooking the dusky, blue peaks of the Cascade Mountains. When I met him he exhibited a quiet modesty, although as he talked about his career I caught glimpses of his notorious feistiness.

For fifteen years Erickson ran the Endangered Wildlife Research Program at the Patuxent Wildlife Research Center in Laurel, Maryland. He was also responsible for directing field studies on rare animals like the Puerto Rican parrot, the masked bobwhite, and the California condor. Few individuals had such a marked influence on the direction of early endangered species work. In a sense, Erickson is one of the fathers of captive breeding in the United States, for the establishment of the program at Patuxent was largely his idea.

But in the mid-1950s Erickson was new to Washington, and virtually unknown. Shortly after his arrival, he was asked to review John Lynch's proposal on breeding whooping cranes. Although there was considerable support for the plan in the bureau, Erickson decided it was too risky, given the poor breeding record of cranes in captivity. He knew of only one instance where caged cranes, a pair of sandhills, had successfully raised young. The concept of captive breeding was a good one, he concluded, but it needed to be refined. He returned Lynch's proposal with the suggestion that the whooping cranes be left in the wild until a series of breeding experiments could be conducted using sandhill cranes as surrogates.

"You have high hopes, you know, when you come up with an idea like that," he said. "You hope everyone will greet it with the same enthusiasm you put into it." To his dismay, service officials did nothing with the proposal that year, or the next. The 1950s drew to a close with the debate over whether to capture whooping cranes still unsettled. It was as if administrators in the bureau, afraid of offending either side, had elected to resolve the issue through indecision.

Disgruntled, Erickson turned to analyzing the data compiled annually on the number of whooping cranes that reached the Aransas refuge from Canada. (The National Audubon Society had distributed hundreds of pamphlets on whooping cranes along the flyway, and fewer birds were being shot.) It had been learned that immature birds did not settle in at Wood Buffalo Park for the summer, but scattered in unknown directions. Erickson suspected that many of them died before they reached breeding age.

In early 1961 John Fitzgerald Kennedy circulated a memo to federal employees. The new president wrote that under his administration innovative ideas were to be encouraged at all levels of government. "Boy, when I saw that, I said that does it." Erickson's face screwed into a frown of determination. "I'd been spending all those years in Washington and feeling like I wasn't accomplishing very much. I sent a very directly worded letter to the division chief. It eventually reached Dan Janzen, the top man in the bureau. I quoted Kennedy's memo; I said it was time to try something new."

An analysis Erickson had completed of annual population counts showed that, year after year, the same few adult cranes were producing chicks. And because so few young birds were living long enough to breed, the flock was probably losing genetic diversity. He urged administrators to approve the breeding experiments with sandhill cranes that he had suggested years before.

Perhaps Janzen, under pressure from the Whooping Crane Advisory Group, had already decided to take action, or perhaps Kennedy's memo indeed forced his hand. Whatever the reason, he referred Erickson's breeding proposal to the Whooping Crane Advisory Group, which quickly endorsed it. By mid-1961 a breeding center for sandhill cranes had been established at the Monte Vista National Wildlife Refuge in southern Colorado, which was on a traditional sandhill crane migration route. Over several years Erickson and a research biologist, Ervin Boeker, put together a small flock of cranes by collecting eggs from wildlife refuges in several states and hatching them in artificial incubators.

Erickson and Boeker found that if they collected eggs instead of capturing live birds, the cranes were less likely to contract diseases at Monte Vista. Like whooping cranes, wild sandhill cranes usually lay two eggs each time they nest, but only one chick survives. As an experiment the biologists decided to try taking a single egg from each of several sandhill nests for hatching in captivity, with the hope that they would not deplete the wild flock. They reasoned that the remaining chick would receive all its parents' attention and be more likely to survive.

At first the biologists encountered problems with incubating eggs and designing a sound diet for the sandhill chicks. But gradually the young captive cranes began to thrive. Careful censuses showed that the wild flock, as predicted, did not suffer from the loss of the stolen eggs. The crane chicks would not mature sexually for three or four years. Until then, the biologists could only speculate about how well the birds might breed in captivity.

In the interim Erickson turned his attention to a more ambitious project. It occurred to him that the Fish and Wildlife Service could establish a program in which biologists raised rare animals and also conducted detailed studies of their behavior, physiology, and territorial needs in the wild. "I viewed it as a genetic bridge," he said. "We could preserve endangered species and learn about them while we also worked on improving their habitat. I saw it as a three-legged stool—propagation, laboratory research, and field ecology—with no leg more important than any other." In 1963 he gave Janzen a detailed proposal for establishing his research center on an old estate on the Patuxent River in Maryland, where the service already had several research laboratories.

It was a period of turbulent change, both in the nation's social institutions and in the way Americans viewed their natural surroundings. Rachel Carson's *Silent Spring* had been published the previous year, and the science of ecology was rapidly evolving. Yet the Fish and Wildlife Service bureaucracy was still inclined to stifle innovation. "There was a lot of feeling that the administration didn't want to do anything that would appear controversial," Erickson said. "But meanwhile the species that so badly needed our help were disappearing."

No action was taken on Erickson's proposal for nearly a year. Then one day in 1964 Erickson went to Capitol Hill to present a photograph of a whooping crane to U.S. Senator Karl Mundt, the ranking minority member of the Senate Appropriations Committee. Mundt was from South Dakota. One of his key aides, Rod Kreger, had become very fond of the whooping cranes that passed through South Dakota during their migration and had asked the Bureau of Sports Fisheries and Wildlife for a photograph of one. When Erickson delivered the photograph, Kreger pulled him aside and asked if there was anything the bureau needed to help save endangered species. "I could think of a few things," Erickson replied dryly.

A few weeks later Kreger invited Erickson to a private meeting. From that point events unfolded swiftly. In November, Secretary of the Interior Stewart Udall announced plans by the Fish and Wildlife Service to construct a research center for endangered animals on its land on the Patuxent River. The following spring Mundt attached an amendment to an appropriations bill that allotted $350,000 for staff salaries, temporary animal enclosures, the construction of two buildings, and other costs. Erickson was put in charge of wildlife research and propagation.

The first animals to be housed at the new Endangered Wildlife Research Center were the sandhill cranes that had been kept at Monte Vista and some Aleutian Canada geese, a diminutive subspecies that breeds only on the chain of islands off Alaska. Soon the program also obtained some masked bobwhite, a small quail that had been extirpated from its range in Arizona but still lived in Sonora, Mexico; a few Andean condors to be used as surrogates for experiments on California condors; and some black-footed ferrets, which were then thought to live only in South Dakota. An official list of endangered animals and fish, first published in 1966, was updated every year. Field biologists set about trying to monitor animal populations, and to learn as much as possible about rare species in the wild. Laboratory

---

researchers studied animal behavior, reproductive physiology, and nutrition. They also ran experiments on disease prevention and treatment, using surrogate species like the sandhill crane.

Erickson, though pleased at the way the program was growing, found himself constantly unable to obtain essential supplies and building materials for pens because of his paltry budget. "We were just operating on a lean purse," he said. "We didn't even have permanent enclosures for the animals until about 1978. All we had were chicken-wire pens and a series of ditches running between them for water. In the winter you had to chop a hole through the ice to provide drinking water for the stock.

"Money was tight every year. That was the most traumatic thing about the job, the uncertainty of it all. When the budget axe began to swing, research was always the first thing to get cut. We were constantly being told that all our funding was going to get slashed, and then at the last minute it would be restored. No one stopped what they were doing for very long to worry about money, though. There was a real *esprit de corps*; the staff who worked with me were absolutely at the front edge of endangered species work."

With the establishment of the breeding center at Patuxent, the question of whether to bring whooping cranes into captivity was finally resolved. Through the years Erickson had come to believe it was foolish not to take some of the cranes into custody. In his opinion, the wild flock was increasing much too slowly. More than seventy immature whoopers had been spotted at the Aransas refuge since the 1940s, but in 1964 the size of the wild flock stood at only thirty-two birds. Young cranes were being lost at a dangerous rate, he contended, and the flock was probably becoming genetically inbred.

Administrators in the bureau and the Canadian Wildlife Service agreed. In 1967, with the reluctant concurrence of the National Audubon Society, a Canadian Wildlife Service biologist removed six eggs from whooping crane nests in Wood Buffalo Park and left eleven eggs for the birds to raise in the wild. The eggs were transported to Patuxent in suitcases padded with styrofoam and heated with hot water bottles. Ten more eggs were stolen in 1968, and ten more in 1969. Although biologists succeeded in getting eggs to hatch in artificial incubators, many of the hatchlings died within a month from bacterial infections and leg abnormalities.

It would take at least four years for the cranes to mature to breeding age. In the interim Erickson and his colleagues continued to try to breed sandhill cranes, at first with disappointing results. Only twenty percent of the eggs laid by the sandhill cranes were fertile. Finally the biologists began inseminating the females artificially with sperm collected from their mates. The fertility rates rose dramatically, to sixty percent, and in time to eighty percent.

Other rare species of birds had bred so prolifically at the Patuxent center that plans were being made for their reintroduction to the wild—some of the

---

first such experiments ever attempted. In 1971 biologists began releasing groups of captive Aleutian Canada goslings in the Aleutian Islands. Most were killed and eaten by bald eagles, and only a few of the survivors learned to migrate to the winter range of the subspecies, the central valley of California. Frustrated, the Patuxent researchers caught some wild Aleutian Canada geese and penned them for several months with the captive goslings. When the birds were freed together, the wild geese flew off by themselves. The subspecies did not begin to gain in numbers until steps were taken to control fox populations on the Aleutian Islands and restrict hunting in the central valley.

The Patuxent researchers also released dozens of captive masked bobwhites into the dry scrub of southern Arizona, only to watch them scatter and disappear. Finally in 1975 a group of captive chicks were moved to the release site and held in large pens filled with desert vegetation. Occasionally they were harassed by people, dogs, and a trained hawk so they would learn to hide. After three weeks they were released. They stayed together and adapted well to freedom. The researchers found it was even more successful to release masked bobwhite chicks with Texas bobwhite foster parents. A Texas bobwhite male could be trapped from the wild and put in a brood box with a clutch of masked bobwhite chicks. (Both male and female bobwhites care for their young. Males were chosen as foster parents because they could be easily sterilized and would not interbreed with masked bobwhites.) Once the male had bonded with the chicks, he was freed so he could raise them in the wild.

Through the years the captive whooping cranes at Patuxent remained nervous and temperamental. Some chicks developed deformed legs because they gained too much weight from eating a commercial bird mash. Others contracted bacterial infections that badly sickened or killed them. (The same infections had caused only minor problems in sandhill cranes.) Even at a young age they displayed extreme aggression toward each other. They were kept in pens with turkey chicks that, because of their shorter stature, enabled the cranes to assert their dominance.

The birds began to reach breeding age in 1971 and 1972. As they matured each one was observed to see if it gave the two-note calls of a male or the three-note, rattling calls of a female. Special stadium beacons were timed to shine before dawn and after dusk so the birds would have as much light each day as at the northern latitudes of Wood Buffalo Park. A few pairs acted like they wanted to breed; they courted and danced and ran across their pens as if eager to fly north. Yet they laid no eggs.

The wildlife propagation program was beginning to come under harsh criticism from Congress and a public that was impatient for results. In late 1973 a Patuxent biologist named Cameron Kepler decided to move four pairs of cranes to separate pens. It had been the custom to keep most of the adult cranes in community pens with about ten birds to promote pairing. This arrangement had worked well with sandhill cranes. Unfortunately, the whooping cranes maintained a rigid social hierarchy that prevented them from courting; the males competed for dominance among themselves, while the more submissive females established a separate pecking order.

Kepler chose the dominant male and female from the flock and placed them in a secluded pen. He paired other cranes as well, according to their status in the flock. The first year the birds appeared too unsettled to breed. In April 1975, however, the dominant pair began building a nest and dancing with each other. Since artificial insemination had worked so well in sandhill cranes, Kepler inseminated the female with sperm from another whooping crane male. A short time later she laid two fertile eggs. One hatched, but the chick lived only six days.

As Kepler watched the other pairs try to breed, he realized something was seriously wrong. The birds bugled together, staged their courtship dances, and tried repeatedly to copulate. But each time a male mounted a female, either he slipped off backwards, as if he was having trouble keeping his balance, or his mate moved out from beneath him before he finished.

Whooping cranes had been living at Patuxent for eight years, and only one pair had bred. Each spring more eggs arrived, taken from wild nests in Wood Buffalo Park. Biologists from the Canadian Wildlife Service were growing impatient. Removing the eggs from the wild did not seem to harm the population; in fact, wild whooping crane young were surviving in record numbers. It was important to hold some cranes in captivity in case the wild flock was destroyed by a disease outbreak or some other catastrophe—but, as Canadian biologists pointed out, the stolen eggs could be sent somewhere besides Patuxent.

Late in the spring of 1975, the Fish and Wildlife Service, at the request of the Canadian government, began to experiment with fostering whooping crane eggs into the nests of wild sandhill cranes. The concept had first been proposed by a Canadian biologist in the 1950s, and it was elegant in its simplicity. If successful it would create a new flock of whooping cranes that migrated between the Grays Lake National Wildlife Refuge in southeastern Idaho and the Bosque del Apache National Wildlife Refuge in central New Mexico.

Grays Lake was the breeding ground for one of the largest concentrations of greater sandhill cranes on the continent. The cranes that nested there had been studied extensively; researchers had even switched eggs from nest to nest to make sure the birds would tolerate such manipulation. That first year, biologists took fourteen whooping crane eggs from nests in Wood Buffalo Park and placed them in the nests of sandhill cranes at Grays Lake. The chicks would be raised by their foster parents and, if all went well, taught to migrate in autumn to the Bosque del Apache refuge, eight hundred miles to the south. The following year they would return to Grays

Lake. Eventually they would pair with other whooping cranes and raise young of their own.

There were potential problems, of course, such as the possibility that the whooping cranes would imprint on the sandhill cranes and never breed with their own kind. But the Grays Lake project was appealing for political as well as biological reasons. It was to be conducted by scientists from Canada, the University of Idaho, the Patuxent center, and the Fish and Wildlife Service southwest regional office. It effectively dissolved what the Canadian Wildlife Service and many American biologists viewed as a monopoly of the whooping crane recovery program by officials at Patuxent.

During the first spring, nine of the fourteen transplanted eggs hatched, and four whooping cranes migrated successfully to New Mexico with their foster parents. Encouraged, the biologists moved more eggs to Grays Lake in subsequent years. But the terrain of the refuge was being drastically altered by a prolonged drought; wetlands that were normally flooded had completely dried by the summer of 1977. The lack of water reduced the area where the cranes could forage and made it easier for coyotes and other predators to reach their nests. Although seventy-five eggs were taken to Grays Lake between 1976 and 1978, only forty-five hatched. Only eleven chicks lived long enough to migrate to New Mexico.

At the Patuxent center, whooping cranes were finally beginning to breed, but they produced fertile eggs only when the females were insemi-nated artificially. By 1978 the Patuxent flock had contributed more than thirty eggs to the Grays Lake experiment. Many did not hatch. When they did, the chicks often seemed lethargic, and their legs and bills were a different color from the chicks that hatched from wild eggs. Roderick Drewien, a biologist from the University of Idaho who ran the Grays Lake project, mentioned the discrepancy to Kepler. "He couldn't explain it, and neither could I," Drewien said. "But it made me start to wonder if there was something wrong with the artificial incubation procedures at Patuxent."

In late 1977 a biologist named Scott Derrickson took over management of the whooping crane flock at Patuxent. Like Kepler, Derrickson was a behaviorist. The cranes were still unable to copulate, and the policy was to inseminate each breeding female three times a week with sperm collected from her mate. About half the eggs were fertile. All were taken from the nest and placed in incubators to protect them from breakage and to encourage the cranes to lay more. But artificial incubation, if done incorrectly, can have pronounced effects on embryo health. Drewien's observations struck Derrickson as important. He wondered if the development of the chicks was being adversely affected because they were not brooded by live birds.

It occurred to Derrickson that the sandhill cranes at Patuxent could be tricked into sitting on whooping crane eggs, at least for part of the month-long incubation period. The sandhill cranes were not high strung; it seemed unlikely that they would break eggs in the nest. In 1978, as the sandhill cranes began to lay, Derrickson switched their eggs with fertile whooping crane eggs. The first chicks to hatch contracted parasites and a bacterial infection. But once the scientists began taking precautions against nest contamination, the results were profound. "We started seeing chicks that were more robust, almost immediately," Derrickson said. "There was a difference even when the egg was left under a parent for as little as five or six days. We never figured out why.

"All of a sudden there was a lot of demand for sandhill crane foster parents. We wanted every egg that was going to be shipped to Grays Lake to be incubated naturally, at least part of the time. So we'd leave an egg under a parent bird for ten days and then switch it with another. It got to be kind of nightmarish, shuffling eggs around all the time."

Derrickson continued to look for other improvements that could be made to the breeding program. The behavior of whooping cranes was one of the most intricate and ritualized of any species of bird. The Patuxent cranes had been hand-raised without parents, either alone or in small social groups with lots of aggression between individual birds. As adults they were sexually inept. Perhaps their upbringing had scarred them psychologically. Perhaps they needed to be handled more gently, and only by people they recognized. Derrickson knew there were ways of overcoming psychological problems in cranes. Another behavioral biologist, George Archibald, had "mated" with a female whooping crane that had become so imprinted on people that she ignored male cranes. For several breeding seasons Archibald slept in a pen with the female at the International Crane Foundation, which he had founded in Baraboo, Wisconsin. He danced with her in feigned courtship and inseminated her with sperm. Finally, in 1977, she laid fertile eggs.

During the spring of 1978, Derrickson spent day after day observing the whooping crane pens from a nearby blind. It was obvious that the birds wanted to breed; they danced frequently and gathered nesting material. But just before a team of biologists arrived to do an insemination, the cranes began to pace nervously, as if they could sense that something was about to happen. After the procedure they remained agitated for several hours.

The inseminations were conducted briskly, since the Patuxent researchers were concerned about handling the cranes too much. More attention was paid to completing the procedure efficiently than to making the birds feel comfortable. Derrickson set out to make some changes. "It just seemed to me that the gentler we were with the cranes, the more likely we were to get good results," he said. "We started fiddling with all sorts of little wrinkles. We decided we would only go in to do inseminations at certain times of day, so the birds would know what to expect. We also used the same crew of four people, and no one else, so the birds would get used to us."

Meant to Be Wild

Derrickson hoped to reduce the psychological trauma of being handled sexually, which appeared to affect the birds profoundly. His methods were scientifically sound. But like George Archibald, he and his assistants approached the cranes with uncommon patience and compassion. Under Derrickson's direction, the insemination team worked cautiously, and members revised their techniques depending on how individual birds responded to them. It might be said that, rather than raping the cranes, Derrickson's team seduced them.

The technique was refined over a three-year period and used on the rare Mississippi sandhill cranes at Patuxent as well as the whooping cranes. Breeding females were inseminated three times a week, as before. They were inseminated again as soon as they laid an egg. "That increased the fertility rate, but it also meant we had to be back on the birds within an hour after they laid," Derrickson said. "That meant watching every pair of birds, every day for months. You're talking about massive burnout on the part of the staff. But there was never a question of whether the birds would be covered; we just did it."

Before the sperm was inserted, one of the team members would rhythmically stroke the female's thighs to heighten her sexual excitement. Afterwards the handlers would continue to stroke her until she reached what could only be described as an orgasm. The strong contractions of her sexual organs seemed to suck the sperm farther into her reproductive tract. Some birds experienced climaxes so strong that they collapsed in the handler's arms.

Other biologists have described Derrickson's work with the whooping cranes at Patuxent as masterful. From 1982 to 1984 the five breeding females in the flock laid a total of ninety-three eggs, more than had been laid in the previous seven years combined. Eighty-one were fertile. The increase was so large as to be startling, both to Derrickson and his colleagues at Patuxent.

Derrickson, though, was still bothered by the inability of the cranes to breed on their own. In 1981 he began letting pairs of whooping cranes raise sandhill crane chicks, with the hope that the experience would sharpen their parental instincts and their desire to mate. But the birds seemed pathetically inept at copulation; the males simply could not stay mounted on the females long enough to finish.

It was a longstanding policy to clip tendons in the wings of the cranes so they would be unable to fly. As he watched pairs fail to breed again and again, Derrickson speculated that, by clipping the males' wings, biologists had destroyed the birds' ability to keep their balance. As an experiment, he began holding full-winged juvenile birds in pens that were fenced on top. In 1984 he placed a four-year-old female in a pen with a seven-year-old male. Both were full-winged and had been raised by sandhill crane foster parents. "They started acting just like birds in the wild," Derrickson said.

"They were doing everything to indicate that they were going to breed normally the next year, including dancing frequently and building a fairly elaborate nest." But that autumn an outbreak of eastern equine encephalitis killed the female of the pair and six other whooping cranes at Patuxent. Most were young females that had only recently reached maturity.

By then Derrickson had decided to take a job as the curator of birds for the Conservation and Research Center, a breeding facility run by the Smithsonian Institution and the National Zoo outside Front Royal, Virginia. Two of the technicians who worked on the artificial insemination team with him also left Patuxent soon after. Researchers at the wildlife center decided not to continue Derrickson's unorthodox insemination methods. The whooping cranes that survived the outbreak of encephalitis did not react well to the change in technique. From 1985 to 1989 the cranes laid only thirty-two fertile eggs.

The fostering experiment also was not going well. After ten years biologists had begun to doubt that whooping cranes would ever breed at Grays Lake. The cranes that survived predation had learned to migrate with their foster parents, but they were inclined to get tangled in barbed-wire fences and to fly into electric power lines, which often killed them. For some reason, many more male whooping cranes survived than females, and the sex ratio of the population was badly skewed. At the end of the spring migration, the female whooping cranes did not return to Grays Lake, but dispersed to other areas. This is a common behavior in wild cranes. Unfortunately, the female whooping cranes wandered far enough from Grays Lake to keep them from finding mates. And a renewed drought had badly degraded the marshes. By 1988 so many wetlands had gone dry that the Canadian Wildlife Service stopped transferring whooping crane eggs to the refuge from Wood Buffalo Park.

In the spring of 1989, Fish and Wildlife Service biologists brought a six-year-old whooping crane female to Grays Lake to see if she would choose a mate and breed. The female had been raised at the Patuxent center; the staff there had described her as too nervous and wild to be easily kept in captivity. She soon started keeping company with one male, but abandoned him for a second male in June. She could fly, but she did not seem to have the strength of a crane raised in the wild.

The pair remained together until the other cranes left on their fall migration. One afternoon in early October the male took off and began circling upward, gaining altitude in long, slow spirals. The female could not keep up; although cranes commonly migrate at three thousand feet, she seemed to have trouble reaching fifteen hundred feet. The birds landed and began to feed. Over the next week the male tried several more times to coax the female to follow him. In mid-October, just before the onset of a snow storm, he flew south alone.

Nearly three hundred whooping crane eggs were taken to Grays Lake between 1975 and 1988, yet in early 1993 there were only ten whooping cranes mixed among the sandhills—seven males and three females. Rod Drewien, who has worked on the project since its inception, believes that the male whooping cranes would breed if they could find willing mates. Drewien hopes to capture one or two of the females, which still spend the summers outside the wildlife refuge, and confine them to a large enclosure on the preserve until they establish bonds with males. He acknowledges, however, that the results of the fostering project have been largely disappointing.

* * *

After more than two unproductive decades of captive breeding, the wild whooping crane, ironically, is now more secure than it has been in half a century. Through all years that the cranes at Patuxent and Grays Lake failed to nest normally, the single remaining population of wild cranes prospered. Over the winter of 1992–93, 136 whooping cranes were counted at the Aransas wildlife refuge in Texas. Conservationists worry that the population could still be devastated by an oil spill or accident involving the toxic chemicals routinely barged along the Intracoastal Waterway south of the flock's wintering grounds.

In January 1993 biologists with the Florida Game and Freshwater Fish Commission settled fourteen juvenile whooping cranes into a large enclosure in the Three Lakes Wildlife Management Area along the Kissimmee River. By placing the birds in the midst of territory used by sedentary sandhill cranes, the biologists hoped to create a whooping crane flock that would live in Florida year-round, and so would not face the dangers of a twice-a-year migration. The experiment, carried out with the Fish and Wildlife Service and the Canadian Wildlife Service, was based on a release procedure that had been used successfully on Mississippi sandhill cranes. The whooping cranes were released in February. Although five were killed by predators within ten weeks, the surviving birds showed no inclination to migrate.

Only eight of the cranes used in the experiment came from Patuxent; the other six were reared at the International Crane Foundation in Wisconsin. In 1989 Patuxent officials transferred nearly half of the center's fifty-four whooping cranes to George Archibald's foundation as a safeguard against infectious diseases. Building on Scott Derrickson's work, Archibald and his staff are attempting to coax the cranes to breed without artificial insemination.

Many endangered species biologists believe that the whooping crane recovery program has been hampered by overly cautious management of the captive birds, such as the Patuxent center's abandonment of the artificial insemination techniques pioneered by Scott Derrickson. The Patuxent program has always attracted close political and scientific scrutiny, which perhaps has made administrators reluctant to deviate from standard research protocol. Moreover, it is controlled by a federal bureaucracy that is underfunded and inefficiently managed, as federal audits have shown.

In contrast, the recovery program for the peregrine falcon was directed from the start by a group of independent scientists who formed a nonprofit foundation in the early 1970s. Although many birds in North America were declining in numbers then, no loss seemed so dramatic, or so easily explained, as the loss of the birds of prey. The once-common bald eagle had grown rare through much of its historical habitat. Populations of ospreys and brown pelicans were diminishing along the Atlantic, Gulf, and Pacific coasts. But the most severely affected were the peregrine falcons that were thought to be the fastest, most powerful flyers in the world, and that, until the 1950s, ranged freely across every continent except Antarctica.

Throughout history peregrines have been admired for their unusual strength and ability. They have extraordinary eyesight and can spot prey from as far away as three thousand feet. When pursuing prey they are capable of bursts of speed of up to two hundred miles an hour. Pursuing a songbird or duck, a peregrine will overtake its prey from above and fall into a dive. It gains momentum with two or three wingbeats, then partially folds its wings, plummets, and kills with a powerful strike of its talons. Everything about the peregrine's physique—its bluntly shaped head, pointed wings, and tapered tail—is built for speed.

The disappearance of the birds of prey began suddenly around 1946, when the chlorinated hydrocarbons used as pesticides, such as DDT and dieldrin, came into wide use. Until then peregrine falcons had not been greatly affected by the settlement of North America. In the 1930s a survey of historical falcon eyries, or nesting sites, in the East had shown that at least eighty percent were still in use—a remarkable number, given the vast alterations to the region over the previous two hundred years. In the early 1940s, roughly three hundred fifty pairs of peregrines were estimated to be nesting east of the Mississippi River and the Ohio Valley. By the mid-1960s, however, the falcons had vanished from the East, and the western population was quickly losing numbers. Falcons were also disappearing from traditional eyries in many parts of Europe.

In 1965 a group of scientists organized an international conference on peregrine falcons at the University of Wisconsin in Madison. Among the conference participants was Tom Cade, then a young ornithologist with a faculty appointment at Syracuse University. An amateur falconer, Cade had owned his first trained hawk at the age of nine. As an adult he had developed more of an interest in wild birds than trained birds, but he held a special fascination for the peregrine.

Cade and other scientists worried that without intervention the peregrine could soon become extinct. "There was a lot of anecdotal evidence that the species was in trouble, even before it was gone from the East," he recalled. "Archie Hagar had been watching fourteen pairs of peregrines in Massachusetts since the thirties, and suddenly in the late forties they began to disappear. He thought it was raccoons getting into their nests. Other people had other theories. In Wisconsin people began comparing notes and realizing that the problem was pesticides, both here and in Europe. We wouldn't find out the exact physiological reasons for a number of years. But it was clear by then that something serious was going on."

Subsequent studies showed that the falcons tended to accumulate high concentrations of pesticides in their fatty tissues. The most prevalent was DDE, a residue of DDT. When birds of prey feed on animals that are slightly tainted with DDE, the chemical builds up in their systems and causes metabolic changes. Eventually females begin to lay eggs with thin shells that can be easily crushed during incubation. "The peregrine populations in the worst trouble were those south of the boreal forests of Canada—the American eastern population most obviously, and the western birds too," Cade said. "The northern subspecies in Canada and Alaska weren't affected as quickly. But virtually every region where falcons nested had some mortality because of pesticides."

In 1967 Cade accepted a position as professor of ornithology at Cornell University. He moved to Cornell with the intention of establishing a center where he and other researchers at the Laboratory of Ornithology could raise endangered birds of prey. As soon as a "hawk barn" could be built, Cade assembled a collection of gyrfalcons, peregrine falcons, prairie falcons, and various species of raptors for breeding.

From the start he harbored hopes of returning the peregrine to the empty eyries of the East. It was, he acknowledged, an ambitious and perhaps unattainable goal; although humans had been training birds of prey for centuries, falcons had been bred in captivity only a few times before. Adult birds caught from the wild seldom adjusted well to confinement. Cade wanted to take some chicks from wild nests and hand raise them, with the hope that they would breed when they matured. But there were no longer any wild nests in the East, and in western states, where falcons were also suffering from pesticide poisoning, wildlife officials were reluctant to let Cade visit eyries to remove peregrine chicks. "There was some concern that by taking chicks for breeding, in the long run we would end up hurting the western population," Cade said. "So we scattered our efforts. We obtained a few chicks from eyries in the Southwest, Alaska, and Canada." A number of falconers also provided the program with peregrines.

In 1971, with the hawk barn stocked with falcons and raptors of various species, the peregrine breeding project officially began. Cade had formed a nonprofit organization called the Peregrine Fund to help cover costs at the facility, and donations were coming in. Graduate students, researchers from Cornell, and several volunteer falconers helped care for the birds, which were kept in chambers with one end open to the outside. Most of the captive peregrines were only a year old, too young to breed. But that spring, researchers managed to inseminate a goshawk and a red-tailed hawk with sperm collected from captive males. As far as they knew, it was the first time artificial insemination had succeeded in birds of prey.

The peregrines at the center did not begin to breed until they reached the age of three. Once they started, however, even Cade was astounded by their fecundity. In 1973 three females laid a total of twenty-four fertile eggs. All but four hatched. (Normally, peregrines lay only three or four eggs in a season. They may breed again if the original clutch is lost.) "We hadn't expected to start really producing peregrines for about five years," Cade said. "But when we got the birds away from the environmental hazards, it was amazing to see how readily they responded to our efforts."

Some of the young birds were seldom handled; others were flown in falconry. What mattered most, Cade found, was that they be treated differently, according to their temperaments. If the researchers were careful, even high-strung birds could be enticed to nest. Frequently, however, the scientists had difficulty matching the peregrines with mates they would accept. "That was a constant problem, one of the quirky things about the business," Cade said. "The females tend to be quite a bit larger, and most of the time they're more aggressive than the males. But we had one male who had belonged to a falconer in California, and he was notorious. He'd literally pin females down. Finally we paired him with a female from Chile that had been confiscated because she was imported into this country without the proper papers. For some reason he immediately accepted her. They became one of our most productive pairs."

The researchers also selected a few male chicks to serve as sperm donors. Each was purposely raised to be sexually imprinted on a human handler. When the bird matured, the handler would visit the falcon wearing a special hat with which the bird could copulate, and the two would "mate." The sperm, collected in the rim of the hat, was then used to inseminate females.

By 1974 the founders of the Peregrine Fund had started to plan for the first release of chicks to the wild. The use of DDT had been severely restricted in the United States, and scientists predicted that the country's birds of prey would gradually recover from its toxic effects. In the interim, the western falcon population could be propped up by fostering captive-bred chicks into the nests of wild birds that were unable to raise their own. In the spring of 1974, researchers in Colorado noticed that a pair of wild peregrines had lost a clutch of eggs—their second clutch that season—from breakage. The scientists decided to put two prairie falcon eggs in the nest to hold the peregrines at the site. The birds accepted the eggs and began to incubate them. As soon as the chicks hatched, biologists swapped them for two peregrine chicks from Cade's captive breeding program. Several weeks later the chicks fledged. It was the first time captive-bred peregrine chicks had been successfully released.

The task of reintroducing peregrines would be trickier in the East, where there was no longer a breeding population. Since the eastern peregrine was all but extinct, the Cornell researchers were faced with creating a new population from whatever captive birds they could breed. Recently the foundation had obtained some falcon nestlings from Scotland and Spain. These birds were different subspecies from the ones that had lived in the eastern United States.

A few scientists outside the program questioned the wisdom of introducing the offspring of European peregrines to North America. (In many cases the release of an exotic species can be devastating to a natural system. The species may compete so effectively against local animals or plants that it throws the entire system out of kilter.) Cade predicted that the Spanish and Scottish falcons would act no differently from the American peregrines; he doubted that many people would be able to tell them apart. "We purposely wanted to create a mishmash population," he said. "We felt that was the best way to guarantee enough genetic diversity for the birds to survive and reproduce in the eastern environment."

Cade and his colleagues hoped to return the peregrine falcon to the East through a technique of falconry known as hacking. When a young falcon or hawk is about four weeks old, it is placed with others in a box high atop a tower or a cliff. Although the chicks cannot escape, the crate contains slits or bars through which they can see out. This enables them to become familiar with the surrounding landscape. After a week or two the crate is opened. The birds are free to begin flying on their own, but they can return to the hack site for food. When they have learned to hunt, the falconer retraps them and begins training them to hunt on command. The scientists planned to hack young peregrines from traditional falcon eyries, from towers they constructed, and from tall buildings in cities, where the birds would have an abundant supply of pigeons for food. Instead of being retrapped once they learned to hunt, they would be allowed to disperse naturally.

Late in the spring of 1974, Heinz Meng, one of Cade's collaborators, settled two peregrine chicks into a hacking station on top of a ten-story building at a college in New Paltz, New York. A week later, when the chicks' flight feathers were fully grown in, he freed them. During the day the young peregrines ventured into the sky above New Paltz, returning to the hacking station to feed on the pigeon carcasses Meng set out for them. For three weeks Meng followed their progress through radio transmitters mounted on their legs. One day the signals disappeared. "We didn't know whether the birds had left the area, or whether something had happened to them," Cade recalled. "They had been doing so well there was no reason to think anything was wrong. Then one morning someone found the wing of a peregrine falcon in a trash dumpster."

At least one and probably both of the peregrines had been killed by vandals. Cade and Meng could take some solace from the apparent success of their first release. But the disappearance of the birds at New Paltz made it clear that reintroduced falcons would face unforeseen dangers.

And now that the Peregrine Fund had proved it could breed falcons, its pending release program was coming under some criticism. Scattered groups of conservationists objected to a proposal by Cade to release the falcons from elevated platforms that could act as eyries if the birds returned to breed. Some of the platforms were to be built along the Atlantic Coast, where there were abundant shorebirds. Although falcons had migrated along the coast, they had seldom been known to nest there; the eastern falcons had been inland cliff-dwellers. By establishing eyries in new locations, the critics objected, Cade's program would change the behavior of the peregrine and possibly endanger nesting colonies of shorebirds. Yet the sites selected for the coastal releases could be easily protected from vandals and owls, which the researchers feared would prey on the young falcons. "We felt that even if we changed the way peregrines selected their nesting sites, we wouldn't be changing the way they hunted or went about their business," Cade said. "But we decided that if people were really concerned about the impact on shorebirds in a certain area, we wouldn't build hack sites there. Because of that, several choice sites had to be abandoned."

In 1975 the Peregrine Fund attempted to release sixteen peregrine chicks from hacking stations set up throughout the Northeast. Each of the stations was staffed by observers who put out food for the chicks and guarded them from predators. A few days after the birds were freed, they began chasing each other playfully, honing their flying skills by making quick turns and dips. Within a few weeks all play stopped; from then on the only thing the falcons seemed to care about was finding prey.

Four of the chicks were freed from an old gunnery tower in the Aberdeen Proving Ground, an Army installation spread across some isolated islands in the Chesapeake Bay. A few weeks after the release, they scattered. One roosted for several months on an office building in downtown Baltimore; another often perched atop a spire on the Chesapeake Bay Bridge near Annapolis. Two settled near Baltimore Harbor, where they could be seen hunting pigeons.

The other falcons were released in more rural settings, including the Taughannock Falls Gorge in upstate New York, where peregrines traditionally nested in cliffs high above a plunging waterfall. One morning an observer arrived at the Taughannock Falls station to find two of the chicks dead. They had been killed during the night, probably by great horned owls. Since the chicks had no parents to protect them from predators, the presence of owls meant that the Taughannock Gorge—and many other traditional eyries—could not be used as hacking sites.

Meant to Be Wild

Of the sixteen peregrines released in 1975, twelve survived until the fall. They were descended from Canadian and Alaskan falcons that spent the winters as far south as Argentina, yet they did not seem to have the same strong urge to migrate long distances. Instead, they behaved like the falcons that had originally inhabited the East; with the onset of cold weather, some flew to warmer states, while others stayed in the region where they were released.

Elated by the rate of survival, the members of the Peregrine Fund staff began preparing for more releases in 1976. Early that spring they were surprised to find five of the year-old falcons roosting near the hacking stations where they had been freed. Cade had predicted all along that young peregrines would be drawn by instinct back to the place where they had first taken wing, and where, if they could find mates, they might settle down to nest. But he had not expected his theory to be proven so quickly.

The Peregrine Fund freed thirty-seven falcons in the Northeast and Wisconsin during 1976. To the north, biologists from the Canadian Wildlife Service released another twenty-one peregrines using the hacking techniques perfected by Cade. For the first time, the Cornell program also attempted to release a pair of bald eagles from a hacking station in upstate New York. It was the year of the American bicentennial, and the national symbol had become nearly as rare in the lower forty-eight states as the peregrine. The Fish and Wildlife Service and the state Department of Environmental Conservation wanted to collaborate with the Peregrine Fund on a release program to shore up the population of eagles in the Northeast. In New York, where as many as forty-one pairs of eagles had traditionally nested, only a single pair remained. "That pair hadn't bred successfully for many years, and someone suggested that we give them some healthy eggs or chicks to rear," Cade said. "The state wanted to make a big entry into endangered species work for the bicentennial, so we also decided to try hacking some eaglets to see if it could be done."

Releasing eaglets is more difficult than releasing falcon chicks, because they are slower to become independent. In addition, as Cade and his staff would discover, eaglets tend to fall off hacking towers before they can fly well. In late June, two nine-week-old wild eaglets from Wisconsin were placed in a hacking station in the Montezuma National Wildlife Refuge near Seneca Falls. Almost as soon as the nest box was opened, the male fell out and flew into some adjacent woods. He refused to return to the hacking tower for food, and he ignored the offerings of fish that an observer set out for him in the woods. Staff members feared that he would starve. But a week later, the male returned to the tower and ate greedily. By late summer both released birds were hunting skillfully. They left the area as the weather began to turn cold.

In only six years the Peregrine Fund had established a breeding program for falcons and devised a successful method of returning birds of prey to the wild. Its techniques had been adopted by federal wildlife officials in Canada and the United States. It had attracted financial support from private individuals, the federal government, and conservation groups like the National Audubon Society. A second breeding center had been completed in Fort Collins, Colorado, in cooperation with the state Division of Wildlife. The center was to be the base of operations for falcon releases and fostering experiments in the Rocky Mountains.

The Cornell researchers continued their breeding experiments and made plans to release peregrines from hacking stations throughout the Northeast and the Middle Atlantic States. Although some of the released birds were descended from Spanish and Scottish falcons, they closely resembled the falcons that had originally inhabited the East. But in June 1977, to his surprise and dismay, Cade received a telegram from the Fish and Wildlife Service ordering him to stop freeing falcons of Spanish or Scottish descent. The telegram noted that President Jimmy Carter had recently signed a presidential order prohibiting the introduction of exotic animals in the United States.

The European falcons were among the most reliable breeders in the Cornell program; several had been mated with American falcons. Within a few days Cade received more bad news. Federal officials had also decided he could no longer release chicks descended from falcons captured from the Pacific Northwest and Canada, which were a different, more common subspecies from the peregrines found east of the Rockies. Only endangered subspecies from North America could be released.

Overnight the service had raised legal issues that, in effect, threatened to bring the restoration of the eastern peregrine to an unfortunate end. Incredulous, Cade appealed to the director of the service for reconsideration of the new policy. He pointed out that the main purpose of the Endangered Species Act was to preserve and restore true species, not subspecies. Federal officials refused to change their stance.

In January 1978 Cade sent a plea for help to some of the world's most eminent ornithologists. The scientists responded with dozens of letters to service administrators explaining why the release of European and Pacific Northwest peregrines should be allowed to continue. Overwhelmed, the service agreed to allow the Peregrine Fund to free chicks descended from the Northwest falcons. The progeny of falcons from other countries could be released occasionally, administrators declared, but only with prior approval. The exotic birds were to be considered part of an experimental population; if any problems arose, they would be trapped or shot.

The Peregrine Fund, freed from federal restraint, has continued to release falcon chicks in the East every year. In 1981 three pairs of wild falcons nested at hacking sites in New Jersey and raised healthy chicks that fledged into the wild. It was the first time in twenty years that wild falcons had bred east of the Mississippi. Twelve years later, more than ninety-nine pairs were breeding along the Atlantic Coast. Unlike the falcons that lived in the East fifty years ago, many nested on skyscrapers in cities and on towers set in salt marshes near the ocean. A portion were descended from falcon subspecies indigenous to Scotland, Spain, Australia, and Chile.

In 1980 the pair of bald eagles released as chicks from the hacking tower in the Montezuma National Wildlife Refuge were discovered nesting in a red maple swamp near Watertown, New York. They have raised chicks there each year since. Hacking techniques for bald eagles have been refined, and state wildlife agencies have released hundreds of wild eaglets from Alaska and Florida. In 1990 thirteen pairs of eagles, all of which had been hacked into the wild, were nesting in New York. And the number of breeding eagle pairs in the continental United States was estimated at two thousand six hundred and sixty—three times the population of 1974.

The Peregrine Fund has released more than three thousand peregrine falcons, in conjunction with its Fort Collins breeding center and the Santa Cruz Predatory Bird Research Group in Santa Cruz, California. In 1991 the organization turned the maintenance of the eastern peregrine population over to state and federal wildlife agencies. The headquarters for the foundation has moved to Boise, Idaho, where Cade and his colleagues have established the World Center for Birds of Prey. There they raise critically endangered falcons and raptors from around the world.

Cade enjoys an international reputation for his work as a breeder of rare falcons, and his dedication to restoring such rare species as the harpy eagle in Central America and the Mauritius kestrel, once thought to be the most endangered bird in the world. But his visionary work with peregrines remains his best-known and most respected accomplishment.

"We were lucky," he said, "in that we were able to get the peregrines that were produced in captivity back into the wild quickly, within one to three generations. It appears that there was no loss of their wild attributes. That, I think, made a tremendous difference."

VIII

LEAVING THE ARK

The Arabian Oryx and the Golden Lion Tamarin

On the last day of January 1982, in a remote region of the Arabian Desert, a herd of ten white antelope ambled through the gate of an enclosure where they had lived for more than a year. The animals bunched in a tight knot and moved cautiously to a feeding trough that had been placed outside the gate. A short distance away a group of biologists and nomadic tribesmen stood watching in silence. As the animals stepped beyond the safe encirclement of the pen to the open desert, a few of the tribesmen fell to their knees, praising Allah.

In this quiet way the Arabian oryx was returned to its rightful place on a stony, hostile plain in Oman. The oryx, a striking animal—white, with distinctive black face markings and long, slightly curved horns—had been hunted nearly out of existence. No one knew of any that still lived in the wild. The freed animals had been shipped to Oman from the San Diego Wild Animal Park, where they were born. To the biologists, the release marked one of the first tests of the theory that an endangered mammal could be restored to the wild using captive stock. To the tribesmen, it marked the resurrection of a cherished symbol, and perhaps a means of preserving their fragile culture.

The Arabian oryx release was one of two wildlife projects conducted during the 1980s that set standards of excellence for endangered species reintroductions. In the second, scientists attempted to save the last struggling population of a tiny, reddish primate known as the golden lion tamarin. Both projects were carefully conceived and executed, and based on a clear vision of what a reintroduction might accomplish. Both were designed to benefit the people of the regions economically and culturally. But where biologists returned the oryx to a desert habitat that was changing comparatively slowly, the tamarin release took place in a Brazilian rain forest that was being radically altered every day. In restoring tamarins, scientists hoped to save not only a species of monkey but an entire jungle.

While every wildlife recovery project has its own peculiar set of complications, saving a population of mammals can be vastly more difficult than saving a population of birds. For one thing, a mammal cannot be switched from nest to nest, as an egg can be; the techniques used to bolster wild populations of whooping cranes, western peregrine falcons, and other rare birds cannot be used with, say, red wolves or black-footed ferrets. Also, many rare mammal species have intricate social customs, so that instead of releasing scattered individuals, scientists must try to release cohesive family groups. Beyond these considerations, the trouble and expense of rescuing a mammal from extinction vary greatly. Predators are among the hardest to preserve; grazers are among the easiest. In that sense the Arabian oryx had an advantage over animals like the red wolf.

The Arabian oryx belongs to a family of horselike antelope and is much chunkier than the petite pronghorn antelope of the American West. Its stocky appearance is diminished, however, by its tapered horns and beautiful pied face. The body is a vibrant white, fading to chocolate legs. In addition to a black blaze on its nose, it has black streaks through its eyes that widen to jagged patches at its jowls. The oryx can run only in short bursts, yet it has remarkable stamina for walking distances. Reintroduced animals have been known to live nearly a year without drinking water.

Three centuries ago, when the oryx lived throughout the Arabian Peninsula, it was admired for its unusual powers of survival, and it was praised in poetry for the beauty of its soft-brown eyes. The species was also known to the ancient Egyptians, who raised oryx and bound the animal's horns together so they would merge into one. These single-horned creatures may have provided the fodder for the myth of the unicorn.

Traditionally bedouin hunters considered shooting an oryx to be a feat of some prestige. The antelope were prized as trophies and sought after for their meat by hunters on camelback. Oryx were persistent travelers, if somewhat slow; hunters often had to ride long distances to find them. All this changed in the 1930s, however, when safaris led by Arab princes began chasing oryx in jeeps. The antelope were easily caught by cars on the stony plains where they tended to live in the summer. Sometimes they eluded hunters by running into the great sand dunes that lay scattered like lakes across the desert terrain. Escape did not always mean survival. The body of the oryx has a delicate system of water retention and temperature regulation that enables it to survive prolonged periods of intense heat. This balance is destroyed when an animal runs more than a short distance. Many oryx fled successfully from pursuers, only to collapse from heat exhaustion and dehydration.

By 1960 the oryx population probably consisted of less than a hundred animals. It was confined to the southeast end of the Arabian Peninsula, to South Yemen and a harsh region of Oman known as the Jiddat-al-Harasis. Herds spent much of their time foraging for grasses and forbs on the flat, stony plains and relaxing under squat, umbrella-shaped acacia trees. In cool weather they sometimes made their way into the midst of a vast sand sea where people seldom ventured. Although oil development was spreading throughout Arabia, the Jiddat was still remote; the only people who ever passed through the region were the nomads known as Harasis and occasional hunters.

In 1961 a hunting party from Qatar surprised several herds of oryx and killed forty-eight of them. News of the slaughter soon reached members of the Fauna Preservation Society, a conservation group based in Great Britain. Enraged members made plans for an expedition to trap some oryx and take them to a wildlife reserve, so the species would not be gunned down to the last animal. The chief game warden of Kenya, Major I. R. Grimwood, offered to lead the capture team.

Grimwood and a small group of men arrived in South Yemen in April of 1962. Rumors were circulating that another sixteen animals had been killed by hunters, which, if true, left less than thirty oryx in the wild. During a month of searching, the party managed to corral only four oryx. One had already been shot, and it died from stress. The surviving three animals were flown to a temporary holding facility in Kenya. Grimwood left Arabia convinced that as few as eleven wild oryx remained.

Once the antelope had been captured, a debate arose about what should be done with them. There were no reserves or national parks within their native range, and the officers of the Fauna Preservation Society were reluctant to confine wild oryx to a zoo. But with only three breeders it made little sense to turn them loose, even on protected range. The rescue effort had attracted wide attention among conservation organizations, including the newly formed World Wildlife Fund. The leadership of the groups could not agree on who should be declared the legal owner of the captured animals, and there was concern that the animals would contract hoof and mouth disease in Africa. Finally, Grimwood arranged to send the oryx to the Maytag Zoo in Phoenix, Arizona. Within two years they were joined by seven more, contributed by the London Zoo and noblemen from Kuwait and Saudi Arabia.

Unlike many disappearing species, the Arabian oryx had been bred successfully by private animal collectors long before its disappearance in the wild. The Fauna Preservation Society harbored few doubts that the World Herd, as the animals in Arizona were known, would multiply bountifully. The first calf was born in 1963. Over the next three years, another five calves were born, but all were male. Biologists wondered ruefully whether the oryx might be bound for extinction after all. The World Herd did not produce a female until 1966, when the seventh calf was born.

By then the Los Angeles Zoo had purchased from Saudi Arabia a pair of oryx and their female calf. Because these animals were unrelated to the oryx held in Phoenix, their addition to the World Herd increased its genetic diversity. The oryx continued to breed prolifically; by 1971 the two zoos held a total of thirty. Six were transferred to the San Diego Wild Animal Park to reduce the chances that an outbreak of disease would threaten the entire herd.

As the World Herd expanded, events were unfolding in the Middle East that would transform the populace and much of the landscape. Great deposits of oil had been discovered on the Arabian Peninsula, and drilling rigs were becoming commonplace in the desert. Even the perimeter of the Jiddat-al-Harasis of Oman was attracting more people. The hunting of antelope with vehicles had been illegal since 1964, but the ban was not well enforced. In 1972 a hunting party killed six oryx on the Jiddat that may have been the final survivors of the wild population.

The nomadic tribes of the region were also constrained by the pressures of development. For centuries the Harasis had roamed the desert in small family groups, tending herds of goats. They lived beneath the limbs of acacia trees, not even using tents to protect against the sun and heat. Frequently a family moved its camp to find better grazing. The harsh climate demanded that they select their camp sites and manage their stock with utmost care. Competent herders had little need for money; most of the time they drank goat and camel milk and bartered for coffee, dates, and fabric.

The oil boom eroded the culture of the Harasis quickly and severely. As development expanded toward the Jiddat, the value of livestock doubled, and the tribesmen learned that there were easier ways to raise goats than turning them loose in the desert. By 1975 a few families even owned pickup trucks with which they could haul water—a luxury that changed their lives considerably. The Harasis moved less frequently in search of good grazing, since they could feed their goats on imported grain. With a steady supply of nourishment, the goats could breed all year and not just after rains, when the grazing was lush. There was more wealth among the tribes, more exotic food, more expensive commodities, more need for cash. There was every reason for the men to abandon their traditional professions, leave their camps, and take jobs working for the oil industry.

In 1974 the ruler of Oman, HM Sultan Qaboos bin Said, asked his adviser on conservation, Ralph Daly, about the possibility of restoring the Arabian oryx to the Jiddat-al-Harasis. After his rise to power in 1970, the sultan had actively encouraged the development of the remote regions of his country. But now, with the oil industry expanding into every quarter, he feared that the indigenous wildlife and culture of Oman were being utterly destroyed. The Harasis mourned the disappearance of the oryx; they had revered it for its beauty and regarded it as an integral part of the desert landscape. If a few oryx herds could be returned to the wild, perhaps some of the men could work as guards against poachers while their families continued to tend stock. The sultan instructed Daly to explore the idea, and he indicated that he would support a reintroduction project financially.

By then there were enough oryx in captivity to chance losing a few in a reintroduction attempt. In addition to the animals in American zoos, several private animal collectors also owned thriving oryx herds. A number of Arab noblemen had started collecting wild antelope in the 1960s, ostensibly for purposes of conservation. To the consternation of the Fauna Preservation Society, such capture expeditions depleted the last tiny bands of wild oryx. By 1980 wealthy families held possession of between eighty and a hundred Arabian oryx—as many as had been produced by the World Herd.

It was impossible for conservationists to tell whether the oryx in private collections were inbred or genetically healthy. The board members of the Fauna Preservation Society and the World Wildlife Fund agreed that any reintroduction should be conducted with animals of known pedigree, and in habitat that had been disturbed as little as possible. Despite the influx of people along its periphery, the Jiddat was comparatively pristine. Its climate and terrain were so formidable that even most Harasis avoided the interior of the desert in summer. After inspecting the region, Daly recommended that some living quarters and holding pens be built in a remote basin known as Yalooni.

In 1979 an English biologist, Mark Stanley Price, moved to Yalooni to plan the logistics of the first release. Stanley Price had previously worked with captive beisa oryx, a more common, sandy-colored species, in Kenya. Arriving in the Jiddat, he found the desert landscape to be full of rich, earthy hues in the slanting light of dawn and dusk, but bleak beneath the glare of the midday sun. Except for the sand sea, the region was flat and rocky, with scattered pans of limestone where grasses and shrubs grew. Between the limestone depressions the ground was strewn with small, shiny pebbles covered with colorful lichens. It was, Stanley Price discovered, a region of extremes; summer temperatures averaged a hundred and ten degrees in the shade, and in the winter temperatures dropped into the forties.

Yalooni was one hundred and eighty miles over rough roads from the nearest settlement. Occasionally Harasis stopped by the outpost, which became known as a place where the desert dwellers could get an occasional meal and assistance with problems. Stanley Price and his small staff purposely cultivated friendly relations with the tribesmen, who would have to share the open rangelands with released oryx. In rainy years grasses were plentiful on the Jiddat, but rainy years were rare. During 1980 not a drop of precipitation fell at Yalooni.

In preparation for the arrival of the oryx, Stanley Price built a row of pens that opened into a two-hundred-and-fifty-acre enclosure, an area large enough for a herd to develop normal social and grazing behaviors. Unlike most conservation projects, the oryx reintroduction was never handicapped by scant funding. Construction of the pens and buildings at Yalooni was funded by approximately $1,000,000 from the sultan's treasury. The expenditure seemed justified to Stanley Price and Daly, especially if the reintroduction enabled some of the eighteen hundred Harasis in the region to continue living in the desert as oryx rangers and goat herders, instead of depending on the oil industry for jobs.

In the spring of 1980, after they had been carefully tested for brucellosis, blue tongue, rinderpest, hoof and mouth disease, and a range of other ailments, five oryx were flown to Oman from the San Diego Wild Animal Park. By the fall of 1981, fourteen oryx had been sent to Yalooni. Most acclimated easily to the climate of the Jiddat.

Slowly, Stanley Price allowed the animals to venture into the large enclosure so they could establish social relations. The animals immediately began browsing on grasses and forbs, which they seemed to prefer over the familiar hay that Stanley Price's staff set out for them. At first the males, vying for a dominant role in the herd, circled each other warily and thrust their horns together in fierce shoving matches. Over time, however, their aggression faded. The animals looked identical to their ancestors; there was no reason to think they would not adjust well to a life of freedom—except for a single question. During their generations in captivity they had been given a constant supply of water. No one knew whether they could readjust to a desert where they would have to go for months getting moisture only from the plants they ate.

The initial release was planned for the cool months of midwinter, when grass was relatively scarce and the herd might congregate around a feeding station rather than immediately scattering across the desert. The animals were to become self-sufficient, but gradually. Especially during dry periods they had grown accustomed to twice-daily feedings of hay and alfalfa. As the release date neared, each morning workers moved the animals' feed troughs a little closer to the gate. On January 31, 1982, they positioned the troughs outside and opened the gate.

The oryx released in the first herd included ten animals ranging in age from nearly four years to eight months. One of the males had achieved a clear position as herd leader, and one of the females was pregnant. They drifted over to the troughs, unperturbed by a small group of observers. After feeding for a few minutes they fanned out to browse on grasses. By night they had wandered a mile into the desert, but they returned to the area around the pen to sleep.

Over the next month the troughs were moved a little each day, until they were two and a half miles from the pen. A dozen Harasis men had been hired as rangers to track the herd and to patrol the plains around Yalooni. Stanley Price knew the animals were growing used to being followed by people in trucks, which might prove to be their undoing if they were ever pursued by hunters. But by getting close to the oryx, the rangers could carefully check individuals for injuries and signs of disease. The reintroduction had been widely publicized and was supported by the local population. Stanley Price doubted that anyone would try to poach the oryx, at least at first.

In March a female calf was born in the wild, the first, possibly, in many years. (Although oryx had not been sighted since 1972, wild herds may have survived in the most remote portions of the Arabian Desert.) The birth was a cause for rejoicing, but also for increased surveillance. Oryx calves spend the first month of their lives lying hidden from the central herd. Even with the constant vigilance of their mothers they are easy prey for ravens, golden eagles, caracals, and the few Arabian wolves that on occasion travel through the Jiddat. On the seventeenth day after the calf's birth, a sandstorm passed over the area where she had been lying. When the storm subsided there was no sign of her or her mother.

The Harasis camped in the region had offered their assistance in emergencies, and now Stanley Price called for their help. For three days men from thirty tribal families combed the desert in search of the mother, with the hope that the calf was still alive. Finally the pair was found, unharmed and well hidden, a short distance from where they had disappeared.

To a large degree the success of the oryx reintroduction depended on the acceptance and good will of the Harasis. Before the release, the tribe had agreed not to herd their animals within six square miles of the pens at Yalooni, although the camp was in the midst of their traditional grazing lands. Once the oryx were freed, the Harasis agreed to keep their stock at least three miles away from grazing herds. The loss of prime pasture was a considerable inconvenience, if not hardship, for the nomads. But if the oryx were to make the transition to freedom, they needed moisture from the lushest, most nutritious grasses and forbs.

After they moved away from the troughs at Yalooni, the oryx appeared to have little need for water. Stanley Price and the Harasis rangers were surprised by how quickly the animals adjusted to their hostile environment. They appeared to recognize even subtle landmarks, such as a pile of camel bones on the desert floor, and most of the animals seemed to have an accurate knowledge of where they were. When an oryx smelled water in the wind, it would walk to the source, cutting a path as straight and true as if it were reading a compass. After nine months without water, two females walked twenty miles to the settlement at Yalooni. Each quickly consumed more than two gallons.

In the early summer of 1984, after an absence of nearly two years, the reintroduced herd wandered back to the vicinity of their old pens. The Jiddat was in the midst of a drought; grazing was scarce, and the oryx probably returned to Yalooni for food. Perhaps to their surprise, they found the area occupied by another herd of eleven oryx released only a few months before. As biologists watched, the two dominant males fought viciously, thrusting their horns together and shoving each other for nearly half an hour. Then the herds split up and formed separate territories.

For the next two years the oryx fed consistently on the grasses Stanley Price's staff set out for them near camp, but their dependence on humans was broken with the coming of plentiful rain in 1986. The heaviest showers fell near the coast of Oman, and the Harasis traveled far south with their herds. The oryx, left without even minimal competition from domestic stock, moved to some richly vegetated plains thirty miles south of Yalooni. "They had a chance to set up territories and interact in a complete absence of people," said Timothy Tear, an American biologist who worked with the project from 1984 through 1989. "They've been independent ever since, and their population is absolutely burgeoning."

Stanley Price left the Jiddat at the beginning of 1987, satisfied that the antelope might permanently repopulate at least a portion of the Arabian Desert. Nevertheless, in a book on the project, he warned that the species would become too inbred to survive unless new bloodlines were added to the free-ranging herds. Inbred calves often die shortly after birth, and Stanley Price worried that genetic deficiencies would badly retard the recovery of the oryx. During the next two years two more herds were released, partly to increase the diversity of the wild population. They immediately mingled with the wild herds and split into smaller groups. "There's a lot of intermixing now," Tear said. "The herd sizes change, depending on what the grazing is like. The females aren't being mated by the same males over and over, the way they were at first. Nobody predicted that the social behavior would change this quickly or be this dynamic."

In 1993, eleven years after the first release, one hundred seventy-three Arabian oryx were living in the wild, and more than thirty Harasis men were employed as rangers or technicians for the project. With their knowledge of the desert the Harasis proved to be unusually good at tracking oryx, especially at finding animals that became lost. The work appealed to them. And while Yalooni was far from their families, their schedules were arranged to allow them to spend at least a week of every month at home. In the custom of Muslim culture, they shared their earnings freely. Because of their affluence, their families and many of their tribesmen were able to continue living as nomads. "The project has helped some of the Harasis retain a modified traditional lifestyle," Tear said, "but I stress the word modified. Too much has changed in the region for their culture to go back to being the same as it once was."

In the spring of 1990, Stanley Price visited Yalooni for the first time since turning the management of the oryx project over to Tim Tear. He found the oryx and their grazing range to be in excellent condition, largely because of an unusual amount of rainfall. "The animals have absolutely prospered," he said. "There has essentially been no mortality."

The return of the Arabian oryx to the wild is known as a classic wildlife experiment because it combined sound scientific technique with a thorough understanding of the people whom it most drastically affected. Through 1992 the Harasis rangers still knew exactly how many oryx were ranging free, and where. They kept detailed records on how the animals interacted, and scientists had been able to chronicle the frequent breakup and formation of herds. The biologists found that subordinate males tended to break away and become loners, rather than staying in herds with dominant males. Females sometimes traveled alone and could be bred by different males, which greatly increased the population's genetic diversity. The biologists also found that when the third herd was released in 1988, the animals were much quicker to explore and try new foods, as if the presence of free-ranging oryx in the desert somehow set them at ease. They integrated rapidly into the wild population. More animals are to be released each year, until the wild population reaches three hundred.

"Since the first day, we've been able to monitor the oryx very closely and watch key social interactions—what happens when a stranger tries to move into a herd, or when two males challenge each other," Tear said. "We know just about every animal—how old it is, what its lineage is, who it has traveled with in its lifetime. In other words, we've monitored the complete development of a wild population from captive-bred stock. I don't know of any other project that has this kind of detailed information.

"The observations are bound to become less precise as the population expands; there will just be too many animals out there to follow so closely. But the main point is this: Reintroduction can work, especially when you involve local people. One helps the other."

* * *

The Arabian oryx was driven to near extinction because its natural defense, its ability to outwalk pursuers, was no defense at all against hunters in jeeps. Such sudden upheavals in the natural balance have become common over the past century in nearly every part of the world. When a species is denied safe haven, either through mechanized hunting, deforestation, or the intrusion of toxins into its food source, it is only a matter of time before its population collapses. Yet too often recovery programs deal only in numbers; they strive to boost populations without preserving the wild conditions that the animals need to survive.

In 1984 a team of scientists from the National Zoological Park in Washington, D.C., released fourteen golden lion tamarins to a forest reserve in the coastal lowlands of eastern Brazil. The tamarin is a petite monkey with a luxuriant orange-red mane that frames its small gray face. Except for its face and long, slender hands, its body is covered with a rich pelage that shines blond in some light and reddish in others. With its striking beauty and inquisitive nature, it is an utterly appealing creature. By publicizing the reintroduction, the scientists hoped to make tamarins a symbol of conservation and, ultimately, to make the preservation of tamarin habitat an international issue. By saving rain forest, they also hoped to save thousands of species—some unknown, some assumed by the general populace to be worthless or even deserving of extinction.

This farsighted approach is a principal reason the tamarin reintroduction has become known as one of the most well-conceived wildlife restoration programs in the world. There are other reasons as well. Through an ambitious education program, the staff of the reintroduction project has won a previously uninformed public to its cause. It has created jobs for local residents and taken steps to restore cleared forest. Researchers have also devised a training program for released tamarins that helps the monkeys learn to forage and move through the strange layered world of the tropical jungle.

In the early 1970s, when the National Zoo first became heavily involved in trying to rescue tamarins from extinction, no one suspected that the species would become such a symbol for conservation. For years, efforts to preserve tamarins were concentrated only in Brazil, where a respected primatologist, Adelmar Coimbra-Filho, had convinced government authorities to ban the export of tamarins and other rare primates. In addition to suffering from the piecemeal destruction of their habitat, the tamarins of Brazil were being captured for medical research and for sale as pets. Three species of lion tamarins, all endangered, were known to be indigenous to the country, including the golden-headed lion tamarin and the black lion tamarin. (A fourth species, the caissara, or black-faced lion tamarin, was discovered in 1990.) The golden lion tamarin once inhabited the coastal forests of two east Brazilian states, Rio de Janeiro and Espirito Santo, but by the 1970s deforestation had restricted it to small, swampy remnants of jungle northeast of the city of Rio de Janeiro. No one knew how many still lived in the wild.

Ironically, the species' position in captivity was even more precarious. In 1972 a consortium of American zoos known as the Wild Animal Propagation Trust conducted a census of tamarins being held in zoos around the world. Of the seventy tamarins located, only a handful were proven breeders. Unless something could be done to boost the birth rate, the species was likely to perish in captivity even before it went extinct in the wild.

The results of the census caught the interest of Devra Kleiman, an ethologist and mammalogist at the National Zoo. Kleiman wondered whether the tamarins' chronic breeding problems might be caused by improper management. Even as late as 1973, many zoos cared for tamarins the same way they cared for Old World monkeys. The animals were grouped together with several adults of each sex. They were fed fruits and vegetables, and only minimal protein. Yet for some time primatologists had recognized that tamarins needed a steady source of protein, such as insects or small animals; otherwise they developed vitamin deficiencies and a form of rickets. "The communication between zoos back then was just terrible," Kleiman said. "For whatever reason, information just wasn't getting out, especially to zoos in other countries."

In 1974 Kleiman was appointed the keeper of the International Studbook for Golden Lion Tamarins. She was responsible for tabulating which tamarins died, which bred, and how many young they produced every year, at every zoo in the world. At that time, studbook-keeping was little more than a matter of listing numbers. "I decided I needed to do more than that, or there wouldn't be any tamarins left to put in the studbook," Kleiman said. She contacted zoos with single tamarins and asked if they could be borrowed for breeding. She inquired among other zoos to see whether they would be willing to trade some animals to increase the genetic diversity of the tamarin families they were holding. She asked the keepers of the primates to use a special feeding protocol worked out by researchers at the National Zoo. And she made a number of shrewd breeding trades that, in effect, gave the National Zoo control over half the tamarins living in captivity.

Kleiman's approach was aggressive and not popular among officials at other zoos. "There was some resistance at first," she said. "But this was happening at a time when the health and the breeding success of the tamarins at the National Zoo was really improving. That gave us a lot of credibility." At the same time, Kleiman began investigating whether the tamarins' breeding had been upset by inappropriate social groupings. Little was known about the species' behavior in the wild; although Coimbra-Filho had initiated some field studies, the jungle was too dense, and the tamarins too spread out, for him to have reached many definitive conclusions. It was thought that tamarins were monogamous and that they lived in family groups. A few studies had also been done on the social behavior of related species in captivity. "A good deal of information was available, but no one had taken the time to put it together," Kleiman said. "There were enough hints to make some educated guesses about the kinds of research that needed to be done."

She began placing two or three tamarins together in separate cages and monitoring their behavior. The monkeys seemed most at ease when an established breeding pair was housed with up to five of its young. Juvenile tamarins helped care for newborn infants, and the experience seemed to enhance their ability to raise young of their own. To her surprise, Kleiman found that females could be kept with their parents only until they started to mature sexually; after that, there was a much greater chance of conflict between mothers and daughters. "There's a great deal of sexual aggression between females," she said, "but we got to the point where we could predict when a family group was getting ready to blow up."

By 1978 Kleiman's studies had initiated major changes in the way tamarins were being cared for at zoos internationally. The results were astounding. Cured of rickets and satisfied with their social arrangements, the tamarins not only bred but bred so quickly that zoos ran out of room for them.

The sudden excess of animals struck Kleiman as an opportunity too precious to be wasted. In 1981 she contacted Brazilian authorities about the possibility of returning some of the monkeys to the wild. Through Coimbra-Filho's efforts the government had been persuaded in 1974 to set aside twelve thousand acres as protected habitat for the species, a patchwork of swampy forest and cleared pasture known as the Poço das Antas Reserve. Nearly half the forest within the reserve had been cut, and it was traversed by a major railroad line. "When I first looked at it," Kleiman said, "I just about threw in the towel. It was in horrible shape. There were squatters. Hunters would get off at a train station in the middle of the reserve, hunt all day, even though it was illegal, and then catch the train home." Early surveys indicated that about a hundred wild tamarins lived within the reserve, but because of deforestation on the fringe of the property, they were isolated from other tracts of suitable habitat. The population appeared stable, but Coimbra-Filho worried that individual family groups were becoming inbred.

In 1983 an American couple moved into a gutted farmhouse on the reserve and set about trying to counter the forces that were pushing the wild tamarins toward extinction. A few years before, James Dietz had completed a doctoral dissertation on the maned wolf population of central Brazil. During the study Dietz and his wife, Lou Ann, lived and worked closely with rural residents. Jim had since taken a job as a research associate under Devra Kleiman at the National Zoo. He returned to Brazil with a grant to study wild tamarins and prepare for a reintroduction attempt in the spring of 1984. Lou Ann, a professional educator, hoped to garner support among local residents for the pending reintroduction through a public education project. To her disappointment, she had not managed to attract much grant money. But she was convinced that she could use the tamarin reintroduction to make local residents more aware of the natural riches being destroyed around them.

On their arrival in Brazil, the Dietzes arranged to hire four of the Brazilians who had worked with them on the study of maned wolves. Before any field work could be started, they had to clean out a ransacked farmhouse and build a simple cinderblock structure for a laboratory. The farmhouse was near the center of the Poço das Antas Reserve, about seven miles from the nearest paved road. "It wasn't a particularly remote site, at least not compared to some other field stations," Jim said. "Once we got out to the asphalt road we were only an hour and a half from Rio. But the dirt roads sometimes got in horrendous condition, especially during the rainy season. There were times when it took a whole day, and all our people, just to get out to the asphalt road to pick up a load of supplies."

Once the project headquarters were established, Jim and his assistants laboriously cut trails through the dense vegetation of the forest so they could map the areas where they were most likely to encounter tamarins. They also searched for suitable places where captive monkeys could be freed. The remaining stands of uncut forest were so thick as to be virtually impassable on foot. Vines and epiphytes covered the tree trunks, and a thick understory of vegetation grew between the trees. The wild tamarins lived in an open layer between the understory and the forest canopy, where they could move about freely on a network of vines. The biologists, mired in the vegetation of the forest floor, became accustomed to hearing the birdlike calls of the monkeys as they passed nimbly overhead.

Through autumn Dietz set up transects to survey both the lands within the reserve and the neighboring ranches. The work proceeded well until late in the year, when the Dietzes learned that a farmer planned to clear thirty-five acres of adjacent forest that contained a family of tamarins. "It happened very suddenly. We had only a couple of days' warning," Jim recalled. "It was right before Christmas, and the farmer had a cutting permit that ran out at the end of the year. There were seven animals living in that area. We hadn't planned to deal with them at all until after the reintroduction. But the farmer wouldn't wait, so we went in immediately to try to trap the animals. We were in there working and trees were literally coming down around us. People were cutting brush out from under us."

Dietz and his assistants set out some small box-traps and settled down to wait as workers continued to fell trees. "We were shocked at the efficiency of the chain saws," he said. Spooked by the noise, the tamarins tried to flee into adjacent woodlands, but they were driven back by rival groups of tamarins. Only five of the animals ventured into the traps. Before the last two could be caught, the farmer's crew set fire to the tract. "We could see them running around in burning branches," Jim said. "We felt pretty miserable."

The biologists managed to salvage the hollow tree where the tamarins had established a nest. They cut a section out of it to be used as a nest box at another site deeper in the reserve. The five captured tamarins were tattooed, and their tails were marked with dye for easy identification. They were closed in the nest box, moved to the new site, and freed. Weeks later Dietz managed to capture the last two members of the family group, an old female and the breeding male. The female was too weak to be released. By the time the male was reunited with his family, his mate had replaced him with another. He left the release site almost immediately.

With the group relocated, Jim returned to his preparations for the reintroduction. In addition to conducting studies of the wild population, the National Zoo had agreed to help the Brazilian government establish a small forest regeneration program, in which trees were planted in certain areas and lime was spread on former pastures. A staff of rangers had also been hired by the Brazilian government to rout out squatters and hunters.

Lou Ann Dietz, meanwhile, was facing challenges of her own. Although she contacted nearly a hundred conservation organizations, she could find few willing to donate funds for educational tools. "Everyone was interested in the reintroduction, but no research had been done on the benefits of an education program," she said. "It wasn't the kind of project an organization like the Smithsonian Institution could sell easily to its donors." Nevertheless, the Smithsonian did make a contribution to the tamarin education fund; so did the World Wildlife Fund and the Friends of the National Zoo. The Wildlife Preservation Trust International, a conservation group based in Philadelphia and the British island of Jersey, donated five thousand posters explaining the ecology of the tamarin and the importance of the rain forest to its survival.

Lou Ann knew she would have to counter some prejudices among the local populace. The Brazilian government had appropriated much of the land within the Poço das Antas Reserve from unwilling sellers, and some of the former owners believed they had been treated unfairly. "We didn't start out with a real good relationship with the local communities," she said. "Nothing had been done to work with the local people; most of them had no idea what the reserve was for. But they were very interested in what we were doing, especially in anything that had to do with the tamarins."

From Lou Ann's account of that first year, one gathers that she pursued her work with little thought for the magnitude of what she was attempting to accomplish. "I never had time to tackle more than one set of problems at once," she said. "Maybe that's why it worked." She set up meetings with local school teachers and with the mayors of the two communities bordering the reserve. She invited the head of a municipal council (the rough equivalent of an American board of county commissioners) to tour the reserve. He accepted with pleasure and brought several other council members along. "After that we put together a meeting and slide show for community leaders," she said. "One of the schoolteachers was so interested she brought eight of her star students. They ended up being the core of my research team."

From the start Lou Ann hoped to conduct a careful survey of residents' attitudes toward tamarins and the rain forest, so she could monitor the success of her efforts. But she lacked the staff for the project until the students volunteered to help. "We put together a survey, as complete as possible, and they went out and interviewed five hundred people. There were those eight students, plus another couple dozen volunteers. They did all of it. Since they were local, they were more accepted than I could ever be, coming from outside and speaking Portuguese with an accent."

The results were in some ways predictable and in some ways surprising. Lou Ann was pleased to discover that many people considered deforestation to be one of the worst problems facing the region. Most of those interviewed recognized pictures of tamarins, but did not know they were endangered, or why. The most startling discovery was that three-quarters of the residents near the reserve watched television—even in rural areas without electricity. "They take car batteries and wire them up to their TVs," she said.

Lou Ann continued to prepare educational programs and materials, including slide shows, pamphlets, and colorful book covers for students. But with the survey completed, she decided to try a more ambitious tack; she asked Brazilian television networks to run public service announcements about tamarins. Although hesitant at first, officials at one of the country's largest networks agreed. "By that time we had a pretty good idea of what kinds of messages we needed to get across," she said. "Forest fire could be a big problem, so we stressed fire prevention. We asked landowners to stop clearing land. Sometimes they would go in and cut trees just to 'clean up' their pastures. We let them know there were good reasons to leave trees standing."

The television messages also discouraged people from keeping tamarins as pets. "That was something I felt I had to word particularly carefully," Lou Ann said, "because it was such an emotional issue. We had to let people know why it was important not to keep a tamarin in their house, even if they loved it."

In the spring of 1984, barely a year after the Dietzes' arrival in Brazil, fourteen captive tamarins were taken to the Poço das Antas Reserve for release. The animals included three breeding pairs and a family group of eight. They had been brought to Brazil the previous winter and held in quarantine at the primate research center run by Coimbra-Filho. All had been chosen for reintroduction in part because they were considered to be of expendable blood; their genes were so well represented in the captive population that it would not affect the species' survival if they died without bearing young.

The researchers had strived to prepare the tamarins as thoroughly as possible for the shock of release. Over a six-month period, a team from the National Zoo—including Kleiman, Dietz, and Benjamin Beck, a specialist in primate cognition—had trained the animals to forage on their own. They concealed food in puzzle boxes designed to resemble hollow branches, or rolled leaves, or other natural vegetation where the monkeys might find fruit and insects. As the tamarins became adept at opening the boxes, the scientists had hidden them more carefully and left some of them empty. The animals were also given whole papayas, bananas, and oranges. Although they had eaten tropical fruit all their lives, they had never been required to peel it or discard the seeds. Frogs and lizards were released into the cages so the monkeys could practice catching small prey. Food was placed so they would have to climb up branches or the frame of the cage to reach it.

In early May the family group was moved to a large outdoor pen on the reserve. Immediately the researchers noticed a problem. Although the monkeys had climbed nimbly around their small cages, they were reluctant to venture up the unfamiliar vines and trees that filled the outdoor enclosure. They were especially frightened of flexible branches that moved beneath their weight. The researchers tried baiting them onto small branches with pieces of banana. In late May they opened a trap door on the top of the pen so the tamarins could leave. "They didn't come out very quickly," Jim Dietz said. "They didn't seem particularly interested in being reintroduced."

When the tamarins did leave the pen, they tended to perch on its top. The scientists, concerned that the monkeys would be preyed on by hawks, hastily dismantled the structure. The tamarins moved into the trees, where another problem became apparent: they were incapable of choosing a sensible path through vegetation. When one spotted some fruit in an adjacent tree, it seldom jumped from branch to branch or scurried along vines to reach the food. Instead it ran back and forth along a branch, growing agitated, or it climbed down and walked to the nearby tree. On the ground the monkeys were in much greater danger from feral dogs and other predators. "They were clumsy. They fell a lot. They got lost from their food supply and had to be rescued," Jim Dietz said. "It was a surprise, because in a small cage tamarins move like lightning."

Three more tamarin pairs were freed on the reserve that summer. One of the pairs, calling loudly through the forest, quickly discovered the location of the family group. The vocalizations and a later encounter between the two groups threw the family into disarray. Some of the animals ran off and become completely disoriented.

Within weeks several of the tamarins had disappeared or been killed. When possible, the researchers trapped those that appeared to be starving or unable to cope with the stresses of freedom. They also tried to coax lost tamarins back to familiar territory with bits of banana. The family group adapted more readily than the tamarins that were released in pairs. The juveniles adapted most easily of all. Over a period of months the youngest monkeys learned to climb on vines and small branches, and gradually they taught the adults to follow them. In February, however, an unknown disease killed five members of the group. By the following June only three released tamarins and a pair of twins, born in March, were still alive.

The reintroduction had been expensive, and a second attempt would probably not be much cheaper. But the scientists had a compelling reason to try again. Over the course of his field work, Jim Dietz had trapped a hundred wild tamarins so he could put radio collars on them or mark their long, elegant tails with rings of dye for identification. Whenever he caught a monkey he also drew a vial of blood for a study on genetic variation within the wild population. An analysis of the blood samples showed the tamarins to have little genetic diversity, with the potential for severe inbreeding. The only way to reverse the trend was by translocating tamarins from other areas or by releasing tamarins that had been raised in captivity. Before a reintroduction could be successful, however, the researchers needed to figure out how to teach captive tamarins to behave like wild animals.

In 1985 two family groups were released on a citrus plantation near the reserve that was owned by the Brazilian state of Rio de Janeiro. One of the groups had been coached extensively in survival skills at the zoo in Washington. For a month it was kept in an outdoor cage with a large nest box and a dense network of small branches and vines. Every four days researchers tore down the tangled vegetation and rebuilt it so the tamarins would have to learn how to move through it anew. Food was scattered around the cage in puzzle boxes or hidden in bromeliads, a spiny plant that grew abundantly on trees within the rain forest. The second family group was kept in an outdoor cage but given food in bowls up to the time of its release. Survival training was expensive and time consuming, and Beck and Kleiman wanted to make sure it was worthwhile; the untrained family was to act as a scientific control.

After release, the trained tamarins began exploring and foraging for food more quickly than the untrained tamarins, which simply sat listlessly. It seemed clear that the untrained animals would not survive without some sort of help. Beck and a Brazilian student, Inez Castro, decided to try fashioning a feeding platform from six lengths of bamboo. They punched small holes in the bamboo, filled the hollow stalks with cut fruit and insects, and plugged the holes with bark or leaves. The tamarins, inspecting the platform, could pull morsels of food out if they learned to look beneath the plugs.

Twice a day the platform was loaded with food and water and hoisted into the trees with ropes. After several days, other bamboo stalks were suspended nearby. As the tamarins learned to check each stalk, the scientists hung a bamboo "trail" through the woods, leading toward a swamp that contained abundant fruit and insects. Each day the tamarins explored a little farther into the jungle. Every evening, however, they returned to the artificial nest box where they had grown accustomed to sleeping. Finally the scientists began hiding food in bromeliads and tree crevices within reach of the trail.

The differences in behavior between the two family groups gradually diminished, although all the animals traveled and foraged more awkwardly than wild tamarins. To Beck, the complex survival training program conducted before the release seemed too costly and time-consuming to be worthwhile. It occurred to him that, trained or not, released tamarins always appeared reluctant to move away from their familiar nest boxes. The postrelease training regimen he developed with Castro had proved effective, and it had not required as much work as building complicated matrixes of vegetation inside cages. The animals could learn to move through the forest after their release, so why not let them?

In the spring of 1986, Beck decided to release a family of tamarins on the grounds of the National Zoo. Parts of the zoo were heavily wooded and could conceivably serve as a training ground for the monkeys. Weeks before the release, a plastic picnic cooler was placed in a cage with the family so members could become accustomed to using it as a nest box. One day the tamarins were anesthetized and fitted with radio collars. After they recovered they were closed in the nest box overnight and moved outside to a large grove of oak and beech. When they awoke the following morning they were free to leave. But for several days they did not move more than a few yards. "We call this the concept of the psychological cage," Beck said. "And for our purposes, it's a very useful phenomenon." The tamarins lingered near the nest box long enough for researchers to begin training them to forage on a platform and "trail" of hollow pipe stuffed with morsels of food. Like their reintroduced brethren in Brazil, they moved clumsily and sometimes fell. As the summer progressed, they learned to climb and run through trees more deftly. Eventually, they began to visit the outdoor cages of other zoo animals, always under the watchful eye of observers.

Beck and Castro's training regimen was modified and used to acclimate tamarins to the Brazilian jungle during releases through 1993. Some of the reintroduced animals were trained on zoo grounds before being taken to Brazil. They were freed on twelve private ranches near the Poço das Antas Reserve because of concerns that the reserve was becoming too crowded.

Trained after release, the monkeys tend to survive for longer periods. But the scientists at the National Zoo have come to accept that captive-born tamarins might never be taught to behave as wild animals. Instead, the hopes for the restoration of the species now lie with the tamarins born in the wild to reintroduced animals. "Tamarins raised in the wild don't have the same problems moving and finding food as tamarins raised in cages," Beck said. "When the released animals breed, that's when we realize the full benefits of reintroduction. It's up to us to get them to that point."

Until the coastal rain forest is no longer under constant threat of destruction, the Golden Lion Tamarin Conservation Program will continue, although its leadership is being slowly transferred from American to Brazilian scientists. Largely because of the education campaign, wealthy landowners now consider it an honor to have tamarins roaming through their forests. The landowners agree never to cut the trees within areas where tamarins are released, or allow the property to be hunted. It is hoped that one day the tamarins within the reserve and those on private land will meet and intermix, as family groups must have done when the forest was unmarred by grazing and agricultural development.

Of the one hundred thirty-seven tamarins freed between 1984 and 1993, forty-three were still alive in early 1993. A few had been living independently in the wild for five years; others still received supplemental food from the field staff. The seventy young being raised by the reintroduced animals all behaved like the offspring of wild tamarins.

It is believed that close to five hundred tamarins now range through the coastal forests. Over the first seven years of the reintroduction program, the research team spent a total of $1,000,000, most of which was provided by private grants. With the population no longer in a state of crisis, researchers have the luxury of conducting rigorous scientific studies that were impossible a decade ago.

Devra Kleiman, who continues to coordinate the conservation program, has undertaken a study of tamarin vocalizations, especially the long-distance call, a high-pitched, birdlike trill that ends in a series of clucks. Kleiman is developing ways of using the call to census tamarins and measure how the behavior of reintroduced animals differs from wild animals. Jim Dietz and collaborator Andy Baker have fitted tamarins in twenty-five native family groups with radio collars and are conducting basic research on their behavior and ecology. "We're not setting out to manage this species; we're setting out to understand them," Dietz said. "That's important in itself. This population is really going to determine what kinds of steps will be taken to save other tamarin species."

In addition to his continuing work on tamarin survival training, Ben Beck has started to experiment with trapping reintroduced tamarins that have lived in the forest for a year or longer and pairing them with captive mates. By associating with veterans of the forest, the captive monkeys learn more quickly how to forage and make their way in the jungle. "We're finding that they don't have to be provisioned for as long," Beck said. "They're given access to food, but they don't take it."

The education project started by Lou Ann Dietz has grown well beyond her expectations. Although she continues as an adviser, the project has been placed under Brazilian direction. "There are six local women on the staff, full time," she said. "They've done plays, musical concerts, presentations for landowners, ecology classes for teachers, a whole range of programs." Many of the students who helped her conduct her initial survey of local attitudes have gone on to college, an unusual accomplishment for children of rural Brazil. In addition, a number of the field assistants who worked with the reintroduction team have gone to college to study forestry, conservation, and biology. "A lot of

Brazilians have received job training through the tamarin project," Lou Ann said. "That's one of its most important benefits."

In February 1990, a fire broke out on a ranch adjacent to the Poço das Antas Reserve. Although it was the height of the normal wet season, no rain had fallen for more than a month. Fed by powerful winds, the flames spread quickly to the reserve, burning sections of cut-over swamp forest and old pasture where trees had recently begun to take hold. No tamarins were injured, nor were their primary habitats damaged. But the fire delayed the regeneration of hundreds of acres by at least six years. "It was very depressing," Kleiman said. "The forest will recover, eventually. But it was a major setback."

It is tempting to assume that the golden lion tamarin, because of the attention and resources it has received, will rebound to healthy numbers. The scientists know better. "There's some challenge to putting tamarins back in the wild," Jim Dietz said, "but we can basically do that now. The real problem is that we may not have enough forest left for the species. The coast is still under incredible pressure from development. I'm afraid we're going to have tamarins left in little islands, if we have them at all."

There are at least two critical lessons to be drawn from the examples of the tamarin and the oryx. First, the amount of preparation needed to return a captive animal to the wild will vary radically, depending on the characteristics and temperament of the species involved. Second, animal reintroductions work best when they are carefully interwoven with the natural ecology of the release site and the social customs of the people who live nearby. Even if the reintroductions had not worked—even if the released oryx and tamarin had proved completely incapable of living in the wild—the projects still would have benefitted the natural landscape and the local populace. It is partly this careful, comprehensive approach that, in comparison, makes many American wildlife projects seem so lacking in wisdom and vision.

With the rain forests of eastern Brazil still being converted to agriculture and urban development, and with the unstable political climate in the oil regions of the Middle East, loss of habitat may still destroy the wild populations of golden lion tamarins and Arabian oryx. Until some way can be found to halt the devastating pace of habitat destruction around the world, scientists have little choice but to maintain substantial collections of rare animals in pens. And with each generation that a species spends in captivity, the chances are compounded that its wild character will be altered. It remains to be seen whether animals kept in zoos for many decades can ever be taught to live as their ancestors did in the wild.

THE ESSENCE OF WILDNESS

The polar bear shuffled along a concrete ledge, his back to spectators. A cool, dim glow lit his large, indoor cage the way the spring sun might glance against Arctic snow, and the ledge where he stood was textured with small bumps like the surface of an ice flow. Separating him from the bustle of the zoo was a chilly, blue pool and, beyond that, a wall of plate glass through which passersby could watch him swim with fluid, powerful strokes.

The bear ignored the pool and the people who watched. Over and over he moved forward three careful steps, backed up three steps, and moved forward. He paced with ponderous slowness, advancing and retreating over a tiny section of the slab as if in a trance. As he stepped backwards he shook his head slightly, like a horse might to fend off a fly. His dance was eerie, disturbing, and ceaseless. Not once during the ten minutes I watched him did he break stride.

The bear was being kept at the Washington Park Zoo in Portland, Oregon, one of the better zoos in the country. His cage had been painstakingly and expensively designed to mimic his native land, and it might have succeeded, except for the fact that it *was* a cage. Polar bears are among the largest and most free-ranging animals—and among the most difficult to keep in captivity. A number of scientists and conservationists believe they should not be placed in zoos at all. The bear I saw in Portland probably invented his halting minuet to relieve his boredom and exercise his languishing muscles.

To me, watching speechlessly, the behavior was startling. Ever since the red wolf was returned to North Carolina, I had been thinking and writing about the differences between truly wild animals and those raised in zoos and breeding centers. Through my work I had traveled to several endangered species projects and visited some of the country's most respected zoos.

I had not found the drab cages and iron bars I remembered from my childhood. During the past decade many American zoos have undergone radical changes in philosophy and design. All possible care is taken to reduce the stress of living in captivity. Cages and grounds are landscaped to try to make gorillas feel immersed in vegetation, as they would be in a Congo jungle. Zebras gaze across vistas arranged to appear (to zoo visitors, at least) nearly as broad as an African plain.

Yet, strolling past animals in zoo after zoo, I noticed the signs of hobbled energy that had found no release—large cats pacing a repetitive pattern, primates rocking for hours in one corner of a cage. These truncated movements are known as cage stereotypes. In extreme cases the constantly repeated motion begins to wear on the animal's health, and it must be put through a lengthy process of rehabilitation. Usually, however, cage stereotypes have no obvious physical or emotional effects. Many animal specialists believe they are more troubling to the people who watch than to the animals themselves. Such restlessness is an unpleasant reminder that—despite the careful interior decorating and clever optical illusions—zoo animals are prisoners, being kept in elaborate cells.

How "wild" can a polar bear, or a wolf, or any creature be after several generations in a cage? This question has assumed growing importance over the last fifteen years, with the ascendance of captive breeding as an important tool for preserving endangered species.

I knew well the rationale for breeding as many endangered animals in captivity as possible. Once a species falls below a certain number, it is beset by inbreeding and other processes that nudge it closer and closer to extinction. If the animal also faces the whole-scale destruction of its habitat, its one hope for survival lies in being transplanted to some haven of safety, usually a cage. In serving as trusts for rare fauna, zoos have committed millions of dollars to caring for animals that will never be placed on display. And in the past decade many zoo managers have given great consideration to the psychological health of the animals in their care. Yet the more I learned about animals bred in enclosures, the more I wondered how their sensibilities differed from animals raised to roam free.

In the wild, animals exist in a world of which we have little understanding. They may communicate with their kind through "languages" that are indecipherable by humans. A few studies suggest that some species perceive landscapes much differently than people do; for example, they may be keenly attuned to movement on the faces of mountains or across the broad span of grassy plains. Their social structures may be complex and integral to their well-being. Some scientists believe they may even develop cultural traditions that are key to the survival of populations.

But when an animal is confined, it lives within a vacuum. If it is accustomed to covering long distances in its searches for food, it grows lazy and bored. It can make no decisions for itself; its intelligence and wild skills atrophy from lack of use. It becomes, in a sense, one of society's charges, an orphaned child, completely dependent on humans for nourishment and care.

Meant to Be Wild

How might an animal species be changed—subtly, imperceptively—by spending several generations in a pen? Sometime after my trip to the Portland zoo, I posed that question to William Toone, the curator of birds at the San Diego Wild Animal Park, which is a breeding center for the California condor. "I always have to chuckle when someone asks me that," Toone replied. "Evolution has shaped the behavior of the condor for hundreds of years. If you think I can change it in a couple of generations, you're giving me a lot of credit."

If all proceeds as planned, the condor will be reintroduced in the next several years—only a moment after its capture, in evolutionary terms. Perhaps Toone was right; perhaps the wild nature of the birds would emerge unscathed, although I was not convinced. But what of species that will spend decades or centuries in confinement before they are released? Can humans, with all our technical skill, hope to preserve the essence of wildness in such animals—an essence we can barely perceive, let alone understand?

The challenge of saving rare animals has always hung on two hinges, one purely physical, one psychological. From the start of the earliest breeding programs, critics worried that caging animals would change them, if not in body then in mind and perhaps in spirit. But concerns about psychological flaws became secondary beside amassing evidence that genetic poverty could damage an animal beyond repair. The consequences of inbreeding can be easily documented; they appear in the animal's inability to produce vibrant young, or its lack of resistance to disease. Scientists believe they may also manifest themselves in the animal's temperament, causing it to be perpetually nervous or easily excited, blurring the distinction between bodily and mental health. Faced with such obvious problems, it is tempting to place all of one's efforts on repairing the physical hinge and hope that the psychological hinge will hold on its own.

In the past twenty years the science of genetics has been expanded and refined until it commands a key position in the campaign to preserve rare animals. This has caused an interesting shift in philosophy and public policy. Thirty years ago the meager funds available for endangered species work were devoted almost exclusively to learning more about animals in their natural habitat. Now, though more resources are available, they must be split between field work and captive propagation. In some programs the opinions of zoo officials and geneticists hold at least as much sway as those of biologists who have dedicated their careers to studying endangered animals in the wild.

There is good reason, of course, to reduce the chances of inbreeding, good reason to fear what geneticists refer to as the deleterious allele. In a healthy, expanding population, natural selection plays an unfettered hand.

Animals that have unusual abilities, such as speed, or sharp eyesight, or some knack for gathering food, tend to survive best and pass their traits to subsequent generations. Slow, sickly individuals cannot compete as well or rear young as prolifically. The fittest survive, and the population grows in vitality. Or so holds Darwinian wisdom in its most simplistic interpretation. But as animals are overhunted or isolated by habitat destruction, the number of genes available to each generation may be halved or quartered. Genetic variants are lost as cousin mates with cousin and father with daughter. Certain traits become increasingly prevalent, while others spill out of the population like seed from a split bag.

The changes that result may be of minor or major consequence. Many Florida panthers, for example, have slightly kinked tails caused by the prevalence of a deleterious genetic variant, or allele. The kink is of little importance, except to show that the fifty or fewer panthers that still live in south Florida are inbred. Of much greater concern is another possible genetic defect that prevents the animals' testicles from fully descending and may seriously inhibit males' ability to manufacture sperm.

Inbred populations often suffer from birth defects, low fertility, and susceptibility to disease. In addition, inbreeding is frequently accompanied by problems in the demographic make-up of wild populations. The sex ratio grows skewed, or the population becomes dominated by elderly animals, with too few juveniles. Those of breeding age begin to have difficulty finding suitable mates, and the birth rate drops. All these factors reinforce each other, until the species is caught in a downward spiral geneticists refer to as the extinction vortex. As the last population falls deeper into the funnel, chances increase that it will be destroyed by a bad storm or a fatal disease.

And yet, in certain circumstances animal populations can recover from severe constrictions in the flow of genes from generation to generation. A few species have been rebuilt to thriving numbers from a handful of breeding pairs. For example, the world population of Laysan teal, at more than a thousand, is believed to have descended from a single pair of birds.

Some animal species, such as lemmings and house sparrows, are less vulnerable to extinction than others. These animals are known as r-selected species; they mature sexually at an early age and produce frequent litters of many young. They also tend to be opportunistic and adaptable to environmental change. Even without the effects of factors like habitat destruction, numbers of r-selected species may fluctuate dramatically from year to year. Their make-up is well suited to rebounding from dramatic, cyclical declines.

In contrast, populations of so-called K-selected species, such as the whooping crane and the grizzly bear, tend to remain stable in the wild, barring outside disturbance. These animals first breed at a comparatively old age and give birth to only one or two offspring a year, if any. They invest great time and energy in rearing each of their young. Their populations do not become inbred quickly; since they reproduce less frequently than r-selected species, there are fewer chances for genetic variants to be lost. But once inbreeding has taken hold, it can be difficult to halt. As their populations lose diversity, individual animals may still appear to be normal. But their ability to cope with change is compromised; they lack the genetic flexibility needed to adapt in a world saturated with people, prone to frequent environmental upheavals, and plagued by new strains of virus and bacteria.

(One of the most interesting case studies of inbreeding concerns the cheetah. In the mid-1980s, geneticists were astounded to discover that the species has almost no measurable diversity; every cheetah so closely resembles every other that skin can be grafted from animal to animal almost indiscriminately. Although as many as twenty thousand cheetahs still remain in the wild, they have a high incidence of abnormal sperm and infant mortality and are unusually susceptible to disease.)

The principal tools for measuring genetic diversity—among them DNA fingerprinting and a more common process called electrophoresis, in which blood proteins are screened for slight biochemical variations—were still in their infancy when captive breeding first came into vogue. The techniques were later honed to a useful, if somewhat imperfect, art. At the same time, geneticists developed complex formulas and computer models to predict how quickly the effects of inbreeding might appear in a population. As scientists learned to peek at the hereditary core of animals, the conventional wisdom about how to save rare species was turned on its head.

By the late 1970s wildlife scientists had begun to suggest that a species generally needed at least fifty animals that were consistently breeding to keep from becoming genetically impoverished in the immediate future. Since not all the animals in a population would breed with equal success and not all gene lines would be equally represented, the actual size of the population usually had to be much larger than fifty. And to guarantee that genetic variants would not be lost from the population over many generations, most species needed a breeding population of about five hundred.

At first some wildlife scientists argued that the breeding population estimates of fifty and five hundred should be used as universal parameters, especially in cases where there was neither time nor funding for geneticists to conduct careful studies of the diversity within a species. It is now more commonly held that no single rule can be applied to all animals. Depending on their behavior and social dispositions, some species might need populations much larger than five hundred to survive intact. The extinct passenger pigeon, for example, may have been doomed as soon as its numbers fell below a threshold of several thousand. With the disappearance of great congregations of pigeons that roosted and nested together, the last pairs apparently did not feel comfortable enough in their surroundings to breed.

But even if it could not be used as a blanket formula, the emergence of the fifty/five hundred rule vastly increased the estimates of how many animals were needed for a species to be considered secure from extinction. In doing so it radically redefined the role of zoos and breeding centers in the preservation of endangered fauna. As a result, within the space of a few years captive breeding evolved from an emergency safety measure to normal practice.

This shift in philosophy had profound consequences. In early breeding programs, such as the first years of the red wolf project, the immediate goal was to restore species to the wild. All haste was made to release animals before they had been confined for more than a few generations. By 1980, however, wildlife managers realized that large numbers of endangered animals would have to be kept in captivity indefinitely. It became common for wildlife conservationists to refer to zoos as modern arks. The notion was simpler in theory than in practice, since no single institution could maintain several hundred rare wolves, or rhinoceros, or bongo antelope. For practical reasons, the captive population of any species needed to be spread among dozens of zoos. Yet, for genetic reasons, each species needed to be treated as one cohesive unit, and mates for individual animals had to be chosen with great care.

The task of managing several hundred far-flung animals as a single population was sobering in complexity. Husbandry techniques had to be standardized, at least roughly, and zoo records had to be translated into a common language, so scientists at one institution could read the genetic and medical history of animals kept at another. In 1971 two Minnesota researchers, Ulysses Seal and Dale Makey, proposed that a computerized record system be set up to keep track of individual pedigrees and medical histories. Seal and Makey hoped to develop a system for collecting information on the numbers and kinds of animals in zoos around the world, as well as their births and deaths, genealogies, and health. Such centralized record keeping would make it easier to manage animal populations genetically, and to decide which animals should be paired for breeding.

Seal's and Makey's brainchild, the International Species Inventory System, opened in 1973 in a small office at the Minnesota Zoological Garden in Minneapolis. The system was funded initially by seed money from private donors, the American Association of Zoological Veterinarians, the Fish and Wildlife Service, and the American Association of Zoological Parks and Aquaria. It soon became self supporting. In the first year, fifty zoos agreed to use the software developed by the organization. By 1991 four hundred institutions were using the programs, and zoos in forty countries were contributing data on the animals within their collections.

Other steps were taken as well to coordinate the efforts of zoos. The International Union for the Conservation of Nature and Natural Resources formed a commission of respected scientists known as the Captive Breeding Specialist Group in 1976. The IUCN is an affiliation of individuals and conservation organizations; it counts among its members the Sierra Club and the World Wildlife Fund. In forming the Captive Breeding Specialist Group, the IUCN established a cadre of wildlife specialists, mostly from zoos, to make recommendations on breeding rare animals for purposes of conservation. The group was given no budget, only some stationery and the power of influence that comes with prestige. Ulysses Seal was appointed to it as chairman in 1980. (In the ensuing decade, Seal, an endocrinologist at the University of Minnesota and the Veterans Administration Hospital in Minneapolis, would emerge as one of the most powerful proponents of captive breeding in the world.)

During the same period, a group of geneticists and zoologists developed a management tool they called the Species Survival Plan, which established captive population goals for different animals. In a species survival plan a team of zoo managers and biologists analyzes the diversity of an animal population and writes a list of goals for future breeding. Through complex mathematical formulas and computer models, the team attempts to predict approximately how many animals are needed to retain ample diversity (say ninety percent of the existing genetic variation) for a set amount of time (say one hundred and fifty years). In many cases certain gene lines are well represented within the population, while others are poorly represented. One purpose of the plan is to smooth out such discrepancies. Occasionally this requires zoos to dispose of animals that are too common, genetically speaking, to provide space for animals with rarer genes.

The influence of species survival plans on the conservation of rare animals was considerable. By the close of the 1980s, plans were in operation for more than fifty species in North American zoos, and European zoos had adopted a similar program. The U.S. Fish and Wildlife Service agreed to incorporate species survival plans into several of its endangered species recovery plans, including that of the red wolf. (The Recovery Plan is the bible of federal endangered species work. A team of biologists and wildlife managers—the recovery team—writes a document identifying the most important goals that the Fish and Wildlife Service should strive for in reviving the species. Theoretically, the service writes recovery plans for every species that is listed as endangered or threatened. In practice, however, plans have been drafted for only about half of the listed species. Recovery plans are as political as they are biological in nature; often what is omitted from them is as significant as what is included.)

The advent of the species survival plan affected only the management of animals in captivity. In the mid-1980s the Captive Breeding Specialist Group began promoting another tool that was more wide-sweeping, known as the Population Viability Analysis. The theory that each species must maintain a certain number of individuals to survive lies at the very core of modern conservation biology. It evolved from concerns that isolated populations are in great danger of sudden extinction, not only because of genetic deficiencies but because of problems arising from demographic imbalances and the unpredictable nature of their environment. It is believed that once a population drops beneath a certain threshold, it inevitably begins to disintegrate.

In a population viability analysis, scientists try to define that threshold through computer models. They also consider how wild animals can be mixed with captive stock to improve the genetic health of the entire species. As described by Ulysses Seal, the analysis is an assessment of risk, nothing more. But its significance for wild populations runs much deeper, for it provides wildlife managers with a potent set of statistics. Using the results of a population viability analysis, geneticists can make recommendations about how many animals should be held in zoos and, based on the quality of the natural habitat, how many should be left in the wild.

In the nearly two decades since the passage of the Endangered Species Act, captive breeding has matured into a carefully regulated business. At the same time, zookeeping has become something of a game of power. Zoo curators buy and trade animals shrewdly, taking care to make sure that their collections are genetically pure—and that they contain the rare, popular species that will attract large crowds or (if the animals are not to be displayed to the public) generous support from donors.

All the while that the genetic hinge has been examined and repaired, a small cadre of scientists has continued to worry about the strength of the psychological traits that must carry endangered animals through lengthening imprisonments. Some dispute the notion of zoo-as-ark as impractical, especially if it means keeping animals in confinement for ten or twenty generations. With no good alternatives, however, there is little for wildlife conservationists to do except to try to alleviate boredom and other maladies of captivity.

And so cages are landscaped and painstakingly redesigned. Attractive though the new American zoos may be, their optical illusions of broad expanses and junglelike coves tend to fool visitors more readily than the animals they hold. At times, even carefully conceived exhibits prove to be totally unsuited for their inhabitants. In the mid-1970s the National Zoo in Washington, D.C., built an enclosure for polar bears that included a large pool and concrete formations that resembled an iceberg. The new exhibit was touted as innovative, lifelike, and humane. But the configuration of the pen was wrong somehow, and it was situated on a south-facing slope that caught the full strength of the summer sun. The bears soon began swimming in continual tight circles. Eventually they were sent to other zoos for rehabilitation.

In some ways the traditional techniques of zookeeping may be exactly the wrong medicine for wild species sickened by the spread of human civilization. Zoo animals are kept in meticulous condition. They are wormed and deloused, and any injuries they receive are promptly treated. Over generations they may lose the resiliency so critical to survival in the wild. A few zoo administrators have begun to talk of purposely infesting captive animals with parasites so they can maintain a natural resistance. This is an important consideration, since parasites place great stress on wild animals, sometimes to the point of death. But because every zoo animal represents a substantial monetary investment, it is unlikely that many keepers will purposely let their displays become mangy or flea-bitten.

One is left to wonder, once again, whether we are diluting the wild essence of animals merely to save the flesh and blood. Whether the captive species we reintroduce to the swamps of North Carolina, the deserts of Arabia, the jungles of Brazil, the mountains of California favor their ancestors even remotely in heart and mind. Or whether they exist solely at our bidding, and only for as long as we prop up their numbers with supplemental food, medicine, and whatever else we decide is necessary for their health.

One week in midsummer I traveled to the farm country of rural Virginia, where the Smithsonian Institution and the National Zoological Park run a breeding center for rare animals. It is an unlikely setting for a zoo, with steep pastures folding against each other like peaks of freshly whipped butter. From halfway up a hill one can see the purple trailings of the Blue Ridge on the western horizon and the roads winding toward Shenandoah National Park. In fields to the east graze animals with exotic stripes and tapered horns: Burchell's zebra, Arabian oryx.

The Conservation and Research Center spreads across 3,150 acres outside Front Royal, with enclosures for ancient species of wild horse and deer, cranes from the Orient, antelope from the Middle East. Small, bright buildings house tiger quolls, tree kangaroos, and tamarins. The purpose of this rural menagerie is partly for research and education, but also to give the animals more solitude and space than could ever be afforded at a traditional zoo. It is home to some of the top captive breeding specialists in the country. My purpose in visiting was to talk to these men and women about the nature of the endangered animals in their care.

Scott Derrickson is an ornithologist with sandy-brown hair and wedges of gray in his sideburns and temples. Until mid-1984 he worked with endangered and threatened cranes at the Patuxent Wildlife Research Center run by the Fish and Wildlife Service in Maryland. He is now the curator of birds for both the Front Royal center and the National Zoo. Among the rarities in his collection are Guam rails, Laysan teals, Bali mynahs, and Manchurian cranes.

In 1987 Derrickson and Christen Wemmer, the director of the Front Royal center, delivered a paper on species reintroduction at the annual conference of the American Association of Zoological Parks and Aquaria. They noted, among other things, that reptiles and amphibians are "genetically 'hard-wired' for conducting their lives." Every component of their behavior seems to be governed by instinct, the way a computer is wired to respond to a stimulus in a predictable way. Birds and mammals are more malleable; some of their actions are instinctive, some are not. The notion that certain wild behaviors could be lost in captivity intrigued me, and I wanted to explore it more fully.

I found Derrickson in a large, cluttered office in the center administration building. We began by talking about wild animals' natural fear of predators. When golden lion tamarins were first reintroduced in Brazil, researchers noticed immediately that the small primates still had an instinctive fear of hawks, although they had lived in captivity for generations. In contrast, black-footed ferret kits, which were being raised at the center, did not seem inclined to hide from badgers and owls. "Very few of the mothers in the ferret breeding program were brought in from the wild," Derrickson said, "and there are questions about whether they know to give alarm calls to signal to their young to hide. Working on that kind of problem is absolutely essential to reintroducing endangered species, but it takes a lot of time. You need a tremendous amount of creativity among scientists."

I asked whether he agreed with William Toone that animals like the California condor cannot be significantly changed by being held in captivity for two or three generations.

"I think it's a valid point for behaviors that are hard-wired, like courtship patterns and some vocalizations," Derrickson replied. "But in captivity you're changing the physical environment and the more subtle social patterns too. When you put animals in captivity you're starting the process of domestication, whether you like it or not."

Over the past fifteen years scientists have learned that there are limits to what captive animals can be expected to retain. Derrickson mentioned the sandhill crane release conducted on the Loxahatchee National Wildlife Refuge in Florida during the mid-1970s. The cranes had been individually hand-reared at the Patuxent center and had grown accustomed to the presence of their keepers. In the wildlife refuge they took to following people down trails. They also had trouble learning to forage for food.

The Loxahatchee study was one of the first reintroduction experiments, and biologists have since begun to use a variety of rearing and conditioning techniques to ensure that birds will act normally after they are released. Still, the study awakened doubts about the ability of captive birds to find food on their own—a behavior that might have been expected to remain instinctive for generations.

"If an animal is given only certain alternatives for feeding in captivity, rather than being exposed to different foods and ways of foraging, then its foraging behavior is probably going to be changed," Derrickson said. "I would prefer to talk in probabilities rather than answer a question yes or no. Are we affecting some aspects of behavior? I would say probably yes, depending on the situation.

"The importance of captive breeding to conservation can't be disputed. We've made considerable progress in the genetic and demographic management of captive populations. But the whole question of behavior management has received much less attention. In part that's because so little is known about the natural history and biology of these species."

Could it be that the way animals learn to maneuver through the world is more complicated than scientists have traditionally believed?

In some cases, Derrickson replied. "There are windows of learning—ages when an animal can pick up a behavior—and then the window closes. The tendency of some birds to imprint on humans is part of that. There have also been some interesting studies done on different species of songbirds. A male hears his father singing at a certain age, and that's when he learns how to put the whole long, complicated song together. Unless he hears it at the right time, he may never learn it."

When scientists attempt to breed animals in captivity, then, they may unwittingly interrupt some process key to their development. And, Derrickson said, by creating conditions in which captive animals will produce young, scientists may unknowingly alter normal learning or socialization patterns.

As an example, Derrickson pointed to the whooping crane program at the Patuxent research center. In the mid-1970s, after the species had been in captivity nearly a decade, whooping cranes finally began to lay fertile eggs. Because the birds never managed to copulate successfully, biologists had turned to artificially inseminating the females. By the early 1980s, however, enough whooping cranes had been bred in captivity for the biologists to begin experimenting with the production of the flock. Derrickson decided to let some of the females try hatching their own young. To his surprise, a few of the birds showed no inclination to sit on the nest.

"They had gotten so used to us pulling their eggs that they would just lay them and get up and walk away," he said. "They'd never been given a chance to express their maternal instincts. We had one bird, EK-2, that would sit down as if she was incubating, but the egg would be a couple of feet away getting cold."

Puzzled, Derrickson and his staff set out to revive the cranes' natural attachments to their young. As soon as a female laid a clutch they replaced it with plaster eggs or the eggs of sandhill cranes. Sometimes the male would stick pieces of straw in the nest as if recognizing that the eggs needed some attention. "Eventually the female might get the idea and sit on the clutch," Derrickson said. "Then we'd put an older egg with a cheeping chick under her to try and really reinforce her maternal instincts.

"We had one bird that finally hatched a chick, but then she wouldn't get off it. She just sat on it and didn't walk it around." He sighed. "We just had to use an awful lot of patience. But we were hopeful that if we let a pair raise their own chick, it would stimulate them to breed successfully without artificial insemination."

Derrickson left Patuxent before the cranes had a chance to mate on their own. A few months later an outbreak of encephalitis killed many of the females that had begun tending their eggs. "Who knows whether they ever would have bred?" he said. "A number of them were mental basket cases."

"I'm not sure there's any behavior that's totally genetically programmed," he added a few minutes later. "It all depends, to a large degree, on the animals and the way they're handled. I think most people who work with captive animals agree that the more the environment can duplicate the wild—both physically and socially—the less likely behavioral deficiencies will occur. But you need to have a knack for working with animals, just like some people have a knack for growing plants. And to some extent we have no choice but to admit that we're shooting in the dark."

"Reintroduction," Larry Collins was saying, "is the pot of gold at the end of the rainbow. It's the bandwagon everyone's jumping on. But why put an animal back out someplace with the same problems that nearly drove it extinct?"

We were sitting in Collins's office at the Front Royal center, philosophizing, as he put it, about the animals in his care. Down a hill from the small house where he worked, clusters of buildings and cages held red pandas, clouded leopards, and golden lion tamarins. Over another hill, past the pens where exotic cranes wailed in trebled, trembling tones, broad pastures of green and gold were dotted with muntjac, onager, and oryx. Collins, the curator of mammals for the center, is a self-professed country boy, genial and bearded with a down-to-earth manner that belies his training as a scientist. He is also pragmatic about the future. He does not expect most endangered species to be reintroduced to the wild in the next fifty years, if ever.

On a daily basis, the staff of the center is much more concerned with immediate tasks—their care of the animals, their research into reproductive biology, and the classes they give for zoologists from developing countries—than with long-range plans for reintroduction. "Philosophically, I see reintroduction as the ultimate goal of zoos," Collins said. "But it's stupid to put animals back out into habitat where they're going to be right back into trouble. It's a waste of your investment. In some cases I think it can even be considered cruel—much crueler than holding them in a well-designed cage."

Collins and I had spent much of an afternoon talking about the stresses animals suffer in captivity. Wild animals, we agreed, live with a constant level of stress from having to find food, avoid predators, protect their young, and endure parasites. But if life in the wild is not easy, I asked, don't its hardships differ from the kinds of pressure animals feel in cages, especially when they are put on display in zoos?

"One thing you have to keep in mind is that animals are highly individualistic," Collins said. "Some react better than others to being on display; some actively seem to enjoy it. Go look at the monkeys at the zoo. They act like they get a kick out of seeing different people. They have a never-ending show going past their cages every day the zoo is open."

In the mid-1970s Collins was put in charge of tending Ling-Ling and Hsing-Hsing, the giant pandas at the National Zoo. One day an animal keeper named Tex struck up a conversation with a man who was watching Ling-Ling lean against a tree. "This guy says to Tex, 'That animal's obviously bored,'" Collins said. "Tex looked at Ling-Ling, then looked back at the guy and said, 'How do you tell the difference between a panda that's bored and one that's content?'

"The point is that we're intimating that we know how animals perceive their world, and that's not safe. It's certainly not scientifically safe. If an animal breeds, produces young, and raises them, then it can't be too stressed out, nutritionally or socially. Beyond that, it's impossible to measure what they're thinking or feeling."

But how long can a species be held out of the wild before it becomes categorically different from its ancestors? How soon will it forget how to hunt or become instinctively acclimated to the presence of humans?

Collins shrugged. "Nobody knows that. It probably varies with the animal. You just need to manage as well as you can to keep them from becoming domesticated. Keep in mind that animals in the wild are going to be evolving over time too; in fact, it's the animals that *don't* change that are most likely to get into problems. If you get an animal that's hard-wired to act a certain way in its environment and then the environment changes drastically, the animal can't adjust.

"You hear people talking about keeping the animals the same so they can be put back in natural areas. There are no natural areas anymore. You hear people talking about putting wolves back in Yellowstone to make it a more natural system. Who are they trying to kid? If wolves ever get reintroduced, it'll be in a heavily managed situation. There aren't any gift shops or macadam roads or hotdog stands in a natural system."

But isn't it likely that an animal's wild traits—the way it perceives its landscape, for example—might be drastically changed in captivity and affect its survival once it is released?

"Sure," Collins said. "It's up to the zoo to be creative enough to halt changes like that. But first you have to know something about the animal's natural history. Most of these species we know very little about."

Scientists have learned a few tricks to relieve the boredom of captivity, and perhaps slow the disintegration of the way an animal perceives the natural world. "You can change its surroundings, subtly but constantly," Collins said. "Its landscape would change; why not its cage? You need to be careful, though; if you change things too much, you can introduce a whole new source of stress."

Some of the most interesting behavioral research has been conducted on animals with severe cage stereotypes. During the late 1980s, Kathy Carlstead, a researcher at the National Zoo, rehabilitated a badly stereotyped black bear by hiding his food under logs and within a haystack so he had to pick through the hay to find it. Before Carlstead began her work, the bear spent hour after hour pacing his cage. Once he was forced to search for his food, however, the pacing all but stopped. Such techniques are called activity therapy. "They're simple and they work," Collins said. "You take a gorilla's food and hide it in a tree so he has to look around for it, or you scatter sunflower seeds around the cage so he has to pick them up. It helps keep his foraging skills intact.

"If you think about it, feeding time is probably one of the times of day an animal most looks forward to. If you just take his food and slop it in a dish, the same kind of food over and over, you're not doing a lot to stimulate him."

He stopped talking abruptly and reached for a glossy yellow helmet sitting on a chair by his desk. "You know what really gives me satisfaction? This." He thumped the side and held it out to me. It was a firefighter's helmet, with a badge bearing a drawing of a leopard, the mascot of the volunteer company based at the center, on its front. "I've been fighting fires for twenty-nine years. In a fire there's one enemy. You have a strong team, and you work together. You have to; otherwise your butt's going to get fried.

"The thing that frustrates me about conservation is knowing what needs to be done and not having any way to do it. We know the rain forest destruction has got to stop, but how do you convince the Brazilian government? How do you solve the global problems—groundwater pollution, the ozone destruction?" He rubbed his eyes and looked at me grimly.

"How do you manage animals to cope with air pollution or bad water? There is no primitive natural system anymore. I think the best we can hope for is a natural system where the flora and fauna are allowed to evolve. But humans have to be part of that evolution. We already are."

I left the Front Royal center the next day with a strange mixture of exhilaration and regret. Before the trip I had feared that my questions would prove too philosophical for the scientists I hoped to talk with, and that I would be greeted, for the most part, with blank stares. I started home uplifted to know that Derrickson, Collins, and their colleagues had puzzled over the same concerns and were searching for solutions. They had spoken of the need for creativity in their work. They seemed to view their science as an uncertain art, a series of portraits begun with paint that changes constantly in texture and color. And yet I found myself returning to the image of the polar bear pacing his concrete slab. It occurred to me that no matter what our intentions, caging an animal robs it of its dignity.

To succeed in this business, Collins had said, you have to keep reaching for the moon, striving for what appears to be unattainable. Driving home, it occurred to me that the goal of all captive breeding—the creation of wild animals that can live in badly altered habitat, without human help—does indeed appear far out of reach, farther even than the moon. None of the scientists I spoke with on that trip or others believe they will live to see the restoration of the animal populations we have decimated in the past hundred years. Where the lack of land does not prohibit such projects, the lack of money will.

What the animals need, more than anything we can offer, is for their wild flocks and herds to be left intact, and their indigenous lands to be left unaltered. Captive breeding is a technical solution to a philosophical problem, the problem of our potential to vastly reshape the natural world. I came away from Front Royal with the utmost respect for the wildlife scientists with whom I had talked, but with a growing dread that unless attitudes can be changed toward animals—unless people in all parts of the world can be taught to cherish the frailty, the ephemeral beauty, and the importance of wildness in its true, unadulterated form—the final goal of captive breeding will never be attained.

Through breeding programs we offer animals time and safe haven, nothing more. The pleasant pastures at the Front Royal center will never be developed or poorly farmed, and poachers will never hide in the wooded groves. I wished with all my heart that the same could be said for the distant velds and snow fields and jungles where animals still run free.

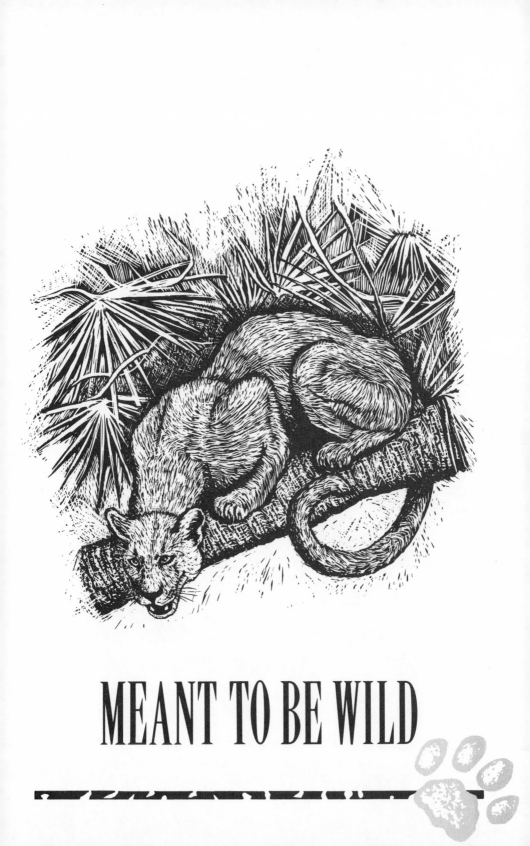

MEANT TO BE WILD

A SINGLE STRUGGLING FLOCK

The Puerto Rican Parrot

At four-thirty on a steamy Puerto Rico morning, a young woman with a single dark braid stepped lightly into some brush on the shoulder of a mountain road. "This is where we need to be careful not to be seen," she whispered. Daylight was still an hour and a half away, but she kept her flashlight off and pushed into a grove of bamboo stalks that rattled together like dried beads. Although it was not likely, somewhere on the dark hillside a figure could be watching, hoping for a clue to the scant, twisted path. Out of sight of the road, the young woman paused long enough to flick on her light. She glanced back at me, nodded, and swung herself abruptly down a muddy incline studded with naked roots.

For more than two months Cathy Blumig had been rising hours before dawn and hiking the remote, rugged trails of the Caribbean National Forest to reach the nesting sites of *Amazona vittata*, the critically endangered Puerto Rican parrot. The trails took her down slippery hillsides, over rushing brooks, and through groves of palo colorado trees that twisted upward like lanky gnomes. Stumbling against boulders and grappling with vines, she passed beneath the spread umbrellas of sierra palm that shaded the stars. Depending on the trail she followed, the walk might take her over an hour; yet in the heavy, humid darkness all she saw of the rain forest were the rocks and roots and mires underfoot, the land crabs that skittered to the edges of the flashlight's beam, the tropical fireflies that hung, unblinking, ten feet above the ground.

Deep in the woods, Blumig would settle into a blind as the morning gathered strength. For more than twelve hours she would keep watch over a parrot nest, recording the movements of the adults until nightfall, when she could leave unseen. Her long hours in the blind would prove invaluable if a predator tried to attack the nest, since the survival of each chick was crucial to the survival of the species. In February 1988, when Blumig and eight other volunteers had arrived in Puerto Rico, a mere thirty-six wild parrots were known to be living in the Luquillo Mountains, their only remaining range.

Blumig was just out of college. She had come to the forest to work as a nest watcher through a program set up by the National Audubon Society and the U.S. Fish and Wildlife Service. On the May morning that I accompanied her, the trails to the blinds were drier than normal—a turn of luck, I thought, as I struggled to keep pace. By five-thirty she had led me to a small stream only a hundred yards from a parrot nest in a gently sloping valley known as West Fork. She turned to me, pressed a forefinger to her lips to signal for silence, and crossed the slick rocks of the stream with a few well-placed steps.

We sneaked to the entrance of a tiny hut barely large enough for two folding chairs. A piece of corrugated metal served as the roof, and the sides were of clear plastic sheeting and burlap that puffed in and out in a light breeze. Several peepholes had been slashed in the burlap to give people inside a fragmented view of the surrounding forest. We climbed in as quietly as possible, peeled off our mud-coated rain pants, and arranged the chairs so we could see through the holes on the north side of the blind.

Thirty feet away was an artificial nest, a curious contraption of wood and plastic mounted on a metal pole. The top section was formed by a hollow log stood on its end. A vine had been strung next to a large hole in the log, and a piece of plastic sewer pipe protruded from the bottom end. In their comings and goings, the parrots landed on the vine, ducked into the hole, and climbed down a wire screen tacked inside the pipe. At the bottom was a wooden chamber where a parrot could comfortably lay her eggs. Fifteen feet up in the air, the West Fork nest looked something like a lopsided smokestack.

The blind was so close to the nest that the birds must have known it was used by some sort of animal. "We try to be real careful not to let them see us," Blumig whispered. "They probably think we're night creatures. They hear us moving around, but always after dark. In the daytime we settle down in here and go to sleep." In truth, she could not afford to sleep a wink. During their shifts the volunteers recorded the activities of adult parrots at four nests scattered through the forest. If a predator threatened attack, the volunteer would clap or whistle to scare it away, but only after giving the parrots a chance to defend their young themselves.

"This pair has a lot of dedication to their nest," Blumig whispered. "I guess I shouldn't anthropomorphize, but it's hard not to, because the parrots all have such different personalities. The male looks like he's spent most of his life commuting on the subway in New York." She grinned. "He's just real beat up and dirty. The female's pretty tough too. Most of the time they're not easily ruffled, but the female's been a little nervous for the past couple of days."

A cool, gray light began to rise, illuminating the vines and bromeliads and ferns that burst from the soil and trees. Although the day was to be sunny, the blind would remain cool, buried as it was beneath the heavy forest canopy. I felt as if I were waiting for a dark curtain to lift and players to appear onstage. I had looked forward to this trip into the forest for weeks, but now, confined to a chair after the strenuous hike, I could not shake a stubborn sleepiness. At five-forty-eight Blumig nudged me. An emerald bird with a thin streak of red above her beak had appeared on the vine, as silent as an apparition. The female parrot. She was bright green except for a white eye ring and the thin crescent of red above her bill; I could not see the iridescent blue that tinged her wings. She had spent the night on the nest while her mate roosted somewhere nearby. She glanced around her, then turned toward the blind and froze, gazing so directly at the structure that I wondered suddenly if she could see through the burlap wall.

This was unusual; normally she ignored the blind. She cocked her head, but her gaze did not stray. Could she sense the presence of more than one of us? Blumig, holding binoculars to her eyes, shifted her weight and grimaced at the slight creak of her chair. The parrot's stark expression struck me as stern and indignant, almost angry. To her we were intruders; we had done something to make her fear for the three eggs we knew to be in her nest. She turned slowly, still looking in our direction. Finally she spread her wings, trembled on her perch, and launched herself, flying low past the hut. A hundred yards away we could hear plaintive squawks as she greeted her mate.

A mere three minutes had passed since she had appeared on the vine. In that time I had gotten a powerful taste of my own clumsiness, of how ill fit I was to enter a world where survival means moving stealthfully and taking offense at every unfamiliar sound. Beside me, Blumig let out a long breath, warm, moist, and full of silent relief.

Until that morning I had seen parrots only in captivity. For the previous four days, in fact, I had been staying at the mountainside compound where the Fish and Wildlife Service breeds Puerto Rican parrots and Hispaniolan parrots, a closely related species that is used as a surrogate to hatch and raise Puerto Rican parrot chicks. Every morning near dawn the hundred birds at the aviary begin to squawk, chirping and rattling for food, banishing the morning peace. With so many, the calls of any one parrot are lost in the harsh, high-pitched chorus that carries down the forested slopes.

But in the wild, the trumpets and squeals of individual parrots are wonderfully conversant, a musical vernacular honed over thousands of years. There are "call-out cries," the soft mutterings of a male summoning the female out of a nest for a foraging trip, and "high squawks," the joyful calls with which the pair announces its presence to other birds. There are trilled warbles and raspy churrs, chortles and honks, and resounding bugles that signal when the birds are leaving an area to hunt for food.

The contrast in the calls is a striking example of the difference between parrots that live in forests and those that live in cages. Parrots are easily domesticated. They can become so attached to their keepers that they will refuse to accept a mate. The wary disposition of a wild parrot disappears after a short period in captivity, making it ill suited for release. Yet during the past fifty years the Puerto Rican parrot has become so rare that confinement offers one of the few hopes for its survival. Since the early 1970s the Fish and Wildlife Service has operated a breeding project for the species in the Luquillo Mountains, a range of dramatic peaks and torrential rains in the northeast corner of the island.

In the spring of 1988, with forty Puerto Rican parrots in the aviary and thirty-six in the wild, the project had achieved a precarious but encouraging balance. Four pairs of wild parrots were known to be breeding. In early spring, depending on circumstances, some of the eggs laid in wild nests might be surreptitiously traded for eggs laid by birds in the aviary. In this way the wild population could receive an infusion of genetically fresh blood. Weak chicks might also be traded for healthy captive chicks. The parrots were so amenable to manipulation, in fact, that they did not seem to mind when biologists put another chick in their nest. A mother raising two young might come home from foraging to find that her brood had mysteriously expanded to three.

Few endangered animal breeding projects have the capacity to swap wild individuals for captive ones so frequently, or so easily. In that sense the Puerto Rican parrot project enjoyed an unusual advantage. But in recent years, releases of chicks to the wild had been curtailed by philosophical differences among federal research biologists, genetics specialists, and the wildlife managers employed by the Puerto Rican Department of Natural Resources. The discord had arisen from a number of troubling questions: Should preserving the wild flock be the most important goal, pursued above all others? Or was it more important, for the sake of genetics, to have a large captive flock? And should the captive birds be completely domesticated to encourage them to breed?

Until the mid-1980s the Puerto Rican parrot project was run by the Patuxent Wildlife Research Center, the research arm of the Fish and Wildlife Service. Among the divisions of the service, the Patuxent center is something of an anomaly. Where the regional offices are responsible for managing wildlife and maintaining national refuges, the Patuxent staff works only on wildlife and environmental research. With its emphasis on scientific experiments and publishing, research differs significantly from the art of management, in which biologists must judge from day to day whether a species is doing well enough to survive. Careful, creative research frequently can provide insight into problems that may not be discerned through casual observation. But critics of the Patuxent center believed that during the early years of the recovery project biologists had wasted precious time and money conducting research when they should have been concentrating on coaxing more captive birds to breed.

In 1984 Jose Vivaldi, then the endangered species coordinator for the Commonwealth of Puerto Rico, began lobbying for the parrot project to be removed from the auspices of Patuxent and placed under the control of the administrators in the Fish and Wildlife Service southeast regional office. Vivaldi charged that the Patuxent researchers had raised only a fraction of the number of Puerto Rican parrot chicks that was possible. The staff members handled the parrots all wrong, he said. For one thing, their strict scientific approach prevented them from treating the birds as individuals and trying different tricks—giving them different size nest boxes, for example—to make them feel more comfortable. And, Vivaldi complained, the Patuxent researchers tried to maintain the wild nature of the birds, instead of using the traditional techniques of aviculture to coax them to breed.

Traditional, in this context, meant commercial. It meant that to boost the number of parrots as quickly as possible, Vivaldi believed the Patuxent staff members should forget they were dealing with wild birds and treat them as pets—naming them, petting them often, and hand-feeding them grapes. According to this approach, a happy parrot is a prolific parrot. But it is not, and will never be, a wild, free-ranging bird.

Before European settlers reached the New World, it is likely that the West Indies were populated by at least six species of macaw, eight types of parakeet, and twelve parrots of the *Amazona* genus—the chunky, rather clumsy parrots so endearing in both their wild and captive incarnations. Three species of parakeet and nine of parrot still survive; the macaws had disappeared by the late 1800s. Most remaining species of parrot are endangered, and the Puerto Rican parrot and the purple-chested imperial parrot of Dominica are the most imperiled of all.

At the time Columbus arrived in the West Indies, the Puerto Rican parrot probably numbered more than a million. (It is the only parrot endemic to Puerto Rico.) It ranged throughout the island and probably thrived at low elevations, where it did not have to cope with the dampness and heavy rain of the mountains. It appears that the decline of the species began in earnest after 1850 as settlement moved up the lower mountain slopes. By 1912 only one percent of the island's virgin forests still stood. The parrots had been pushed to the highest elevations; any that lingered near farm fields to feed on grain or seeds were shot.

As the parrot population declined, the odds increased that it would be utterly depleted by a storm or an outbreak of disease. In 1898, and again in 1928 and 1931, major hurricanes passed over Puerto Rico, destroying buildings and roads and stripping trees of foliage. Unable to feed in the decimated forests, the parrots put aside their fear of humans and moved down the mountain slopes to forage.

In 1937 rangers for the U.S. Forest Service counted about two thousand parrots in an informal survey in the Luquillo Mountains. The count was conducted during the first stage of an ambitious road-and-trail-building program that opened the national forest to unprecedented use by humans. The Forest Service also began logging selectively to create more space for commercially valuable timber. Among the trees chosen for cutting were some of the largest of the palo colorado, a softwood species that grows in twisted shapes and often forms deep cavities in its trunk.

World War II had just ended, and the West Indies were starved for energy. Palo colorado wood, considered worthless for lumber, was valued as a source of heat. From about 1945 to 1960 local men known as *carboneros* felled hundreds of palo colorado trees and set them on fire, smoldering them until the wood turned to charcoal that could be sold as fuel. The *carboneros* hiked fifteen miles or more into the forest, where they camped and tended their fires until the charcoal was ready to be hauled down the mountains. For a hundred-pound bag of charcoal they were paid about ten cents. Not surprisingly, they supplemented their earnings whenever possible by stealing parrot chicks from nests. Parrots were popular as pets and in some places as food; a live chick brought between $2.00 and $5.00.

During 1953 and 1954 counts of parrots in the Luquillo range located only about two hundred birds. By the mid-1960s the Forest Service had stopped selectively cutting timber, and *carboneros* were no longer allowed to fell palo colorado trees, in which, it turned out, the parrots liked to nest. But the species continued to falter.

In the autumn of 1968 a Patuxent biologist named Cameron Kepler arrived in the Luquillo Mountains to study the failing parrot population. Kepler was being funded meagerly by the Fish and Wildlife Service, the Forest Service, and the World Wildlife Fund. Congress would not pass the Endangered Species Act for another four years. The pesticide DDT had not yet been banned, nor had the Environmental Protection Agency been formed. Public sentiment for saving endangered animals was growing, but it was not prevalent, especially outside the mainland United States.

As his first task, Kepler arranged with biologists from the U.S. Forest Service and a local college to conduct a census of the parrots. The birds tended to flock together in early autumn and could be seen easily from ridge tops. One clear day the spotters posted themselves through the forest and began counting parrots. Only twenty-four appeared.

"The optimist in me always wanted to believe there were more birds out there," Kepler said, "but there was no way to know. We spent the first year trying to find out as much as we could. We went to overlooks in remote valleys. We set up transects and went through them looking for parrots. We talked to people who had lived in the forest for years. They'd say, oh, there are parrots in the next valley over; you need to try there. So we'd try there and find nothing."

The following April, when they should have been tending nests, eighteen parrots were seen flocking together in the West Fork valley. Over the summer Kepler and some Forest Service biologists put up a number of artificial nest boxes, hoping to attract parrots to sites that could be easily watched. The parrots ignored them; instead, the boxes were taken over by pearly-eyed thrashers and swarms of honey bees. Kepler turned his attention to other tasks, such as poisoning the rats that had been introduced to Puerto Rico and that preyed heavily on parrot eggs.

He also began looking, largely in vain, for natural parrot nests. He located a man who had once worked as a *carbonero* and nest robber, and who agreed to take him to the parrots' traditional nesting trees. Most were empty. By 1971 Kepler had located five nests within two low valleys. He noticed that breeding pairs sometimes fought ferociously over tree cavities that seemed to him to be poor sites for raising chicks. "Every time we looked at a nest we found a new problem," he said. "Some had no dry places for the birds to lay their eggs. Some had rain literally pouring into them. The adults and chicks would come out of the holes all mucked up with water and wet wood, really in bad shape. We didn't know exactly what was wrong, so we didn't know what to do."

Given the precariousness of the parrot population, it seemed wise to establish a captive flock for breeding. For two years Kepler searched fruitlessly among zoos and pet traders for Puerto Rican parrots. Finally he located two females that had become so attached to each other that they repeatedly tried to mate. The only way to obtain a male was to capture one from the wild flock. In early 1972 biologists trapped two birds, an adult believed to be a male and a juvenile bird, in mist nets set above the forest canopy. They were loaded into shipping crates and put on a flight to Miami, where they were to be quarantined before being taken to the Patuxent center in Maryland. A few days after they reached Florida the young bird died of stress.

That winter Kepler transferred to a job managing the whooping crane flock at the Patuxent center. The parrots were still in rapid decline; by the spring of 1972 the population stood at about fourteen. Although he struggled to maintain a degree of optimism, Kepler knew the species was not likely to survive. His sentiment was shared by his colleagues in the Forest Service and the Puerto Rico Department of Natural Resources.

Perhaps the time of the parrot had passed. Perhaps the ecology of Puerto Rico had been altered too radically for the species to continue to live in the wild. One problem was particularly perplexing. Since the 1950s, the pearly-eyed thrasher, a large, aggressive, and traditionally rare bird, had become abundant in the forest. Thrashers competed with parrots for nest cavities. Whenever they found a parrot nest untended by adult birds, they pecked the eggs or killed the chicks. In 1973 Kepler's replacement, a Patuxent biologist named Noel Snyder, decided the parrots needed some help defending their nests and began shooting invading thrashers. At one nest Snyder shot twenty-six thrashers within a few weeks.

Obviously a different solution was needed. Snyder had already started investigating the possibility that good nesting trees were scarce in the forest; the previous summer he had begun an exhaustive survey of potential nests in the valleys where parrots had been known to breed. (Over a number of years Snyder and some field assistants would inspect twelve hundred trees, usually by shinnying up their trunks.) A substantial number of trees, mostly palo colorados, contained hollows that seemed like suitable nest sites. Yet the parrots always bred in the same few cavities that were unusually deep and dark. Apparently the others were too shallow, too wet, or too close to the ground for the birds to feel comfortable.

It occurred to Snyder that the thrashers might leave the parrot nests alone if artificial nest boxes were set out for them nearby. But what kind? No one knew much about the breeding biology of pearly-eyed thrashers, or about the size of the tree hollows they liked to use for raising their chicks. Intrigued, Snyder and a Forest Service biologist named John Taapken began tinkering with different designs for thrasher nest boxes. They built skinny, deep nest boxes and wide, shallow boxes. They built straight vertical nests, nests with angled chutes, adjoining nests with a deep hollow on one side and a shallow hollow on the other. Finally, after setting out nearly a dozen different designs, the biologists discovered that thrashers were reluctant to venture into deep cavities where the bottoms could not be seen from the entrance. They preferred to lay their eggs in nests that were only about two feet deep, much shallower than most of the nests used by parrots.

In 1974 Snyder and Taapken mounted a shallow wooden nest box on a tree a few yards from an active parrot nest. By the middle of the spring a pair of thrashers had moved into the box. They paid little attention to the parrots, and they defended the area vigorously against intruding birds. Over the next two years the biologists refined the nest box design and deepened the cavities where the parrots had traditionally laid eggs. By the time Snyder left the project in 1976, the pearly-eyed thrasher had ceased to be much of a threat, and the wild parrot flock had finally started to grow. Eight chicks fledged that spring, bringing the number of wild parrots to twenty-two.

"When you're trying to save endangered animals, what helps you are the strange, rare events that open your eyes to all sorts of problems you didn't know existed," Snyder would recall later. "We were lucky; we happened to notice that thrashers wouldn't go into deep nests. But we wouldn't have seen that if we hadn't been watching parrots and thrashers closely every day. Intensive field research is boring to a lot of people. But it's the only way to find out why the wild population is failing."

Snyder had once had a graduate student named James Wiley. In 1973, on Snyder's recommendation, Wiley was hired as a research biologist for the Puerto Rico Department of Natural Resources. Three years later when Snyder left the island to work at the Patuxent wildlife center, Wiley became the leader, and some say the eventual savior, of the federal parrot project.

Wiley was a quiet man, shy around strangers but with a dry wit that made him popular among his colleagues. For thirteen years he and his wife, Beth, lived in a drab cinderblock building that served as the aviary. Built in the 1940s, the structure had been used previously as a National Guard training barracks. "By the time the Fish and Wildlife Service got it, it was pretty trashed out—no windows, no running water," Wiley said. "But it was as hurricane proof as we could ever hope." The aviary had been built on a steep mountainside up a winding, dangerous road in a forest frequented by thieves; it was not unusual for hikers to return from trails without their wallets and jewelry. A ten-foot chain-link fence had been built around the building to discourage break-ins. Whenever the Wileys left the grounds, they carefully locked themselves in and out.

The birds were kept in a second-story room with concrete floors and cages ten feet high. Three large windows lined one wall. Although adequate, it was a cheerless home for the eleven Puerto Rican parrots that had been collected by 1976. Most had been taken from wild nests as chicks. A half dozen Hispaniolan parrots, captured from the Dominican Republic by smugglers, were kept in outside cages. They had been discovered and confiscated at the San Juan airport.

The Hispaniolan parrots settled into their life in captivity with no trouble. The Puerto Rican parrots seemed much more high strung. Cut off from sun, rain, and wind, they acted alternately bored by their lack of freedom and threatened by the workers who entered their cages to clean and deliver food. Some pairs began to pluck each other's feathers. "We left them alone as much as possible," Wiley said. "We specifically wanted to treat them as wild animals. We left the windows open so they could hear the calls of wild parrots flying by, and we played them tapes of wild birds calling in the forest. We gave them as many indigenous foods as we could— bark to chew on, sierra palm, leafy plants—at least so they would grow up recognizing them as good to eat."

Most of the birds in the aviary were either too young or too old to breed, but they had been separated into pairs after a chromosome test distinguished males from females. In 1977 a pair of birds began to act as if they were courting; they lounged in the wooden nest box built on the side of their cage, and they screeched aggressively whenever a person approached. Four eggs appeared in the nest, an unusually large clutch. "They were doing everything perfectly," Wiley said. "We were all feeling pretty cocky about it until I noticed that the male had laid an egg." The chromosome test for the "male" had been faulty. Placed together in a cage, the two females had bonded sexually and had both laid eggs—which were, of course, infertile. Although Wiley separated them, they ignored other potential mates and continually called to each other across the aviary.

In addition to running the breeding program, Wiley spent vast amounts of time studying the wild population. He was employed by the Patuxent center, but he did not see his role as limited to research; his one objective was to preserve the parrot through research, management, or whatever conservation techniques were needed. There were four breeding pairs in the wild, and he intended to guard their nests closely. When possible he sent Forest Service workers out to observe the birds, but he also depended heavily on volunteers. "Anyone who showed up was commandeered," he said. "I had just about every member of my family and Beth's family working at one point or another. Even so, it was like fire fighting; we only managed to cover the nests about sixty percent of the time."

He developed a reputation as inexhaustible. Most days he rose before three o'clock in the morning and worked past ten at night. He had been an athlete all his life and needed only twenty minutes to walk trails that took most people more than an hour. He tried to make up for a severe lack of resources by working quickly and tirelessly. "For all intents, I was in a foreign country, and the Puerto Rican parrot wasn't a priority as far as the American people were concerned," he said. "We managed to get the equipment we absolutely needed by being resourceful. Frugal."

Despite his efforts, the number of wild birds rose and fell erratically. Eight chicks fledged from wild nests in the spring of 1980, but within a year seven of them had disappeared. The wild population consisted of only about twenty birds, the same number as when Snyder had left four years earlier. The captive flock at last began to show promise in 1979, when the first Puerto Rican parrot chick hatched. It was immediately placed in a wild nest. Two more chicks hatched in the aviary in 1980, and two more in 1981.

From the start the Hispaniolan parrots raised young with little trouble. But many of the Puerto Rican parrots seemed completely inept at courtship. Of the six pairs in the flock, only two produced fertile eggs. Three of the females had no mate at all, and attempts to inseminate them upset the entire flock.

If the Wileys were dedicated, they were also isolated from the world of commercial aviculture. In 1980, with the field program taking nearly all of Jim Wiley's time, Beth Wiley volunteered to manage the captive flock without pay. During the breeding season she arose several times a night to hand feed Puerto Rican parrot chicks, which were considered too valuable to be left in the nest with their nervous parents. Wiley worried that the captive program was suffering from a lack of professional guidance. He began asking officials at Patuxent to hire an aviculturist for the project. They responded by putting Beth Wiley on the payroll in 1982.

Occasionally, officials from the Department of Natural Resources questioned whether it was healthy to keep the captive parrots inside. The Wileys saw no reason to move the birds to outdoor cages, where they might be vulnerable to storms, predators, and vandals. "At that point we didn't have any information that keeping the birds outside would make a difference," Wiley said. "Many people breed birds well inside; many breed birds well outside. It's a debate that rages among aviculturists."

Jose Vivaldi of the Puerto Rico Department of Natural Resources first became openly critical of the parrot project in 1983. Until then he had accepted Wiley's explanation that the parrots were too high-strung to be easily raised in captivity. But one day he received a letter from Ramon Noegel, an animal breeder who lived near Tampa, Florida, and who wanted a permit to collect a few iguanas in Puerto Rico. Noegel's letterhead listed him as a breeder of parrots. "So I called him up," Vivaldi said. "I was told he had the biggest parrot breeding center in the world.

"I went to see him. He picked me up at the airport and took me to a house, his house, on a suburban street. I had expected a big compound with security and fences and gates, but there was none of that. There were all these birds in outside cages. And they were calm and quiet."

Noegel claimed to have been breeding some species of West Indian parrots since the mid-1970s. All his birds had names. Their cages, half covered by a roof and half exposed to rain, seemed spacious, clean, and comfortable compared to the indoor cages at Luquillo. Each had a special slot through which keepers could slide trays of food, so that they needed to go inside only for occasional cleanings. Noegel took great pleasure in talking to his birds and feeding them treats like grapes. Vivaldi returned to Puerto Rico slightly stunned, "and then I began my holy crusade."

Vivaldi insisted that the Wileys move the parrots outside and adopt Noegel's techniques for breeding. He also suggested that Wiley forget about preserving the wild flock and concentrate solely on breeding. Every year Wiley would smuggle chicks from the aviary into wild nests to replace infertile eggs and weak young birds. And every year, Vivaldi complained, at least a third of the chicks disappeared. It seemed like a waste to put parrot chicks in the wild; they should all be kept at the aviary to increase the breeding stock.

Vivaldi had other misgivings as well, and now he voiced them loudly. He charged that the parrot project was squandering money on research that was incidental to the recovery of the species. He claimed that the Luquillo forest was too damp and too high in elevation for a Puerto Rican parrot flock ever to prosper there. Since the late 1970s the recovery team for the species had been exploring the possibility of raising and releasing parrots in a forest reserve in Rio Abajo, a drier, mountainous region in central Puerto Rico. In 1985, at Vivaldi's urgings, the Fish and Wildlife Service agreed to pay for the construction of a second aviary at Rio Abajo that would be managed by the commonwealth. Vivaldi intended to hire a commercial aviculturist to run the aviary according to Noegel's teachings.

Wiley did not place the same faith in Noegel's claims. "You hear that Ramon Noegel has bred this many parrots and produced this many chicks in a year, and it sounds impressive," he said. "But when you ask to see his breeding records, things start to get a little fuzzy." Vivaldi's criticisms of the project bothered Wiley. But he was far more troubled by what he perceived as a lack of support for the parrot project from administrators at Patuxent.

In the spring of 1984 Forest Service workers had spotted two parrots displaying courtship behavior in a section of the national forest not normally used by nesting pairs. To Wiley the discovery presented a unique chance to study how parrots selected a nest site. By watching the new pair closely, perhaps he could figure out how to lure parrots into artificial nest boxes that could be easily protected from predators. If he could provide more good nest sites, more parrots might begin to breed. He asked for funding to hire some additional field staff temporarily. The request was denied. "I was told I had to choose between hiring people at the beginning of the season to watch this new pair or hiring them at the end of the season to watch the chicks as they were starting to fledge," he said. "There was no choice. When the birds first learn to fly, they're very vulnerable to predation. They have to be watched." He complained bitterly of the decision in a memo to Patuxent officials.

He began to wonder whether his bosses at Patuxent cared about preserving the wild flock. They complained that he was not being accountable; several times he had been told to cut back on the amount of time he was spending in the field so he could deal more efficiently with administrative tasks. "I was supposed to sit behind a desk, take orders, and get all my information about the parrots from field technicians," he said. He resisted. Without going into the field he could not maintain a continuous familiarity with the wild parrots; nor, he believed, could he make good decisions about how the flock should be managed. He knew the parrots well; he recognized individuals by their idiosyncrasies. He wanted to keep close track of them so he could piece together information about the social make-up of the flock. A number of biologists, including Vivaldi, had suggested that the parrots be banded or otherwise marked for easy identification. Wiley feared that leg bands and wing tags would destroy the parrot's natural camouflage and make them more susceptible to predation by hawks.

The wild flock seemed to be expanding, slowly. The captive flock did not. Of six potential breeding pairs, only two pairs had managed to lay fertile eggs.

Vivaldi's complaints were increasing in intensity, and administrators in the Forest Service were beginning to question whether the recovery program was making enough progress to justify their continued support. Wiley agreed that the breeding project needed help from a professional aviculturist, but he could not, he told Vivaldi, force Patuxent to hire one.

In 1984 Vivaldi approached administrators in the southeastern regional office of the Fish and Wildlife Service with his complaints. (The regional branches of the service are not affiliated with the Patuxent wildlife center, with which they often compete for funding and power.) Vivaldi's move, calculated to bring changes to the parrot project, succeeded masterfully. By asking for assistance from the southeast region, he provided the fodder for a major skirmish over control of the program.

Wiley was taking some chancy steps of his own. In an article published in the journal *Bird Conservation* in 1985 he wrote, "Serious problems with the Puerto Rican parrot program threaten the future survival of the species." He asserted that there was neither ample budget nor staff to run an adequate breeding facility and field program, and that development of an artificial insemination technique for the lone female parrots at the aviary had been delayed by a lack of funding. "Apparently," he concluded, "it is easier for federal government agencies to ignore the problem of wildlife species in an associated country than it would be in a state with full representation in Congress or where public awareness and concern are greater."

The tone of the article was too critical to be ignored by officials at Patuxent. In the summer of 1986, with little warning, Jim Wiley was transferred back to the mainland to conduct field research on the California condor. Administrators, questioned about the move, said Wiley had lost all perspective; he had become too close to the project to effectively do his job.

"I certainly had lost *their* perspective," Wiley said with a characteristic wry grin. He was seated at a booth in a small Mexican restaurant in southern California, near the wildlife refuge where he was running experiments on the release of condors to the wild. His brow was deeply creased below a receding head of dark, curly hair. He wore a sweat shirt, a blue bandana knotted casually around his neck, and a pair of baggy cotton trousers held up by colorfully striped suspenders. He was at once as likeable, and as intensely energetic, as he had been described to me by others.

"I tried to fight the transfer," he said. "It didn't work. And the project has changed a lot. Right now no one knows what parrots are out there in the wild. They've lost that thread, that continuous thread that Noel Snyder and I kept going over the years.

"We still don't know how the birds select a nest site. We don't know the incidence of 'divorce' and repairings, or of what happens when a mate is killed, whether a widow will choose a new mate. . . . No one wants to put much emphasis on natural history anymore. Plain old nineteenth-century biology has gone out of vogue, but it's always been the most important thing in determining what the animals need. You can say, sure, we'll figure out what they need after we're breeding them in captivity, but if the wild flock completely disappears, you'll *never* know what they need.

"Call it ecology or behavioral biology or whatever you want, but it's the natural history that continues to be important." He paused. "In this era, though, everyone seems to want to forget all that and concentrate on captive breeding."

Seven A.M. at the aviary. All night the forest had reverberated with the soothing, two-beat song of the coquis, the copper-colored tree frogs so abundant in the rain forests of Puerto Rico. In the hazy hours of dawn their high, clear calls sounded like water dripping over rocks, but with the rise of light they had trickled away beneath the morning clamor of the parrots. Now, with fog nudging the mountains, all sound had stilled, and the parrot compound had taken on the cold, musty feel of an outpost in a remote portion of the Third World.

Upstairs in a tiny kitchen crowded with desks and incubators, a pale, blond woman named Betsy Anderson readied a plate of cut fruit for five Puerto Rican parrot chicks. Anderson was a biological technician who lived at the aviary. She unlatched the side of an oblong wooden box that served as an incubator and swung it open, exposing the chicks, which blinked in the fluorescent light. Emerald feathers spottily covered the month-old chicks, but the youngest, at about three weeks, had only blue-gray quills that barely hid the bumpy skin of its neck and head. It lolled sloppily against the others, its beak open to show a fleshy tongue. Already it had a white circle around its eye and the endearing, quizzical expression of a parrot.

It was the spring of 1988, and the parrot project had changed. The aviary was well-equipped and adequately staffed. The wild nests were guarded every day by student volunteers. A professional aviculturist had come to work on the project as a volunteer, and plans were being made to hire her. In the yard, parrots fed and preened in brand-new, spacious cages equipped with slots through which keepers could slide trays of food. Each cage was half covered by a roof and half exposed to sun. Each was built almost exactly to specifications written by Ramon Noegel.

The five chicks in the incubator box ignored the fruit and looked at Anderson, as if hoping for another choice. "That's it, guys," Anderson said, quickly closing the box. "We're weaning them, and we have to be sort of heartless about shutting them in there with solid food." Four feet away, another incubator held a tiny, naked bird with bulbous blue eyes. It was only three days old; a few damp, white feathers clung to its flesh-colored wings. Anderson lifted it out, cradled it carefully in a paper towel, and lowered it into a shallow porcelain cup. It stretched its wobbly neck and squeaked, begging for food. Anderson extended a syringe toward the chick and squirted a small bit of cereal mix into its mouth. "Little overanxious there," she laughed, as the chick, seeking more, strained upward until it nearly fell over. "That's good; a strong begging response is a sign that he's doing okay."

In the modern incarnation of the parrot project, the aviary employed four people. While money was still not plentiful, no one had to pay for parrot food out of their own earnings, as the Wileys had done. For fiscal 1988 the Patuxent wildlife center had allocated an annual budget of $677,200 to the parrot project, twelve times the budget in 1981, when Wiley was running both the field program and the aviary alone. (Between thirty and forty percent of the annual budget was allocated for administrative overhead and was not received by the field program.)

Although the Patuxent center had greatly increased its commitment to preserving the parrots, its control over the project had crumbled. In 1985, after talking with Jose Vivaldi, officials from the southeast regional office began suggesting to Fish and Wildlife Service administrators in Washington that the parrot project was being poorly run. After Wiley's transfer, Puerto Rican officials formally requested that the project be placed under the authority of the southeast region. In letters to Frank Dunkle, the director of the service, the officials complained that Patuxent scientists were using antiquated breeding techniques and mishandling the project. Dunkle granted their request.

Politically speaking, the realignment of power represented a substantial victory for Vivaldi. The focus of the project had shifted from basic research to management, and from maintaining the wild flock to building up the number of birds in captivity. Since 1986 all but two of the chicks hatched at the aviary had stayed there instead of being fostered into wild nests. The captive parrots, however, were still being handled very little. Although Noegel's ideas had greatly influenced the way the birds were fed and housed, several other commercial aviculturists had been consulted about breeding methods. On their advice, Fish and Wildlife Service administrators had decided that the captive Puerto Rican parrots should not be coddled. Designs were still being refined for the new aviary in the mountains of Rio Abajo to the west. (The facility was to have opened in early 1987, but its construction had been delayed repeatedly by administrative problems within the Department of Natural Resources.)

That morning Vivaldi had stopped by the Luquillo aviary on an informal visit. He and I sat at a picnic table a few yards from a cage of squawking Hispaniolan parrots. Overhead, blue-bottomed clouds collided and split, dropping a few desultory sprinkles. "I come up here now, and I feel good about things," Vivaldi said loudly enough for Anderson, passing by, to hear. "I would not have said that three years ago."

Vivaldi still believed no research should be done on the captive flock. He had objected to a study of artificial insemination being conducted on Hispaniolan parrots by a Patuxent researcher, who collected sperm from male parrots and injected it into females. Vivaldi believed the experiments were too costly and disruptive. "It upsets the birds, and I think we could breed parrots with traditional methods for much less money," he said. "What do we need to save the parrot? This is the only research we should be doing."

He looked at me with a small, tight grin. It struck me that he believed the Puerto Rican parrot suffered no ill that could not be remedied by the right brand of aviculture, preferably Ramon Noegel's brand.

I decided to play devil's advocate. "The Hispaniolan parrots don't seem to have any trouble breeding here," I said. "Maybe the Puerto Ricans aren't breeding as well because they're too inbred."

"They have never been handled correctly. So we do not know if that is true or not."

I asked whether he believed it was important to maintain the wild flock, even if it meant fostering some captive chicks into wild nests. He crossed his legs, grinned again, and shrugged. "Of course the wild flock is important. I would not want to see it disappear. But if we breed enough parrots we can release a new flock at Rio Abajo. I believe Rio Abajo is a better place to have parrots anyway. It's drier."

I thought of the female parrot staring angrily toward the blind where Cathy Blumig and I had sat the day before. There was no resemblance between her demeanor and that of the placid, well-fed parrots in the cages set around the yard of the aviary. A parrot raised in captivity might learn to behave with the same wariness, the same wildness, as one that had grown up in the forest. But it would take time, and perhaps some training, to acclimate released birds to the dangers posed by predators. To me, it seemed safer to make sure the wild parrots at Luquillo survived, so chicks from the aviary could be fostered into their nests and could grow up with wild parents. Yet Vivaldi was suggesting that if they all disappeared, scientists could simply put together a new flock somewhere else.

Vivaldi's opinions were extreme, and not shared, I discovered, by Jim Wiley's replacement. Down the twisting mountain road from the aviary, Marcia Wilson sat at a cluttered desk in an office near the small town of Palmer. As the new project leader, Wilson was responsible for the day-to-day operation of the aviary and the field program. She also had to negotiate a delicate line of diplomacy between her bosses at Patuxent, the southeast regional office of the Fish and Wildlife Service, and the project's collaborators in the U.S. Forest Service and the Puerto Rican Department of Natural Resources. Wilson leaned back in her chair and sighed. "I'm finding out that there's a lot more to this job than pure biology," she said.

Until late 1987 Wilson had worked at the Charles Darwin Research Station in the Galapagos Islands. During the 1988 breeding season, she had experienced swings of buoyant optimism and deep disappointment as the wild parrots raised chicks, or abandoned nests, or disappeared. Now, late in the breeding season, she was inclined to be stingier with her hopes. "You come into something like this, and you think you can see at least a small light at the end of the tunnel," she said, "and then suddenly it's like there's another rockslide ahead; the light disappears. But we are making progress.

"It really is important for us to bolster the number of birds in the captive flock. We need to increase the size of the flock here, and we need to have some to send to Rio Abajo, whenever it's finished. But, to me, by far the most important thing is keeping the wild flock alive."

Through the student volunteer program, the wild parrot nests were being guarded more scrupulously than ever before. Occasionally, when the breeding birds were off foraging, their nests were inspected to make sure the eggs and chicks were healthy. That spring one nest had developed a bad leak, and a pool of muddy water had formed in the bottom. Biologists substituted plaster models for the real eggs, which they then took to the aviary for hatching. A few weeks later the plaster eggs were traded for two healthy chicks.

During the previous ten years the number of wild parrots had doubled to thirty-six; but in 1988, as in 1978, only four wild pairs were known to be breeding. I asked Wilson if it was possible that more birds were nesting than scientists had discovered.

"Sure. We answered that question this year; we found a new nest," Wilson said. "I don't think there are many more, though; the number of wild parrots just isn't increasing fast enough. That's the most frustrating thing about the whole project. A third of the chicks are still being lost in their first year, mostly to hawks, we think. And we're not sure how to change that."

Unless the number of breeding pairs increased, Wilson acknowledged that the wild population was likely to grow slowly. "We keep hoping the flock will get to a certain size and then really start producing a lot of chicks," she said. "And I guess that is a possibility. But I think it's more likely that we'll keep on the way we have—losing some birds, gaining others. Eventually, you've got to think the population will stabilize. All we can do in the meantime is try to save the birds in as many different ways as we can."

Despite a promising start, 1988 did not turn out to be a particularly good year for the Puerto Rican parrot. Eight chicks were added to the aviary flock, but six others died of infections or various mishaps. Several captive parrots laid eggs, but two-thirds of them were infertile. The newly discovered wild nest was destroyed, possibly by rats. In December biologists counted only thirty-four wild parrots, two less than the previous winter.

In the spring of 1989 the wild parrots successfully raised nine chicks. That summer biologists sighted forty-five wild parrots, including at least two that had not been spotted in previous counts. Marcia Wilson knew that several of the young birds would not live through the winter. Still, it seemed that the wild flock was finally recovering.

The mood of the staff was much improved, although Wilson suspected that more changes in policy lay ahead. In June an influential group of wildlife scientists had traveled to San Juan for a meeting on the future of the Puerto Rican parrot. Among them was Ulysses Seal, the internationally renowned captive breeding specialist from Minnesota. Seal had been asked by the southeast regional office to do a population viability analysis—an analysis of how likely a species is to go extinct—for the Puerto Rican parrot and the rare Puerto Rican plain pigeon.

Over the previous three years Seal had become a central figure in federal endangered species programs. He had already conducted population viability analyses for the black-footed ferret, the Florida panther, and the red wolf. The list of participants for the Puerto Rico meeting included biologists, nutritionists, aviculturists, geneticists, veterinarians, and members of the Captive Breeding Specialist Group of the International Union for the Conservation of Nature and Natural Resources. It also included two field biologists who had come to be considered renegades by the Fish and Wildlife Service, Jim Wiley and Noel Snyder.

In the years since he had left Puerto Rico, Snyder had become convinced that the wild population of Puerto Rican parrots, since it was slowly increasing, should be preserved virtually at any cost. (This was an interesting and somewhat ironic position. Until 1986 Snyder had worked on the recovery program for the California condor, which was quickly declining in numbers. He had lobbied heavily for the entire condor population to be captured for breeding.) He had quit the service and was running a program to restore the thick-billed parrot to the southwest United States. He, Wiley, and Cam Kepler had recently published a book that was respected as the definitive work on the Puerto Rican parrot.

Snyder knew that whatever conclusions were drawn at the meeting would likely become the guiding policy of the parrot project. Since Wiley's transfer he had worried that too little was being done to increase the number of parrots breeding in the wild. Now he worried that Seal and the scientists in the Captive Breeding Specialist Group did not know enough about the biology of wild Puerto Rican parrots to make sound judgments about the management of the species. When the meeting convened, he and Wiley were seated in the audience, along with Marcia Wilson, Jose Vivaldi, and Hilda Diaz-Soltero, the field supervisor from the southeast regional office of the service. (Vivaldi and Diaz-Soltero were married.)

A population viability analysis is, essentially, a tool for calculating risk.

A group of scientists writes a scenario: Suppose there is a population of parrots, most of them of breeding age, in a forest that is occasionally swept by disease and storms. How inbred are they? How long is each one likely to live? By varying factors like the number of parrots, their pedigrees, their breeding success, and the frequency of catastrophic events, the scientists come up with an educated guess on how long the species will survive. From there they try to calculate how many parrots are needed to maintain a healthy population for several centuries. The process is more complicated than can be easily explained, but the purpose is not. By massaging a set of data, the scientists try to predict what nature is most likely to do.

For three days members of the Captive Breeding Specialist Group led discussions and ran computer models to consider the plight of the Puerto Rican parrot. At the end they concluded that the species was not substantially inbred, at least not yet. To keep the population healthy, they said, the captive flock should be expanded as quickly as possible. It should also be divided among three separate facilities, two in Puerto Rico and one in the mainland United States, so it could not be devastated by a single storm or outbreak of disease.

A report issued later by the group noted that Jim Wiley had hastened the growth of the wild population by fostering chicks from the aviary into wild nests. Seal and his colleagues believed that too many of the fostered chicks had come from the same two breeding pairs; those pairs were now overrepresented, genetically, in the wild population. Nevertheless, the scientists noted that Wiley had bolstered the number of free-ranging birds to the point that the flock was now likely to survive a major hurricane. If the accelerated rate of growth could be maintained, the wild population would become increasingly stable. There was good reason, then, to supplement it with chicks hatched in the aviary. But for the time being it was more critical, for genetic reasons, to increase the number of birds being held in captivity.

The group recommended that chicks from the aviary not be released to the wild unless the captive population had increased by at least six over the previous year. If two of the adult parrots at the aviary died, for example, eight chicks would be held in captivity the following spring—two to replace the dead birds, plus six more. If the captive flock produced more chicks than that, half of them could be considered for release to the wild.

Since 1980 the aviary had raised an average of between four and five chicks a year. Under the group's guidelines, no chicks would be fostered into wild nests until the captive birds began to breed much more consistently. Even then only the "excess" chicks—and, Snyder feared, only the money and resources that could not be readily spent on the breeding program—would be dedicated to the wild flock.

Snyder reacted angrily. In a series of comments attached to the group's report, he warned that the wild population was still too small to be assured of surviving a major hurricane. If the wild population disappeared, he wrote, it would not be easy to create a new one with birds raised in captivity. "Captive breeding is out of control," he said. "It's become a panacea. I support it for some species. But I just don't see that a species as high-strung as the Puerto Rican parrot is ever going to breed as well as some other species will in captivity. Putting all our efforts into increasing the size of the captive flock is exactly the wrong strategy."

It was the same familiar quarrel that had bedeviled the parrot project for years; but with the involvement of the Captive Breeding Specialist Group, it moved into a new arena with a more powerful set of players. It was, in fact, the same familiar quarrel that has bedeviled nearly every endangered species project, beginning with the whooping crane. In the case of the Puerto Rican parrot, events in the fall of 1989 would show, dramatically, just how much was at stake.

In mid-September a tropical depression formed off the coast of Africa and embarked on a wobbling course toward the islands of the Caribbean. As it gained strength the staff of the parrot project tracked its progress with alacrity. When its wind speed built to eighty-five miles an hour, meteorologists named the storm Hurricane Hugo. "For some reason little sensors went off in my head," Teri Sorenson said. "I just had a gut feeling that we were going to get nailed by this one."

Sorenson was the aviculturist hired by Patuxent in 1988 to manage the Luquillo aviary. She had no research background—in fact, she had no college degree—and was therefore an oddity among her Ph.D. colleagues. But she had an unusual knack for handling captive parrots, which made her a valued member of the staff.

As Hugo drifted closer to the eastern shore of Puerto Rico, Sorenson stockpiled water and food and readied some small cages that could be stacked in a room of the old aviary building. On Sunday, September 17, she moved the parrots inside. By nightfall all was secure; she could do nothing more but close herself inside the cinderblock building and listen to the rising wind. A graduate student was staying at the aviary temporarily; the two women settled in for the night, hopeful that the metal plates and hurricane shutters they had put over the windows would hold.

The winds at the center of Hurricane Hugo were a sustained one hundred forty miles an hour, with occasional gusts to two hundred. It was the worst hurricane to threaten Puerto Rico in sixty years. As it approached the coast from the east, it began to drift slightly north. Its impact on the island was more of a glancing blow than a direct hit, but in a few hours it transformed the forests of the Luquillo Mountains from a lush mélange to a desert of leafless snags.

Marcia Wilson had taken refuge with her husband in their house outside the forest. At 9:00 P.M. the power went out as the front edge of the storm eased against the island. By 3:00 A.M. a deafening wind had taken hold. "No one could sleep, it was so loud," Wilson said. "I could hear the roof on our patio and an aluminum awning ripping away, piece by piece. We were in a house made of cement, so we knew we'd probably be okay. And at least we had been able to do something to protect the parrots in the aviary. But I had the most sickening feeling, because there was nothing we could do to help the wild birds."

In the darkness of the aviary Sorenson had the odd impression that she was inside a bird cage draped with a cloth cover. All night she could hear debris battering the building. She busied herself moving desks and equipment away from windows where water was spraying in. A door on the south side of the building blew out, leaving one room exposed to the rain. "I didn't have time to be scared; I was busier than hell," she said. "I figured that the wild birds were probably taking a beating, and that I might have all that was left of the Puerto Rican parrot under my care."

The wind continued past dawn and into the full light of day. Around two o'clock in the afternoon, as the gusts began to diminish, Sorenson and the graduate student cautiously opened the front door. Before the storm, the yard of the aviary had sloped into a bank of trees and bamboo. Now, however, there was no vegetation; they could see all the way to the ocean. "My first thought," Sorenson said, "was, oh my God, the poor wild parrots."

In the days immediately following Hugo, parrots were spotted in groups of two and three, acting confused. Wilson, hiking up what remained of the mountain road, was astounded by the devastation. "It was like a bomb had gone off," she said. "There was nothing but dead snags and brown as far as you could see. I remember noticing lots of hummingbirds down low, looking for flowers. There weren't any. And then I'd hike around a curve and there would be a little valley of green where none of the trees had been touched." Other birds were staying close to the ground too, searching for food and hiding from the hawks that seemed to pepper the sky.

By Friday, Wilson had received confirmed sightings of five parrots. Some of the birds were spotted in the foothills, where they had wandered, apparently in search of food. There seemed to be no pattern to their movements. Wilson noticed that they were acting furtively and calling to each other very little. The standing trees, burned by salty rain, turned even browner.

On September 27, Wilson and Hilda Diaz-Soltero organized the first comprehensive count of parrots after the storm. Lookouts posted throughout the forest spotted twenty-three birds. A second count a week later turned up only thirteen. Fearing that the parrots had begun to starve, the biologists enlisted fifty volunteers for a major census in mid-October. Twenty-two parrots were sighted. The next week observers counted only eight.

The counts remained low through November: ten birds one day, six another. The parrots had always dispersed in the fall, so Wilson was not particularly worried. The trees began to leaf out. All at once the forest looked like a North American landscape in the throes of spring. Wilson saw no reason to do anything but clear the trails, repair the nest sites, and wait for the breeding season. At least one pair of breeding birds was known to be alive; it was likely that more were hiding in remote parts of the forest. But in December, Ulysses Seal and administrators in the southeast regional office of the service suggested that it might be best to trap whatever parrots could be found and hold them for several months, until the trees recovered their foliage and began fruiting. They noted that if the wild population had dropped as low as the counts indicated, it would probably not survive. Snyder condemned the proposal as ludicrous and predicted that more parrots would appear in time for breeding season, as they had every year. Wilson completely opposed capturing the birds and asked that a decision to trap the parrots be delayed at least until February, when breeding pairs would start looking for nests.

In January biologists counted twenty parrots, and the proposal to capture the wild flock was put aside. To Wilson's relief, the parrots stopped acting so secretive; now they squawked and squabbled loudly. With the onset of the tropical spring, parrots began inspecting the nests where they had raised chicks in years past. "We can't tell if they're the same birds or new pairs yet," Wilson said. "We're not optimistic that we're going to get any chicks at all this year. I'm just amazed they're alive and acting like they want to breed."

In April, unusually late, three pairs of wild parrots—two that had nested the previous year and a new pair—settled down to breed. Two of the nests failed because of unknown causes. A healthy pair of chicks fledged from the third nest, surprising Wilson and leaving her encouraged about the resilience of the wild population. The birds in the Luquillo aviary also began to breed late, but produced five chicks. Two were fostered into one of the failed wild nests. Under the circumstances, all parties agreed that the wild population needed to be supplemented, at least for one season. The chicks were taken back to the aviary when the wild parrots failed to care for them.

Autumn counts in 1990 turned up only about twenty-two parrots, the size of the flock when the Wileys arrived in Puerto Rico in 1976. But the population estimates were based on periodic censuses; researchers well knew that more parrots might be living out of sight in the dense jungle foliage. Indeed the flock seemed to stabilize quickly, considering the extensive damage. Six pairs—five of which formed after the storm—bred in 1991 and 1992. And their chicks survived in record numbers.

Instead of devastating the population, the hurricane seemed somehow to have boosted its productivity. Perhaps Hugo had stimulated the trees of the forest to produce more fruit during the critical season when parrots need more nourishment to lay eggs and care for young. Or perhaps it killed enough older parrots to give younger birds a chance to breed. Several nests were laid in new areas of the forest, areas that before the storm had been frequented by older, non-breeding birds. Whatever the reason, by the spring of 1993 the wild population had reached thirty-five adults, and at least ten chicks were expected to fledge by fall.

In 1993, after twenty-five years of labor to save the parrot, what did the Fish and Wildlife Service have to show for its efforts? It had a single struggling wild flock, the existence of which was its most notable accomplishment. It had the Luquillo aviary, where the parrots were still not reproducing well, presumably because of the humidity of the facility's mountainous elevation. It had a new aviary at Rio Abajo, which was finally completed in 1990 after years of bureaucratic entanglements within the Puerto Rican government. To make sure the facility was free of disease, only Hispaniolan parrots were raised there during the first two years. And it had a total of forty-eight adult Puerto Rican parrots living in captivity. There was still no planned third breeding site, as recommended by the Captive Breeding Specialist Group; there were still no plans to create a second wild flock. Although the bickering between different factions had abated, bitter differences remained as to whether more emphasis should be put on preserving the wild flock or maintaining the species' gene pool in captivity.

Ulysses Seal and his colleagues had urged that some of the parrots be taken to a zoo in the United States, where they would be safer from tropical storms. But the move had been vigorously opposed by Snyder, who argued that the parrots were much more likely to contract diseases at a mainland zoo than at an isolated breeding facility in Puerto Rico. After much debate, the secretary of the Department of Natural Resources decided that none of the birds would leave the island.

To some degree the Puerto Rican parrot project in the 1990s is in a position to be envied by other endangered animal recovery programs. It has a wild population that is increasing and that can, if policy permits, be supplemented easily with captive chicks. It has an ample budget and staff. It has at its disposal the expertise of dedicated scientists and animal breeders. Tragically, it has not traditionally had the kind of management that can respond quickly to the sudden changes that tend to beset wild animal populations. It has lacked flexibility and vision, and its progress has been impeded by constant political jousting.

On my last morning in Puerto Rico, I hiked with a student volunteer to a lookout above the mountainous valley known as East Fork. A fog had settled its skirts over the Luquillo forest, but as we climbed the last slope the gauzy clouds dissipated, stretched thin by the early sun. We teetered up a set of iron spikes driven into the listing trunk of a palo colorado tree, ascending to a platform fifty feet above the ground. As we took our places, a splinter of green detached from a nearby tree and glided in front of us. It was a male parrot, flashing the malachite of his flight feathers, a color as deep and true as the Caribbean Sea. We watched him with binoculars until he disappeared behind a ridge.

We had climbed to the lookout to count parrots, and we were not disappointed. As the morning lengthened we heard raucous chortles from the crown of a tree to our left and saw movement between limbs where two parrots were foraging for fruit. In front of us another pair skimmed the canopy of the valley, their emerald backs exactly the color of the leaves. Scaly-naped pigeons skated beneath us on the wind, and a black swift bobbed across the sky. A bananaquit landed almost within reach, opened its crimson-lined mouth, and sang. But our loyalty was to the parrots. The sight of each one brought a single, reverent exclamation: "There!"

During three hours on the lookout we spotted five Puerto Rican parrots, a significant percentage of all alive on earth. A century before we might have seen dozens or hundreds, but the ecosystem that supported them no longer exists. Eventually the species might adapt to the new landscape of Puerto Rico—parrots have even been known to thrive in cities—but not without a measure of help that is wisely, more carefully administered than any we have given so far.

It was midmorning; the tropical sun struck my face with rude power. I shouldered a day pack and looked once more through limbs to where a parrot sat, idly scratching its neck with its foot. I grabbed the top iron spike and swung my weight down the tree trunk, moving toward home.

THE SOUL OF THE CONDOR

The young bird stood unsteadily on a sloping sandstone ledge, a huddle of cocoa feathers against the streaked yellow rock. A harsh east wind boiled up the sheer slopes of the mountains and riffled the dull feathers on the bird's back, pulling through them like fingers through hair. From my vantage in a blind a half mile away the bird looked bedraggled and ill at ease. Although she would come to depend on wind, she had not often encountered such breezes during the three weeks she had wandered free on a pinnacle of crumbling stone in the Hopper Mountain National Wildlife Refuge of southern California.

Beyond the pinnacle lay the dusty flatness of an agricultural valley, three thousand feet straight down. The mountains rose so steeply from its floor that I could imagine stepping off the hillside and soaring, arms spread, above the quilt of green and russet fields. But whereas I was gravity-bound, the young bird was not. Turning to face the wind, she opened her wings tentatively, flexing them as a gust caught her and took her off the rock, tossing her only a few feet in the air before setting her down. She flapped again and jumped, only to come crashing back; but on the third try her body lifted and hovered over the rock, rising on thermals as if pulled upward by a rope. As she climbed, her great wings stretched to a span of ten feet, and she dangled her legs awkwardly as if using them as a rudder. Nearby, ravens circled, small as jays beside a hawk. The young bird floated undisturbed as wind shook the oaks on the mountain, letting gusts play through her primaries, the fingerlike feathers at the tip of each wing. No other bird could fly as she flew, a condor in the California sky.

It was January 1989, the beginning of what a certain cadre of biologists hoped would be the year the condor returned to the mountains north of Los Angeles. Nine months before, the fledgling had hatched in a nest box at the San Diego Wild Animal Park. If she survived the winter, by summer she might range as far as she chose, passing thousands of feet above the foothills of the San Joaquin Valley, riding thermals north to the favorite roosts of condors in years past. If she chose. In all likelihood she would seldom stray far from the rocky pyramid that served now as her nest, since whatever food she needed would be set out nearby. Neither would she live in the wild long enough to breed. She was an Andean condor, transplanted from South America, and by 1992 she would be captured and removed from the mountains, lest she pose a threat to the purity of the California condors to be released on the wildlife refuge.

The fledgling—Y-1 as she was known among biologists—brought her wings down a single time, arching them in a slow flap that made me think of her as lightly embracing the wind, encircling it, steadying herself on its back. Her altitude dropped slightly, but she compensated quickly and rode higher, swaying a hundred feet above the sandstone in a slow, graceful curve. For all appearances she was in complete control, yet she had been flying only a few weeks, and never in such wind. Below her the grassy hillside dropped and rose in an eruption of gray mountain spines that extended east to the horizon. To the southwest, out of sight behind a ridge, lay the Pacific; to the southeast, the dull orange smog of Los Angeles. The condor began working her way back, descending in slow curls, pointing her primaries down and stretching her legs toward the ledge where she had taken off. But the wind would not let her go, not yet. She teetered comically in the air two yards above the ledge; rather than flying she looked like she was hanging from a string. Finally she dipped low enough to touch the stone and began to flap vigorously, shaking the wind from her wings. She hunched herself against the breeze and walked slowly behind a rock, out of sight and out of the sky for the evening.

I straightened my back and relaxed my neck, which had been craned at an odd angle during the twenty minutes the bird remained in the air. Her performance had kindled a feeling of expansion in my chest, a warmth that radiated through me, titillating me, defying speech. I had never seen a condor in the wild, nor was I likely to see many others. Y-1 was the most advanced of seven Andean condors that would fledge at Hopper Mountain in early 1989, and that would act as models for the release techniques to be used eventually on the California condors that lived only in captivity.

Y-1 would spend the night resting beside two other chicks released on the sandstone roost, one from the Buffalo, New York, zoo and one raised at the Patuxent Wildlife Research Center, the Fish and Wildlife Service research branch in Laurel, Maryland. In a week four more chicks would be let out of a large cage several miles away. If all went as planned, the seven birds would form a loose flock that foraged within the wildlife refuge, feeding on carcasses set out for them. Their silhouettes above mountain peaks might send shudders of appreciation through the people that spotted them, as the sight of Y-1 in flight had sent a buzz of awe through me.

But there were some through whom the sight of a condor would send a small surge of pain, the way the glimpse of a person with a familiar

---

Meant to Be Wild

mannerism might remind one momentarily of a beloved friend who has died. To these people the California condor perished forever in 1987, when the last wild bird was brought into captivity. Nothing, not a successful breeding program, nor the possibility of a future release, would ever lighten the loss of the species as it once had lived, untagged and unmolested by humans.

The condor, symbol of heaven and death. Certain Indian tribes believed it could fly to the gods; others refused it religious significance because it scavenged among corpses for food. As a symbol of rare wildlife its significance to Americans has also been mixed. In the history of endangered species work, no program has attracted such public attention, or generated such ill will.

Largely because of its size and powers of flight, the condor has been admired, and even revered, by the peoples of the West for hundreds of years. It can soar for an hour without a wing flap, circle a mountain in a matter of minutes, cruise comfortably from sea level to altitudes of nine thousand feet. Passing overhead, its wings emit a steady whine, almost like whistling wind. In the air no bird rivals its beauty and skill.

Stories of the condor's agility in flight comprise a kind of modern folklore among field biologists. In 1976 William Toone, now the curator of birds at the San Diego Wild Animal Park, spent weeks searching unsuccessfully for condor fecal samples before coming upon a bird. "I had walked all over the mountains without even seeing a condor, much less getting close enough to get a fecal sample," he recalled. "I couldn't understand why there was such a fuss about condors; I thought it was all kind of stupid. Finally one day I decided I was going to climb the highest peak in the area and sit there until I saw one.

"I started up a ridge, and just as I was getting close to the top a condor flew over. He was maybe fifteen feet above me, flying pretty fast. He didn't move a muscle that I could see, just dipped his primaries a little to let some of the air slough out, but he stopped dead. He craned his neck down to look at me and we eyed each other for—God, it seemed like an hour. Then he adjusted his primaries and took off, spiraling straight up until I couldn't see him anymore. It was like he flew to heaven. From that day on I said, count me in fellas, I'm a believer."

On the ground or a perch the California condor appears, in the opinion of many beholders, grotesque. Its plumage is dark brown and dull, except for the long, white triangle on the underside of each wing. Its naked head and neck, a blend of startling oranges and reds, is lumpy and reptilian. A ruff of shaggy, black feathers rings its neck. After a meal of decaying calf, its hooked beak may be spattered with remains and the reddish crop may be distended, like an oversized blood vessel. Yet field biologists describe the species as appealing and full of character. Unlike an eagle or a hawk, its face can be expressive. With crow's-feet lining its eyes and inflatable air sacs in its cheeks, it is capable of striking many different poses, looking regal one moment and comical the next.

In 1987 the Fish and Wildlife Service captured the last free California condor, a male adult known as AC-9 (Adult Condor 9), and took him to the San Diego Wild Animal Park. AC-9 joined twenty-six other condors being held at the wild animal park and at the Los Angeles Zoo, the only two breeding centers for the species. In captivity AC-9 would never again feel the upward surge of a thermal or spot a stillborn calf on a grassy hill beneath seven-thousand-foot peaks. But neither would he consume a lead bullet that might lodge in his system and kill him.

The problem of lead poisoning, more than any factor, put to rest a debate of unusual acrimony over whether the foundering condor should be removed from the wild or left alone. The argument had fostered contempt, even hatred, between people who in other circumstances might have held each other in high regard. It had derailed careers and started a lawsuit between the Fish and Wildlife Service and the National Audubon Society, the joint operators of the now-defunct Condor Research Center in Ventura. And it had left the field biologists who struggled to save the species frustrated and emotionally overwrought.

By 1989, however, the debate had long been silenced. On my way to California that January, a number of people asked me why so much time and money had been devoted to preserving a species that seemed so outmoded, so unable to survive in the modern world. I had scarcely known what to answer; indeed, on occasion I had wondered why myself. Sitting on the hillside and watching Y-1 soar in forty-mile-an-hour gusts, I began to understand how the beauty of a bird in flight might capture a person's heart, and how that person might cling unreasonably to a plan he believed to be crucial to the survival of the bird. How a person would mourn the loss of a species so ancient, so large yet defiant of gravity—a species that could soar without effort, the way I had always dreamed of soaring myself.

Until 1984 biologists believed North American condors had always been restricted to a range that reached from the high country of the Pacific Northwest south into Mexico and east to Texas, with some scattered populations in Florida. Fossilized bones of *Gymnogyps californianus* and a similar species, *Gymnogyps amplus*, had been found in California, Oregon, New Mexico, and in cliff caves of the Grand Canyon, but not in numbers to indicate that condors had ever been plentiful. Scientists concluded that the birds may have been most abundant during the late Pleistocene Epoch and that they thrived best in warm climates. Then, in 1984, 1985, and 1986, three condor bones were excavated from a swamp in Genesee County in upstate

New York. Dated to eleven thousand years, the bones were too scattered to tell whether they belonged to one bird or three.

The discovery of condor remains within a hundred miles of the reach of glaciers was somewhat of a shock. It suggested that at one time the birds had been able to withstand penetrating cold, and that during the Pleistocene they conceivably could have inhabited most of North America. If so, the extinction of the great Pleistocene mammals—mastodons, mammoths, giant beavers, and ground sloths—would have thrown the species into a state of crisis. With no ready source of food in the East and Midwest, the range of condors would have shrunk to the coasts and mountains of the West, where they could scavenge the carrion of sea mammals, elk, and antelope.

If indeed the condor has spent the last ten thousand years in steady retreat, the question of how much should be done to preserve it becomes more complex. Since the mid-1980s a small group of paleontologists has argued that, as a species, the condor is outmoded, and that the latest drop in population is merely the final step in a natural process of extinction. This theory has never won wide support. But even if scientists were to uncover more evidence in its favor, arguments that the condor should be allowed to die out are not likely to hold much sway with federal officials. The condor recovery project is among the most expensive attempts to save an endangered species in the history of the United States, and perhaps the world. Since 1970 the federal government and the state of California have spent millions on condor preservation, much of it in administrative costs that, because of bookkeeping, cannot be easily tallied. In fiscal 1989 alone, the budget for the project, with twenty-eight California condors in captivity and seven Andean condors in the wild, was about $750,000. Another $750,000 was budgeted in each of fiscal 1990 and 1991. Between forty and fifty percent was spent on administrative overhead.

Despite an uneasy sentiment among conservationists that the money devoted to saving the condor might be more wisely spread among several species, there is little chance that the condor project will be scaled back. The program is a classic example of how government policy toward wildlife can be taken to an extreme. By struggling to preserve it at such high cost, federal administrators have made the condor into a symbol for all endangered species. They have gambled that the condor's plight will draw attention to less appealing rare animals, and that public concern for the condor will generate support for other conservation projects. In the process they have backed themselves into a confrontation with nature they cannot afford to lose. Were the condor to go extinct, in the future biologists might find it difficult to attract support for endangered animal recovery projects.

It should be noted that the amount of money earmarked for condor preservation seems large only in comparison to the meager budget set aside for most endangered species. Supporters of the condor project say the real problem lies with the federal government's unwillingness to commit adequate resources to protecting the nation's natural treasures. They also point out that, regardless of whether it had been slipping toward extinction on its own, the condor began to suffer serious losses soon after white settlers arrived in the West.

Historical records show that in the 1800s condors were shot for sport and, according to some accounts, baited into corrals from which they had too little room to take flight. In the late 1800s and early 1900s their eggs, pale blue or green and the size of large avocados, became popular among collectors and sold for as much as $250 apiece. During the Gold Rush, condor quills reportedly were used as vials to carry gold dust. Although no exact counts are available, a significant proportion of condors also may have been poisoned during the latter half of the nineteenth century, when settlers commonly laced carcasses with strychnine and arsenic to kill grizzly bears and wolves. In addition to its intended victims, the poisoned bait attracted untold numbers of other mammals and birds.

By the 1930s the condor was known to exist only in the mountains that line the southern third of the San Joaquin Valley—the coastal range to the west, the Sierra Nevadas to the east, and the Transverse Mountains that form the terminus. Condors could still be spotted from the ridges and towns that lay along their flight routes, but their appearances were increasingly rare. Little was known of the biology of the species, other than that breeding birds laid a single egg in the spring, and that pairs usually nested only every second year. Laws had been passed against shooting condors and collecting their eggs. But, with such a low reproduction rate, the population did not seem likely to grow quickly, if at all.

In 1939 the National Audubon Society hired Carl Koford, a graduate student at the University of California at Berkeley, to conduct a detailed field study of the condor over several seasons. Koford's monograph, published in 1953, is still considered the definitive work on the biology of the species—its feeding habits, mating rituals, and rearing of young. Reading the study, perusing its thick weave of data and observation, one senses that the years Koford devoted to studying condors were among the most inspired of his life.

From March 1939 through June 1941 Koford spent nearly every day in blinds watching condors, compiling thirty-five hundred pages of field notes. His observations were expanded in subsequent years, and in the 1950s he became recognized as the world authority on the species. He often cautioned that condors should be studied from a distance and disturbed as little as possible. His skills as an observer and chronicler of detail remain unquestioned, but his views on the fragile nature of the species are still debated vigorously.

Koford believed the condor to be critically endangered, although even as late as 1950 the birds could be spotted from certain vantage points on virtually any day of the year. His monograph estimated the number of condors at sixty, based on sightings of between thirty and forty-two birds at once from observation posts in the mid-1940s. There was no valid evidence that the population was in decline, he wrote, but neither did it seem to be on the rise. With only sixty birds, a third of them immature, he believed the success of every nest to be essential to the survival of the entire population.

Most of Koford's monograph is written in the detached, factual style characteristic of scientific research papers. Koford noted that the decrease in the condor population could have genetic consequences, and that, with fewer birds to scan the countryside, the species might have increased difficulty finding food. In a final section entitled "Conservation," his tone grew more emphatic. Condors should not be bred in captivity, he insisted, until all strategies for maintaining the wild population have failed. "It is extremely doubtful that a condor raised by hand and lacking the experience gained by being raised in the wild could survive for long if released," he wrote. "Release of captive condors might well introduce zoo diseases into the wild population. . . . The beauty of a California condor is in the magnificence of its soaring flight. A condor in a cage is uninspiring, pitiful, and ugly to one who has seen them soaring over the mountains." Instead, he advocated the protection of all wild nests, the passage of federal laws protecting the species, and an ambitious education program for people who visited condor habitat.

Koford did not restrict his lobbying efforts to his monograph. In the early 1950s he learned that ornithologists from the San Diego Zoo intended to capture a pair of mated condors for the establishment of a breeding program. The zoo had obtained trapping permits from the California Fish and Game Commission and the federal government. Before the capture team could be successful, however, their permits were revoked. Representatives from the Audubon Society and the Sierra Club, arguing that the loss of a pair of breeding birds would adversely affect the wild population, had convinced members of the state legislature to block the trapping. The conservationists' opposition was led by Koford.

Years later, biologists working on the condor recovery project would complain that if a breeding program had been established in the 1950s, there might have been an ample number of birds to release in the late 1970s, when only about thirty wild condors remained. Koford's critics scoffed at the notion that the capture of two birds would have affected the wild population, and they speculated that he had greatly underestimated the number of condors alive during his study. His count of only sixty helped halt the capture of birds for the breeding program at the San Diego Zoo and hastened the establishment of the Sespe Condor Sanctuary, a mountainous tract of thirty-five thousand acres. A number of field biologists speculated that Koford had purposely given the lowest possible estimate of the condor population to heighten concern about the species' continued survival.

Koford's friends describe him as a meticulous old-school scientist and a man of deep compassion. He was soft spoken, personable, fond of dressing in flannel shirts and jeans. He disliked the technical hardware of modern biology, preferring to spend long periods patiently observing wildlife over tracking animals with radio transmitters. His easy personal style was appealing to local activists, including members of the Sierra Club, the Friends of the Earth, and other groups that in the 1970s began to pay close attention to the recovery program for the condor.

By 1978 the direction of the program had shifted from simple field observation to aggressive management. A number of large tracts of land had been preserved as wildlife sanctuaries. A recovery plan had been written, calling for the maintenance of a minimum population of fifty birds, and a recovery team had been formed of eminent biologists. Government scientists believed that between forty and fifty condors remained, but that— unlike previous years—a frighteningly low percentage of the birds were immature. A study from 1966 to 1976 by Fish and Wildlife Service biologist Sanford Wilbur concluded that it was only a matter of time before significant numbers of condors died of old age.

Wilbur believed that the entire population was being sustained by only a few condors that mated and bred consistently every two years. The effect on the population would probably be minimal, he wrote, if several young condors were trapped for breeding. True, no one had ever bred California condors in captivity, but wildlife scientists had experienced little trouble in breeding Andean condors, a very similar species. Since the birds would not mate before they reached the age of six or seven, researchers should have ample time to refine techniques for reintroducing the birds into the wild.

The publication of Wilbur's study coincided with the release of a report by a panel of experts appointed by the American Ornithologists' Union and the National Audubon Society. The panel concluded that the condor population had become too precarious to sustain itself and urged the federal government to begin trapping condors for captive breeding and radio telemetry studies as soon as possible.

All through the years that federal biologists amassed the evidence to justify captive breeding, Koford continued to warn that to handle the condor excessively—to disturb its nests, mount transmitters on it, or haul it from the wild—would be to drive it to extinction. In an impassioned article written for the newsletter of a local Audubon Society chapter in 1979, he argued that Andean condors tend to be much hardier than California condors and to have much greater tolerance for changes in habitat. It is doubtful, he wrote, that breeding California condors in captivity will prove to be easy. The article raised strong objections to the government's plan to steal eggs from condor nests in hopes of forcing mated pairs to "recycle" or "double-clutch"—to produce a second egg the same season. Such manipulation "imposes on the female physiological stresses which may shorten her reproduction life and decrease her lifetime egg production," Koford wrote. The striking color of an individual's head and its white wing patches were probably important in determining its social rank, he added. Marking birds with bands and wing tags could disrupt the social organization of a flock.

Koford offered a long list of questions about condor biology that he insisted be answered before any birds were radio tagged or removed from the wild. How frequently did condors feed on deer, which had diminished in numbers since the 1960s? Did the prevalence of Compound 1080, a poison commonly used to control populations of ground squirrels and coyotes, affect the health of condors? How could more birds be coaxed into the western part of their range, which was being used by only a few?

"Do we want to replace wild condors with cage-bred, hand-raised birds?" he finished. "A wild condor is much more than feathers, flesh, and genes. Its behavior results not only from its anatomy and germ plasm but from its long cultural heritage, learned by each bird from previous generations through several years of immature life. . . . If we cannot preserve condors in the wild through understanding their environmental relations, we have already lost the battle and may be no more successful in preserving mankind."

Koford died of cancer a few months after the article appeared. Had he lived, perhaps he would have modified his stance as the number of wild condors dwindled to twenty-five, and then to fifteen. With his death his pleas would be amplified by dozens of supporters, many of whom vowed to carry on his crusade against a wildlife establishment they would come to view as unreasonable, self-promoting, and, to a large degree, evil.

At about one-thirty on a warm, sunny afternoon in June 1980, a small group of men approached the top of a sandstone cliff set among the bleak desert crags of the San Rafael Wilderness Area. The team included Jeff Foott, a noted wildlife photographer, and three ornithologists: Bill Lehman and John Ogden of the National Audubon Society, and Noel Snyder, the Fish and Wildlife Service biologist whose innovative experiments with nest boxes had helped save the Puerto Rican parrot. Snyder worked for the service's Patuxent Wildlife Research Center, the branch responsible for the condor program.

The men had picked their way up the backside of the cliff on foot. Seventy-five feet down its face, accessible to humans only by rope, a pair of condors were raising their chick, hatched from an egg laid on the floor of a small crevice. By examining the chick the biologists hoped to tell whether its parents were breeding well or experiencing reproductive problems. They also hoped to gather data on how quickly condor young developed in the wild—data that would be an important gauge for chicks hatched and raised in captivity.

The previous day the team had entered another nest and taken measurements from the only other condor chick produced that spring. This second chick was somewhat older and, Bill Lehman would discover, feistier. As the other men waited, Lehman and Foott rappelled down the cliff to the small opening to the nest. Lehman was to handle the bird while Foott captured the procedure on video film.

Inside the tiny cave, Lehman found the chick hunched in a corner, an angry black bundle, lanky and rather reptilian-looking. It hissed and bit at him, boxing him with its wing tips. Lehman dodged the blows and grabbed the bird by its head, pinning its wings together and pushing it into a small backpack to be weighed. Over the next thirty-five minutes he measured the wing span of the bird, the length of various primary feathers, the length of its beak. He had handled other wild birds, but never vultures and certainly never California condors. When he released the chick, its head began to wobble weakly. A few seconds later it collapsed, dead from stress.

Afterward, Lehman's handling of the chick would be criticized by many conservationists as unnecessarily rough, and his intrusion into the nest cave as needless. An autopsy showed the chick to have been in fine health, although (as the champions of John Ogden and Noel Snyder would argue) it may have been unusually susceptible to stress because of an intangible genetic weakness. As it was, the dead chick became a symbol of all that might go wrong when scientists try to handle rare animals.

Within days the California Department of Fish and Game revoked the permits under which Snyder and Ogden, the principal investigators, were conducting their research. In the furor surrounding the incident, members of the Sierra Club and the Friends of the Earth charged that the permits had not allowed the biologists to handle eggs and chicks in the first place—a contention disputed by the research team.

"Obviously we made a mistake, a bad one, but it gave us some experience for working with the rest of the birds," Noel Snyder said. "We found out the hard way that handling late-stage condor chicks is very difficult, very stressful, and we stopped doing it. For some reason it doesn't seem to make as much difference when the birds are younger. That doesn't excuse the loss of that chick. But if you're sensitive to the situation, as you should be, you lose one bird and you don't repeat the mistake."

Depending on whom one asks, Snyder is either a man of deep sincerity and compassion or an autocrat driven by ambition. I had read much about him, mostly in articles that had been highly critical of his role in the condor project. Beginning in 1980, Snyder worked closely with John Ogden at the Condor Research Center. Ogden had been responsible for conducting a major radio telemetry study on the species, while Snyder concentrated on studying its breeding biology. Snyder also pushed relentlessly for captive propagation, believing that without it the condor was doomed. My reading had prepared me for a man with a certain sharpness to his manner, a technocrat with little patience for the more contemplative side of endangered species work. When finally we met I was caught unawares by what I found: a burly man with a soft face, a patchy gray beard, a quick laugh, and green eyes that often seemed kind.

As a young man, Snyder had harbored a passion for classical music and had become an accomplished cellist. Among the several degrees he holds is a bachelor of music from the Curtis Institute of Music in Philadelphia. Long before he began work in California, however, he had given up music to devote more time to the preservation of endangered species. By all accounts he is a man consumed by his work. He had shown unusual dedication during his work on the Puerto Rican parrot recovery project, and in California he attacked his job with equal vigor. "Noel accomplished more than nearly any field biologist in a short time, in terms of monitoring condors, because of his incredible level of energy," John Ogden would tell me when I visited him at his new job studying wading birds in south Florida. (Past fifty now, both Ogden and Snyder have moved on to other species, other pursuits.) "Noel focused himself on the condor entirely. But he also got very involved with it emotionally, and it tore him up. It was . . . ," Ogden paused, reflective, and combed his hand through a thick, white beard. "I think it's fair to say it was extremely difficult on all of us. I tell people I'm really twenty"—a laugh—"but I got all this gray from working in California."

In sorting through the divisions that plagued the condor project during the early 1980s, I expected to find the story of a clash between clear ideologies, with Noel Snyder and the people who favored captive breeding of the condors in one solid camp and the conservation groups that opposed capturing the birds in another. But the tale is more complex than that, for Noel Snyder shares more philosophy with Carl Koford than his enemies might admit. Both men handled condors extensively, and they agreed that the species needed to be studied closely in its natural environment. Both logged thousands of hours watching condors in the field. And both believed so strongly in their abilities to prescribe a cure for *Gymnogyps californianus* that they fought bitterly against anyone who disagreed with them.

But Koford and his allies had established a history of experience with the condor that Snyder did not have. For several years the California chapters of the Sierra Club and the Friends of the Earth had lobbied unsuccessfully for restrictions on hunting, oil and gas leasing, road building, and the use of Compound 1080 in key areas of condor habitat. Snyder and Ogden arrived in

California in 1980 prepared to lay the groundwork for a captive breeding program and a major telemetry study. "The conservation groups had already decided what was killing the condor," Snyder said. "Our question was, where's the evidence, the data, to show that shooting is a major problem? Or 1080? There wasn't any." His attitude struck many people as arrogant.

"I think Noel and John first got themselves into trouble by coming out here and playing down the importance of what other people already knew," said Mark Palmer, a Sierra Club staff member who was then the chairman of the organization's California wildlife committee. "We would have meetings with them, and they would sit back with smiles on their faces when we told them some of the things we had observed. The people in those meetings had been watching condors for years. But their opinions didn't count because they didn't have precise data to back up what they knew to be true."

Palmer had become involved in the condor controversy in the late 1970s, when he served as the chairman of a special Sierra Club task force set up to examine the federal proposal for a captive breeding program. "We never had any problems with the idea of captive breeding," Palmer said, "as long as it was done on a fairly limited basis, and as long as steps were taken to improve the habitat and the health of the wild birds. But it's clear now that a limited program was never their intention."

In contrast, the Friends of the Earth, a more radical conservation group, steadfastly opposed all attempts to capture condors. "We didn't feel there were enough birds to split the population into a captive group and a wild group," said David Phillips, who until 1984 worked as the wildlife program coordinator for the Friends of the Earth. "There had undoubtedly been a decrease in the population from shooting and eggshell thinning caused by DDT. But DDT had been banned, and the birds had started reproducing again. There were only about twenty-five birds out there, but half of them were trying to breed. The situation seemed ripe, to us, for a gradual increase in the population. And we knew that wouldn't happen if they started pulling birds."

The views of the anti-capturists might not have affected the recovery program if the biologists at the Condor Research Center had exercised complete control. But the decision on how to manage the condors lay with several administrative bodies, including the California condor recovery team appointed by the Fish and Wildlife Service. In theory, the recovery team held final authority over the project. In practice, however, the ultimate power was split between the National Audubon Society, the service's Patuxent research center, and the California Fish and Game Commission, a state agency that generally regulates hunting and fishing but also oversees the management of endangered species within California. It was this body that revoked the research permits held by Snyder and Ogden after the death of the chick in 1980.

The action was a serious blow to Ogden's and Snyder's field research, since it prevented the biologists from doing anything besides observing the birds from afar. They had worked out a strategy needed to save the California condor, with backing from the recovery team, and they were anxious to get started. They were also anxious to rectify the problems they believed had been caused by Koford, whom they saw as having prejudiced public opinion by promoting the condor as a fragile, mystical bird.

"Carl was wrong," Ogden said. "It's a shame, but he was wrong. He bothered condors extensively himself. He went into nests and handled young; he even took a photographer to a blind that was very close to a nest and let him blow a police whistle to get the condors to sit up and look out. Then he maintained that the birds were very sensitive and easily disturbed. As a result, in my opinion, there was very little monitoring of the condor in the fifties and sixties when we might have had a chance to find out a lot about the population. The prevailing wisdom was to leave them alone."

Ogden and Snyder could not convincingly dispute Koford's theories, since no one knew much about the temperament of the species, or about the movements and feeding habits of individual birds. The condor was still somewhat a creature of legend when, in October 1980, the two researchers went to northern Peru to work with biologists who were releasing captive-reared Andean condors into a wild flock. The range of these particular condors appeared to be limited to a dry, rocky coastline where marine mammals often washed up dead. During the trip, Snyder and Ogden trapped several condors and mounted button-sized, solar-powered trans-mitters in the crooks of their wings.

"We thought we had gone to Peru to work with a small subpopulation, and that we'd be able to radio a lot of birds that were coming down only on a little patch of coast mountains to feed," Ogden recalled. "East of those mountains was the Sechuran Desert, fifty to seventy miles wide and virtually rainless. A wasteland. One day we got in a plane and started following a radioed bird and it kept going and going, all the way to the foothills of the Andes. We couldn't believe it. It turned out that birds were crossing back and forth over the desert, some of them almost daily. It was the first indication any of us had of how far condors could fly, and how fast."

To Ogden and Snyder, the revelation strengthened the case for a telemetry study of the California species. How else could biologists learn enough about the condor to save it? How could they determine the kind of terrain it most favored, or learn its foraging patterns, or recover dead birds in time to discover what had killed them? Nevertheless, they could not convince their critics that the value of tracking California condors out-weighed the dangers posed by handling the birds long enough to fit them with radio tags. From 1980 through 1983, whenever the state Fish and Game Commission met to discuss the condor, David Phillips and other activists were present to testify against both the telemetry studies and the need for captive breeding. The activists questioned every aspect of the research proposals—the ability of biologists to trap condors safely, to fit them with radios that would not affect the birds' flight, to track them reliably through the rugged mountain terrain.

The opponents also criticized Snyder and Ogden for dismissing the population estimate of sixty condors made by Koford in the 1950s. "A number of people felt the population figures were being manipulated to make it look like condors were declining very rapidly," Palmer said. "If there really had been substantially more than sixty birds in 1950, like Snyder and Ogden were saying, and the number had dropped to twenty-five by 1980, there's much more reason to step in with radical measures like captive breeding." Finally, the conservation groups claimed that the population was breeding and maintaining itself fairly well.

"At the time we were extremely frustrated, because many opponents of the program had no field knowledge of the species," Snyder said. "We were dealing with a very emotional group of people who believed in a mythical condor. We resented their power to prevent what we considered to be essential actions."

In retrospect, it seems that what the condor most needed was for the people dedicated to preserving it to agree on a course of action that could be adjusted as the situation changed. Instead, the direction of the project wavered between two bitter extremes. The clash of philosophies between the conservation groups and the Condor Research Center focused on the very concept of wildness. To Palmer, Phillips, and their allies, a condor raised in captivity—one that had never jumped into space and soared above cliffs and chaparral and conifer-covered slopes—was not the same as a wild bird. Neither was a condor that had been marked with radios and wing tags. Some activists, including Phillips, argued that what the birds needed was simply enough pristine wilderness to survive. The Condor Research Center biologists countered that the pressure of development was simply too great; the birds would have to adjust to the presence of hunters and oil and gas rigs and other problems caused by people. In the mid-1980s Snyder even began to suggest that captive-bred condor chicks could be "trained" to forage in a small portion of the species' original habitat, to make sure they fed on carcasses that were not contaminated with Compound 1080 or lead pellets. "... It could be argued that preservation of the existing traditions of the wild population should be avoided, as these traditions may have some fatal flaws that have been contributing to the species' plight," he wrote in a 1984 report. "For example, it is possible that continued traditional use of foraging areas is exposing the remaining individuals to high risks of mortality."

"They were making the problems the condor's own fault, rather than dealing with the problems of *human* overuse of the species' traditional habitat," Phillips said, traces of exasperation in his voice even after many years. "By saying, 'Oh, the birds are flawed,' they didn't have to deal with toxins, or restricting oil and gas leases, or banning hunting and off-road vehicles."

However much the condor biologists may have been hobbled by dissension from the conservation community, the blow that would prove ultimately most crippling would be the divisiveness within their own ranks.

In 1982, after extended discussion, the Fish and Game Commission issued permits to allow the Condor Research Center to trap condors, fit them with radio tags, and draw blood from them, the only certain way to determine their sex. Between twenty and twenty-five condors remained in the wild. Each year several birds disappeared, killed by unknown causes.

The commission's action might have marked the beginning of an era of great productivity for the center, had the relationship between John Ogden and Noel Snyder remained congenial. But in the two years since they had begun to work together, their spirit of cooperation had grown brittle. To some degree the problems were the result of differences in personal style. But they also arose from a basic disagreement about the pace at which birds should be brought into captivity. Both men believed it was still possible to sustain the condor in the wild. Snyder, however, feared that many of the wild birds might die off before an adequate gene pool had been preserved in captivity, and he lobbied for establishing a breeding program as quickly as possible. Ogden wanted to move more slowly. While in favor of captive breeding, Ogden thought it was more important to maintain a sizeable wild flock so biologists could study the behavior of the species and learn why so many birds were disappearing. The disagreement, though slight at first, blossomed into bitterness, distrust, and resentment.

The first obvious fracture in the biologists' relationship appeared with the trapping of the first condor, which landed on a carcass on Columbus Day 1982 and was caught by a net shot from a small cannon nearby. According to Audubon biologists, Snyder immediately suggested that the bird be taken into captivity. Ogden disagreed. The matter was settled when Charles Fullerton, the executive director of the Fish and Game Commission, sided with Ogden. The bird was radio tagged and released. (Snyder would later deny that he had ever favored bringing the bird into captivity.) A short time after the incident the two biologists began arguing over the publication of a paper on a method of identifying individual condors through patterns of molting feathers.

"It was tragic, because when Noel and John stopped speaking to each other, the lack of communication filtered down to everyone else in the program," said Jesse Grantham, an Audubon biologist who worked on the project from 1980 until 1986. "Noel had his group of people watching nests, and John and I had our group of people doing telemetry work. Neither side knew what the other was doing. And there was absolutely no talking going on between them."

Nevertheless, the Condor Research Center managed to proceed with its plans to establish a captive population. In August 1982 the biologists removed a young fledgling from its nest after they became convinced that its parents were neglecting it (a claim disputed by David Phillips and other activists). That fall the biologists trapped and radio tagged three condors by luring them to strategically placed carcasses and snaring them in cannon nets. A fourth bird, a juvenile, was trapped with the hope that it would prove to be a female and could be taken into captivity as a mate for Topatopa, a seventeen-year-old male that had been held at the Los Angeles Zoo since 1967. The juvenile bird turned out to be a male. But it had been caught just before the onset of bad weather and heavy wind; with gusts of up to sixty miles an hour, the biologists were reluctant to release it. After holding it for several days they noticed that it had begun to gape and pant heavily. "We went back and looked at some old photos, and we could see that the bird was gaping even in the wild when it wasn't under stress," Grantham recalled. "There was no doubt in my mind that there was something wrong with it, and that it should be taken in.

"Noel told me he was worried because there was a lot of pressure from the conservation groups to release the bird. . . . There was even talk of stationing federal marshals around the building where we were holding it because of fears that someone would try to steal it."

By the close of 1982 three condors were living in captivity, Topatopa at the Los Angeles Zoo and two others at the San Diego Wild Animal Park. The following spring the research center began experimenting with double-clutching to try to force the condors to double their rate of reproduction. Eggs and chicks were to be stolen from nests and raised in captivity by condor hand puppets so they would not imprint on their keepers.

As with every component of the program, the decision to remove condor young from the wild had been debated furiously before the Fish and Game Commission. The people in favor of the proposal had included not only Snyder and Ogden but administrators from zoos that hoped eventually to obtain condors for their collections. Recently Snyder had begun conferring closely with zoo officials—largely, his critics speculated, to shore up his eroding political base with the Fish and Wildlife Service. Snyder found increasingly that his bosses at the Patuxent center ignored his recommendations, even though they had hired him to be their field expert on the condor. Patuxent administrators had come to regard him as impulsive, strong-willed, and difficult to supervise. By 1983 Snyder's relationship with his bosses had deteriorated beyond repair.

Despite the personal differences that had surfaced between Snyder, Ogden, and other biologists at the Condor Research Center, all agreed that stealing condor eggs and chicks was a sound way to increase the number of captive birds. Since the parents would probably lay a replacement egg for each one stolen, the number of young produced each year might be vastly increased. The anti-capturists might not have objected to the proposal if the replacement egg had been left in the nest. But during 1983 and 1984 the biologists brought into captivity ten eggs and three chicks—the entire class of young for both years. The condors double-clutched just as predicted; in 1983 one pair even produced a third egg. But not one was left in the wild. The biologists had decided it would be safest to rear them all in captivity, since the mortality rate for condor young was fifty percent. And by stealing every egg and chick, the biologists hoped to force the adults to mate the next season—a full year before they normally would have bred again.

To the anti-capturists, the notion that the condors would not be affected by the constant surveillance, the intrusions into their nests, and the repeated loss of their young seemed ludicrous. "If you remove all your productivity from the wild, of course your population is going to be affected," Phillips argued. "Their whole argument became a self-fulfilling prophesy; the condor population was decreasing, so they needed to bring all the wild birds in." Phillips also worried about the stress the condors suffered from losing egg after egg. "They were never allowed to raise a chick, so there was no positive reinforcement for them," he said. "Why should they continue to try to breed?"

"That's just hogwash," Snyder said. "We never saw any decrease in courtship or breeding; in fact, we saw a great increase. The only thing that stopped those birds from breeding was death."

Between breeding seasons Snyder and his field assistants worked on examining nest sites, while Ogden, Grantham, and several other Audubon biologists concentrated on trapping condors for radio tagging and medical examinations. Although the trapping had gone well in 1982, week after week in 1983 the team caught nothing. "We still hadn't handled condors enough to know what kind of bird we were dealing with," Ogden said. "We still thought, to a degree, that the condor might be smarter than we were. We kept trying to hide the net better, hide the cannon in a less obvious place."

By the summer of 1984 only about sixteen condors were alive in the wild. Birds were still disappearing for unknown reasons; although the biologists speculated that they may have been killed by shooting, collisions with electric wires, or poisoning, they had little evidence to confirm their suspicions.

It is not possible to recount all the difficulties and political battles that hamstrung the recovery program during the mid-1980s. With the relationship between Snyder and Ogden frayed beyond repair, every turn of events spawned more resentment and suspicion between the two research teams. Snyder pressed for expanding the number of birds in the breeding program, but he also proposed that a few year-old birds be released from captivity into the wild—once their gene lines were adequately represented in the captive flock. Ogden continued to have qualms about depleting the wild population before more could be learned about individual behavior. "I just thought we'd be losing more than we'd be gaining in terms of protecting condor habitat and answering all the questions we still had about the species in the wild," he said.

Jesse Grantham, concerned that the wild birds were feeding on carrion contaminated with lead or toxins, began lobbying for an around-the-clock feeding program, where carcasses known to be free of contaminants would be placed within obvious sight of condor roosts and nests. "I thought we should have been babysitting birds," he said. "I thought we should have made sure that we knew where they all were at every minute, and that every time a bird came down on a carcass it was one of ours, and it was clean. Why weren't we all talking about the pros and cons of a supplemental feeding program? That's exactly the kind of thing that should have been discussed openly between the two sides, but wasn't looked at seriously until it was too late."

The winter of 1984–85 would prove to be the worst in the history of the recovery program. That fall fifteen condors were known to be alive, including five breeding pairs. Within a few weeks Ogden and his staff managed to trap five condors and fit them with radio transmitters or replace old transmitters that were wearing out. "We never figured out why we managed to trap birds that year and not the year before," Ogden said. "But we came away from handling them convinced that the condor was not a creature of some mythical intelligence, that in fact it was rather simple. One morning our truck broke down and we were delayed while we worked on it. A condor sat on a bluff and watched us, then watched us put out the carcass and walk into a blind very close by. Then he landed on the carcass and we caught him."

The trapping was the last field work Ogden completed on the condor project, for that fall he moved to Florida. The following January, when the staff of the research center began monitoring condors for signs of courtship, they could find only nine. Six birds, all of which had been expected to breed that spring, had disappeared. One had been wearing two radio transmitters. The only clue to the birds' disappearance did not surface until April, when the bird wearing the transmitters was discovered in the foothills of the Sierra Nevadas, too weak from lead poisoning to feed or fly. It died before veterinarians could treat it.

Nine condors, in Noel Snyder's opinion, were not enough to maintain a wild population. Convinced now that every bird should be taken into captivity, Snyder worried anew about the genetic health of the species. A series of tests on the captive condors had shown that they were already low in diversity and probably could not survive without serious inbreeding.

The only way to save the condor, Snyder believed, was to add the bloodlines from the wild birds to the captive stock.

Even so, the capture of the entire wild population was a radical proposition, and most of the Audubon staff members at the research center opposed it. The presence of condors in the mountains had forestalled a number of development projects, including a generating plant with more than five hundred windmills that had been planned along a major condor flight path. Once every condor had been placed in captivity, the Audubon staff members feared there would be no limit to the development within the species' historic range.

Grantham also worried that the project was drifting away from its original goal of preserving the condor throughout its habitat. "We were being told we deliberately needed to wipe out the historic traditions of the condor," he said. "That flew in the face of everything that we as conservationists and biologists believed in. And there were huge gaps in our knowledge about the birds. We learned very little about what might be called its culture—for instance, about why some birds flew north into the Sierra Nevadas at certain times of the year, or why they all ended up on the Hudson Ranch in August and September. Maybe biologically none of that was important. But from a philosophical standpoint, it was important to me that condors continued to roost in the same trees and forage in the same areas that they used ten thousand years ago."

Grantham argued that a few condors should be allowed to remain in the wild, even if they had to be given Andean chicks to raise in place of their own young. The traditions of the birds would remain intact, and biologists could continue to learn about the dangers that faced condors in the wild. Snyder viewed the suggestion as well-intentioned, but missing the point. "I think the question we need to ask is whether those traditions were worth saving, or whether they were really part of the problem," he said. "We had condors foraging in areas where hunting is allowed. They were feeding on carcasses that contained lead pellets. No one would have suggested a complete closure of condor habitat to hunting, because the backlash wouldn't have been worth it. People out there would have shot condors just for retribution; make no mistake about it. It was not a good political situation. So wouldn't it have been counterproductive to try and save those traditions? Did we really want to fight the kinds of battles we were bound to lose?"

In the spring of 1985 the biologists trapped three more condors, all from family lines that were not well represented in the captive flock. Six were still free. In June the state Fish and Game Commission voted to remove all condors from the wild and asked federal administrators at the Patuxent Wildlife Research Center for their concurrence. Instead, the Patuxent administrators suggested that three birds be released from captivity—in direct opposition to Snyder's recommendations. The wild birds could be safely maintained, they reasoned, if biologists provided them with an abundant supply of calf carcasses known to be free of lead and toxic contaminants. Snyder protested that birds raised in the wild could never be trained to feed consistently on a source of food provided by humans. Nevertheless, the Patuxent administrators' position was ardently supported by the top staff members of the National Audubon Society.

The state Fish and Game Commission would not agree to the release of any captive birds. But all that summer and fall the staff of the research center hauled animal carrion to the remote mountain slopes where six condors still roosted. In early autumn the Audubon biologists began trying to trap several birds for radio tagging, including a female known as AC-3 (Adult Condor 3). AC-3 already wore a radio transmitter, but its battery had worn down and needed to be replaced. The biologists also wanted to draw blood from her to check for elevated levels of lead and other toxicants. "The problem was that under the terms of our permit we needed to have a veterinarian there when we caught her, and the vets were only available on certain days," Grantham said. "So if she tried to feed when we didn't have a vet on hand, someone would flush her off the carcass."

In retrospect, Grantham believes the biologists made a critical mistake by preventing AC-3 from eating meat they knew to be free of poisons, because somewhere that fall the condor devoured an animal that had been shot. When finally the biologists trapped her in late November, she seemed healthy and alert. They changed her radio transmitter, drew a vial of blood, and released her. A few days later, veterinarians reported that her blood contained dangerously high concentrations of lead.

AC-3 had since flown with her mate into Bittercreek Canyon, an area of sloping sandstone cliffs in the foothills surrounding the San Joaquin Valley. Grantham noted that the bird's crop was full, an indication that she had fed recently, but her mate seemed hungry. "Usually they stayed right to-gether," he said. "In fact, we had high hopes they would breed that year, and when courtship begins the birds are inseparable. So it seemed unusual for her not to be ready to feed when he was." The male landed on a carcass the biologists had put out, ate heartily, and flew from the canyon, leaving AC-3 behind.

For several days AC-3 remained in the same vicinity, flying little. A system of high pressure settled over the area; the wind died, and there were no thermals the condor could ride to other foraging areas nearby. Still, AC-3's behavior struck the biologists as unusual. One morning Grantham and Dave Clendenen, a biologist on Snyder's staff, hiked into the canyon to look at AC-3 more closely. The condor let Clendenen approach to within a few yards before flushing. "We both came away feeling that something was seriously wrong," Grantham said.

The following day the biologists agreed to take a helicopter into the canyon in hopes of getting close enough to AC-3 to catch her. But the aircraft frightened the condor into an area of sheer cliffs that was nearly inaccessible to humans. A veterinarian studied her through a spotting scope and decided she was healthy enough to be left alone. Within a few days, however, AC-3 had begun roosting on the ground. "Condors just don't do that," Grantham said. "We went into the canyon and put out a carcass. She flew toward it, then started walking toward it, almost crawling. A vet and one of our biologists walked up to her, and she just tucked her head down and let them grab her.

"Her whole upper digestive tract was paralyzed by lead poisoning. Meat was putrefying in her crop, and she was starving to death."

AC-3 was taken into captivity in early January. Veterinarians treated her with medication to reduce the lead content in her bloodstream, but her digestive system never recovered. Ten days later, as surgeons prepared to implant a tube in her stomach to feed her, she died.

Of the great flocks of wild condors, only four males and a female remained. Abruptly the Patuxent administrators reversed their position and called for the capture of all wild birds. The poisoning of AC-3 had changed the minds of several members of the Audubon staff as well. If a supplemental feeding program could not prevent condors from ingesting lead, perhaps nothing could. But administrators for the National Audubon Society did not agree. That winter the society filed suit against the Fish and Wildlife Service, seeking an injunction against the capture of the last birds. Among other things, the society hoped to force the service to restrict the authority of zoo officials, whom society administrators believed had gathered too much control over the recovery project. The injunction was granted, then overturned in federal court the following June.

Snyder and his allies had clearly won. During 1986 and 1987 the biologists from the research center had little to do but sit in blinds or pits and wait for condors to alight within range of their traps. In April 1986, Dave Clendenen removed the last egg ever to be laid in the wild by a California condor. Its sire, AC-9, had been weighed and measured by biologists shortly after his hatching in 1980—the day before another chick was weighed and measured to death. AC-9 had been known to biologists all his life. He seemed to like people; sometimes he showed off his aerodynamic skills by flying circles above groups of bird watchers. As Clendenen removed the egg from a nest cave high in a cliff, AC-9 soared by the opening, as curious of humans as always.

Another year would pass before AC-9 himself could be captured. In the spring of 1987, though, his was the sole silhouette of a condor in the California sky. On Easter Sunday, AC-9 landed to feed on the remains of a calf in a pasture, near a strangely shaped group of shrubs. The blast of a net from a cannon brought his life of freedom, and the era of the wild condor, to an undignified close.

A hard south wind tousled the stiff knots of chaparral, stirred road dust into ragged funnels, and rattled the contorted oaks that spilled down grassy mountain slopes. It was a Santa Ana wind, the gale of the southwestern desert, as bitter and cutting in the Transverse Mountains as a freshly sharpened saw. On a narrow, precipitous road a battered government pickup eased over a series of speed bumps, then resumed its ponderous climb, threading up switchbacks with patient steadiness and an incessant mechanical whine.

An hour earlier, just before dawn, I had left the home of James Wiley in the valley town of Camarillo. A few last stars hung in a sapphire sky, and silhouettes of houses appeared as the darkness began to dissolve. Even in the dimness I could sense the manicured lushness of the California coast, the residential gardens and orange trees and windrows of eucalyptus. Wiley, a man of few words with strangers, was mostly quiet as he turned off a busy highway and wound northeast through canyons to the road that led to Hopper Mountain. We were out of the range of irrigation, and all greenness had vanished. The hillsides, folded one against another with triangulated steepness, were covered with stiff bunches of chaparral and sage. What color existed was subtle: the gray-green of the sage, the orange-buff in a ridge of exposed rock, the bone white of sycamores along a stream. We were on our way to a hillside where young Andean condors roosted, an aerial distance from Camarillo of about twenty-five miles. A condor might have flown it in half an hour, but the drive up the meandering mountain roads would take three times that long.

It was January 1989 and a new era had unfolded, marked by the absence of political battles. The California condor had not lived in the wild for nearly two years. The immediate fate of the species lay firmly in the control of the Los Angeles and San Diego zoos. The jointly operated Condor Research Center had been dissolved. Authority for the recovery project was now delegated to the Fish and Wildlife Service western regional office in Portland, with the Patuxent research center taking a secondary role and making plans to divest itself of the program entirely. Jim Wiley had replaced his old friend Noel Snyder, who resigned from the service in 1986.

A few of the field biologists who had worked under Snyder remained on the project, mostly for the chance to test the release of captive Andean condors to the wild. Controversy was no longer a hindrance to their work; the specifics of the release experiment had been negotiated with state officials to the last detail. Only female condors would be set free, to prevent them from breeding. They would be removed from the wild within three years to eliminate the possibility that a South American species would become established in California condor habitat.

The pickup traversed a shale-covered slope and creaked around a hairpin turn, moving laboriously up the gravel road. Dust seeped into the cab, settling over the dash, the seats, and us. A thin, rusty pipeline followed the road for a while, then draped down a cliff and across the canyon like a suspension cable on a bridge. All around were other trappings of the petroleum industry—yellow and green oil pumps set on flat pads of ashen soil, a thin silo marked with the words "PRODUCED WATER," a wall of squarish fans to burn off excess natural gas. We had entered oil country even as we entered the Hopper Mountain National Wildlife Refuge.

Wiley pulled on a pair of sunglasses, anticipating the sharp rays that hit the windshield as we crested a ridge. The pitch of the hillsides softened, and cropped grasses covered the rocky soil. The slopes had started to green after a recent rain; their mixed mat of gold and jade gave the mountainside a pastoral appearance. A half dozen cattle, standing beside the road, observed our approach with dull incredulity. Wiley stopped on the lip of a hillside that fell to the south and rose again in a yellow tower of stone. "Let's take a look from here," he said.

The breeze slapped at our faces as we stepped off the road and picked our way a few yards down the hill. We dropped to the ground and half reclined, our backs pressed hard against a rock to reduce the angle at which the wind could assault us. Wiley, his hands steadier than mine in the gusts, kept his binoculars trained on a small, black chip that soared over the stone pyramid. Although I knew it to be a condor, the bird was too distant for me to have any appreciation of its size. It hovered unsteadily for a moment, then dropped gracefully and landed out of sight. "Not bad," Wiley said, "for a beginner."

At forty-six, Wiley was as lithe as a dancer, with unusual strength in his arms and legs from decades as a competitive cyclist. He frequently rode as much as a hundred miles on his days off, looping through the foothills outside Camarillo. Like Noel Snyder, he was a man of compulsive work habits. In California he was working as the chief research biologist on the condor project, directing field work and collaborating with ornithologists from the Los Angeles Zoo on the Andean condor release. The field technicians who worked under his direction knew him to be even tempered, taciturn, and quick witted, with a dry sense of humor.

The release had been designed to test whether young condors could be kept out of oil and gas fields and gamelands where they might feed on carrion contaminated with lead shot. If successful it would be a large triumph indeed. It would show that biologists could transform the condor from a maverick bird to one that seldom strayed from a small foraging ground. As the young condors began to explore, the biologists would try to coax them toward Hopper Mountain by setting out calf carcasses at progressively greater distances from the sandstone roost. With no parents to teach them, would the birds discover the meat? "That's part of the experiment, to see whether they'll learn to key in to other carrion feeders—eagles, ravens, and vultures—that can help them find food," Wiley said. "If they can't, there's a question of how well they'll do."

We returned to the truck and drove north toward a site where the field crew had erected a roost box for four additional Andean condor chicks that were to be released the following week. If all proceeded as planned, Y-1 and her two nest mates would follow the trail of carcasses to the second roost, where they and the new birds would form a cohesive flock. First, though, all would have to learn to fly. Y-1 and Y-3, the youngest bird roosting on the sandstone pinnacle, had both been taking short flights for weeks. But Y-2, a chick raised at the Patuxent center, seemed unusually reluctant to leave the ground. "Her development has always been slower than the other two," Wiley said. "She's more timid. Her reaction when we moved her into the roost box was to tuck her head and stay huddled in a corner instead of taking any kind of defensive posture. We're afraid she may be holding up the other birds."

The second group of chicks was being kept in a roost box on an exposed ridge several miles away. Wiley turned off the main road and stopped to open a gate across an overgrown jeep trail. Oaks lined the narrow road, their leaves curled and brittle, their branches pearly from wind. Occasionally the slick, red limbs of manzanita poked from the waving masses of shrubs. The prickly undergrowth—sage, bitterbrush, and fine grasses—had the intricate texture of a tapestry. Chiseled mountains surrounded us in every direction. "This is as far back as you can get in the wilderness," Wiley said expansively, "at least in California." Several miles down the trail, he swung the truck into a small parking place and led me on foot down the muddy, stone-strewn road, as comfortable and agile on mountainous terrain as he was anywhere on earth.

We stepped into the brush to reach a hidden trail. A short distance away was a plywood blind built eight feet above the ground on thick timbers. The blind caught the full force of the wind, which vibrated against it and hid the noise of our approach. We removed our boots quietly and climbed a ladder to a trap door in its floor.

The interior was dark but for the beam from a small flashlight. I could feel a layer of foam padding beneath me and could barely see the crouched shape of Dave Clendenen, who was monitoring the behavior of the chicks. "Look here," Wiley whispered. He lifted a flap of material to expose a small one-way mirror set into the wall of the blind. Through it I could see an outdoor cage where a young Andean condor perched on the lip of a water tub. A few yards away sat a second chick, its feathers tickled by breeze. The cage was at the same level as the blind. To leave it, the birds would have to fly.

This was my first encounter, close up, with condors. I was prepared for them to be large, but even so I was startled by the size of the six-month-old

The Patuxent Wildlife Research Center is in rolling land just outside Washington, D.C. Photo by Ray Erickson, U.S. Fish and Wildlife Service.

One of the first whooping cranes taken into captivity from the wild. Photo by Ray Erickson, U.S. Fish and Wildlife Service.

To test imprinting behavior, biologists raised whooping crane chicks with sandhill crane puppets. Photo by David Ellis, U.S. Fish and Wildlife Service.

Wild whooping crane eggs are transported to breeding centers in a special incubator built inside a suitcase. Photo by G. Smart, U.S. Fish and Wildlife Service.

Two-week-old peregrine falcon chicks, hatched in captivity. Photo courtesy of The Peregrine Fund, Inc.

An adult male peregrine falcon. Photo by Dr. David H. Ellis, Institute for Raptor Studies.

A hacking tower in Crown Butte, Montana, for releasing captive-bred peregrine falcon chicks. Photo by Bill Heinrich, The Peregrine Fund, Inc.

Biologist Teryl Grubb rappels down a 500-foot cliff in Arizona to reach a peregrine falcon eyrie. Photo by Dr. David H. Ellis, Institute for Raptor Studies.

An adult male Puerto Rican parrot. Photo by James W. Wiley, U.S. Fish and Wildlife Service.

Wild Puerto Rican parrot chicks are usually fitted with radio transmitters before they fledge. Photo by G. Lindsey, U.S. Fish and Wildlife Service.

An Arabian oryx on the Jiddat-al-Harasis desert plain. Photo ©
Tim Tear and Deborah Forester.

A wild oryx herd grazing. Photo © Tim Tear and Deborah Forester.

Timothy Tear, center, and his colleagues release a captive oryx into a holding pen at Yalooni. Photo © Tim Tear and Deborah Forester.

Once courtship begins in late winter, wild California condor pairs become nearly inseparable. Photo by James W. Wiley, U.S. Fish and Wildlife Service.

In flight a California condor's shape—the broad tail and the
reaching, almost rectangular wings—is unmistakable. Photo
by James W. Wiley, U.S. Fish and Wildlife Service.

An adult golden lion tamarin. Photo by Jessie Cohen,
National Zoological Park, Smithsonian Institution.

Devra Kleiman has con-
ducted extensive research
on the golden lion
tamarin's behavior. Photo
by Jessie Cohen, National
Zoological Park, Smith-
sonian Institution.

Juvenile black-footed ferrets in their nest box. Photo by LuRay Parker, © Wyoming Game and Fish Department.

In Wyoming's breeding center, ferrets are kept in table-top cages. Burrows made of plastic tubing hang beneath. Photo by LuRay Parker, © Wyoming Game and Fish Department.

A wild black-footed ferret. Photo by LuRay Parker, © Wyoming Game and Fish Department.

Researchers prepare to weigh a ferret caught from the wild population near Meeteetse. Photo by LuRay Parker, © Wyoming Game and Fish Department.

At night in the shine of spotlights, the wild ferrets looked eerie, "like something from E.T." Photo by LuRay Parker, © Wyoming Game and Fish Department.

Biologists treed this male Florida panther in 1981 and fitted
it with a radio collar. Photo by Chris Belden, Florida Game
and Fresh Water Fish Commission.

Biologists Marnie Lamm and David Maehr assist veterinarian Melody Roelke, right, as she examines a wild
panther. Photos by William Greer, Florida Game and
Fresh Water Fish Commission.

A wild western cougar from Texas before its release in north Florida. Photo by Chris Belden, Florida Game and Fresh Water Fish Commission.

chicks. They stood more than two feet tall and weighed twenty pounds. Rather than birds they reminded me of young bears, partly because of their dark plumage and naked, mud-colored heads. As adults their ruffs would become milky white, unlike the California condor, and their heads and necks would turn brownish-gray with rosy highlights. As chicks, however, the two species were virtually identical. The bird on the water tub began flapping her wings, which, even partially extended, looked far too wide to fold comfortably again onto her back. She jumped into the air and threw her head back in enthusiastic frolic, then landed and walked calmly toward the entrance of the plywood roost box.

Through a peephole on the other side of the blind, I could see the third and fourth chicks sitting quietly in the roost. At four months of age the youngest chick's feathers were still a downy gray, and she was slighter than the other birds, although nearly as tall. She stood next to the plywood wall, only a few feet from me but unaware of my presence.

Behind me Clendenen was keeping track of the birds' actions with a special scanner gun, the kind used in department stores to record prices. On a clipboard he held a list of possible behaviors—feeding, preening, frolicking, and sleeping, among others—each of which had been assigned a bar code that could be deciphered by the scanner. If a bird preened for ten minutes, Clendenen would skim the scanner over the corresponding code and time. The information was retained in the memory of the scanner, to be loaded into a computer later.

By analyzing such detailed observations, the biologists hoped to determine whether the chicks were developing normal patterns of social behavior. Some of the birds had been fed and cared for by hand puppets until they were old enough to feed themselves. The others had been left in the nest with their parents. Had all the birds developed socially at about the same rate? Or had the chicks raised by puppets been deprived of social interactions needed for them to learn normal patterns of behavior? This last question was important, since all California condor chicks bred in captivity would be reared by puppets. Parent birds would not be trusted to raise their own young, since they frequently broke the eggs by sitting on them or rolling them around with their feet.

So at set intervals, a biologist checked the four chicks still confined to the cage on the windy ridge to see if they were feeding, scratching, fighting, jumping exuberantly into the air, or sulking in corners by themselves. To see, in effect, if they were getting along. Under normal circumstances such interactions would not have been an issue, since in the wild condor chicks are raised alone. But this was not the wild; this was an attempt to create a species of bird that could survive in modern California. It struck me as odd, and a little sad, that the four chicks were being watched so closely to see if they conformed to a human notion of what was appropriate behavior.

A half hour later, as Wiley and I walked back to the pickup, I asked whether he was bothered by the idea of training condors to depend entirely on people for food. "If it's the only way to have condors in the wild, then it's worth doing, even if it makes the birds different," he replied. "There's no way to maintain condors and let them range freely, at least not as freely as they once ranged. Land uses have changed too much. Condors have depended for the last century on cattle ranches as a source of carrion. But the day of the large ranch is over; people can't afford to hold that much land in one piece anymore. Rangeland is being converted to crop agriculture or developed into office parks," he smiled grimly, "which seem to be taking over the world.

"I don't think it will be a problem to keep the birds in an area where they can be managed. I hate that word. I hate the concept. But it's an unfortunate fact. If we want to have condors in the wild at all, they're going to have to be managed. And they're going to have to be managed within a smaller area than their historic range."

Midmorning. A crystal light poured over distant gray ridges and washed down the grassy side of Hopper Mountain to the knoll where five people toted saws, ladders, ropes, and large sheets of plywood along a trail. It was the yellow sun of January, weak from its winter slant but bright enough to warm the south-facing slopes. The Santa Ana winds had ceased, and the dizzying, blue sky revealed no hint of the season, nor of the cold weather forecast for later in the week. It was a good day, Wiley and his field assistants agreed, to build the first of several elevated platforms where carcasses could be set to lure the condors from the pinnacle roost toward Hopper Mountain.

The platform was to be constructed eight feet up in a holly oak on the very edge of the knoll, just above a steep drop. The splayed branches of the tree could cradle a sheet of plywood large enough to hold the corpse of a calf. Viewed from the air, it would be an obvious perch for carrion-feeding birds.

In addition to Wiley, the construction crew included Clendenen and Jack Ingram, a professional roadie for rock-and-roll bands in Los Angeles. Although Ingram had never been employed by the condor project, he had logged thousands of hours as a volunteer since the early 1980s, and he and Clendenen had become close friends. Their relationship, like their commitment to the recovery program, was founded on a deep spiritual appreciation of the condor and its place in the landscape of California.

Enrique Zerda Ordáñez, a biologist from Colombia who was visiting the project, had come along, as had I. We ferried the materials to the oak from a pickup parked on a ridge, then paused to examine the spread of branches and discuss how the platform should be braced. Clendenen and Ingram climbed into the tree while Wiley stood below, handing up tools and making suggestions about trimming branches. Zerda Ordáñez and I watched nearby.

Beyond the oak, on the far side of a narrow valley, a thin beard of conifers spottily covered a mountain ridge. To the south, the sandstone pinnacle jutted abruptly from a rolling hillside of grass and clay. I had asked to accompany Wiley that day partly with the hope of seeing a condor in flight again, although I knew it was possible the birds would never leave their roost. Y-1 and Y-3 had flown early that morning, and Clendenen had told me the birds tended to be less active in the middle of the day. Still, the air was crisp and fresh, the breeze moderate but steady enough to keep a condor aloft. Standing on a sheltered mountainside, drenched in sunlight and with a spectacular view, I was willing to wait a long time.

The men working on the platform obviously did not need any help. Zerda Ordáñez and I sat down with binoculars to scan a cluster of oaks for interesting birds. Occasionally Zerda Ordáñez called on a two-way radio to another Colombian biologist who was spending the day in the blind near the pinnacle: "¿Qué pasa? Where are the condors?"

"Sleeping behind a rock," came the reply.

Hours passed. I finished writing a long entry in my field journal and began watching an acorn woodpecker perched in a pine snag. The bird seemed to be hunting for insects or perhaps a cache of nuts; it moved methodically clockwise around the trunk as if wanting to leave no inch uninspected. I had almost forgotten about the distant roost box when a terse call came over the radio: "Condor up."

It was Y-1, circling lazily above the roost. Her large, yellow wing tag glinted in the sun. I called to the men in the tree, who paused long enough to glance toward the pinnacle but continued to work. Zerda Ordáñez was off on a short walk. The condor stopped circling and hovered in one spot, then began to rise, riding a thermal straight up. I followed her with binoculars until she was only a speck against long wisps of clouds.

Zerda Ordáñez returned, and I began trying to show him the tiny figure in the sky. But once I had taken my glasses from my eyes to point in Y-1's direction, I could no longer find her. On a hillside below us, two bluebirds fluttered gracefully in midair, hawking for insects. Otherwise the sky was empty.

"Hey, Enrique," a call came over the radio, "do you see the condor up there?"

Wiley and Clendenen, down from the tree, scanned the clouds with their binoculars. Y-1 was radio tagged, but until now she had never flown far enough from the roost for the biologists to need to track her. Wiley picked up the radio. "What kind of reading do you get on her?" he asked the biologists in the blind.

"She's somewhere up to the north."

"Which way's north?" I asked.

Wiley looked at me blankly. "Toward us," he said, surprise in his voice.

All work forgotten, the five of us hiked up the ridge toward the pickup, holding our binoculars ready. Only two days had passed since I had first watched Y-1 land clumsily on the pinnacle. Wiley, Clendenen, and Ingram climbed the ridge with long strides, anxious to reach the pickup where they would have a full view of Hopper Mountain. The dark form of an eagle crested the ridge and we froze. "Golden," Ingram said. "I've seen that happen before, though—you come up a hill and there's a condor right in front of you."

Wiley and Clendenen were the first to the top. Above Hopper Mountain, far enough away that her wing marker was barely visible, Y-1 soared on outstretched, steady wings. Her legs were tucked neatly against her; she had lost the need to use them as a rudder. No one spoke for many minutes. "She's really got her wings," Ingram said finally, awe in his voice.

"Is this sooner than you expected her to fly?" I asked.

"Oh, yeah." Wiley and Clendenen answered together. "This is real quick, faster than we expected by many days," Clendenen added.

The condor arched her wings, seesawing for a moment before regaining her balance. Her silhouette against the thin, trailing clouds had the unmistakable shape I had seen only in books—the broad tail and the reaching, almost rectangular wings.

"God, she looks like a California," Clendenen said.

"The question now," Wiley said, "is whether she can orient herself and find her way back."

I did not want to think about what might be in store for Y-1 if she could not retrace her way to the pinnacle. But even as Wiley spoke the condor moved slightly to the south, hovering above a hillside where a dozen cattle grazed. "Look," Ingram said, "she's checking to see if anything down there's dead."

We seated ourselves beside the truck to watch the condor return to her roost. The dark slash of her shadow, rough-edged and moving with almost frightening speed, crossed a hillside below us. She flew directly to the pinnacle and beyond, sailing over the valley with its quilt of farm fields, circling back against a rocky ridge. She swept out over the valley a final time before completing two tight circles above the stony pyramid and setting down shakily.

Her maiden flight had taken forty-five minutes. Without speaking, Ingram and I rose and began to applaud, clapping loudly in the dull, gold light of late afternoon, clapping with admiration and joy for a debut we had never expected to see.

"I think in five years," William Toone was saying, "we're going to be drowning in California condors. I would stake my career on it. We're going to have so many birds in captivity we'll wonder where to put them all."

We were sitting in a picnic area overlooking a shallow pool in the San Diego Wild Animal Park. Toone, the curator of birds, had escorted me to the pool to show me the assortment of stunning water birds—Dalmatian pelicans, white spoonbills, sacred ibises, Demoiselle cranes—that lived in the constant bustle of one of the largest zoos in the country. "I have three hundred species in my collection," Toone had sighed when I met him, "and everyone wants to talk about condors."

By Toone's own calculations, 1989 was to be the year scientists proved that California condors could be easily bred in captivity. Twenty-eight condors were alive. Most had been brought in from the wild as eggs or chicks between 1982 and 1985. In 1988 a seven-year-old male condor and his mate, a bird of unknown age, had produced the first chick ever conceived and hatched in captivity. Now, with more condors reaching sexual maturity, Toone was especially optimistic about the season that lay ahead. "We have four pairs of birds already courting here, and another in L.A.," he said. "We've never had any problem breeding Andean condors in captivity. It shouldn't be any harder to breed Californias."

A tall, animated man in his early thirties, Toone was self-assured and outspoken. His staff was in charge of caring for the condors at the wild animal park, and he seemed to personify the pragmatic, hopeful attitude that now pervaded the recovery program. Virtually nothing was going to stop the condor from breeding successfully, he told me cheerfully; and he did not believe that a condor in a cage was much different from one in the wild. "The condor has this kind of mystical significance to people because of its powers of flight," he said. "But it's just a bird. That's all. It's no more or less significant than any other bird. The one thing that's different about it is that it's become a flagship for endangered species."

In captivity the condor was being given the utmost in technological care. Video monitors were trained on every portion of the cages that held the great birds and their potential mates. As soon as an egg was laid, it was removed to an incubator. "Condor eggs have only about a fifty percent hatch rate, where we can get about a ninety-seven percent hatch rate with artificial incubation," Toone said. "As long as there are so few birds, we're going to pull every egg."

As a condor chick pipped out of its shell, it might be aided by technicians wielding surgical tweezers. The recorded calls of vultures would be played to mimic the sounds it would hear in the wild. It would never see more than the face of its "parent," a pigskin and fiberglass hand puppet built to resemble the head of an adult condor. For meals it would receive minced baby mice and the yolks of chicken eggs instead of carrion regurgitated by its mother or father.

I would not be allowed to see the condors, since they were being kept in cages away from public view. Strict security measures were being taken to keep the birds isolated, mostly to make them suitable for release to the wild, but also because of fears that someone might try to harm them or free them. A few years before, members of the radical conservation group Earth First! had obtained a film of a condor thrashing its wings against a chain-link fence. The footage, taken from within the security area at the wild animal park, had given the impression that the bird was trying to beat its way out of its cage. "It was a simple example of cling-flapping, something all condors do," Toone said. "There are pictures of them hanging onto cliff sides and trees and cling-flapping. These condors do not have a hankering to escape. They're just not that enamored of flying. They fly to look for food, and once they're full, they sit themselves down."

I was inclined to disagree, having watched Y-1 soar only a few days earlier. All the food Y-1 could have wanted had been set out on the sandstone pinnacle, yet she chose to fly as far as Hopper Mountain, several miles from her roost. When I mentioned this to Toone, however, he merely shrugged. "Sure, the birds are going to explore," he said. "I'm not saying they're just eating machines. Looking at condors in captivity is like looking at people in the hospital—there's no question that they should be out walking around. But the character of a hospital patient doesn't completely change just because he's hooked up to a bunch of IVs and tubes.

"The one big question mark is whether this population is so genetically weak that we can't save it," he added. "Some of the early genetic work that's been done on condors has shown it's a very inbred population. But it's also a very healthy population, for all appearances. I personally don't think the genetics issue will toll the death knell for the species. Who knows? In a couple of generations we may start seeing condors with birth defects. But I'd bet against it.

"The California condor—maybe I shouldn't say this so emphatically, but I will anyway—this bird is going to be a dreamboat. In five years we'll have more than we know what to do with. In ten years we'll have condors back out in the wild, because we've got the people and the dedication and the know-how to make it happen. There's just nothing I can see from this point in time that's going to put up any roadblocks big enough to stop us."

By June of that year, Toone's optimistic predictions were in the process of coming true. Four condor chicks hatched that spring, including one conceived by a pair of six-year-old birds at the Los Angeles Zoo that were presumed to be too young to breed. Several pairs double-clutched after their eggs were stolen for artificial incubation. Although many of the eggs were infertile, Toone and his colleagues remained hopeful that the birds would breed more successfully as they matured.

In the Transverse Mountains, Y-1 continued to explore the sky above Hopper Mountain, until one day she discovered the second group of condors on the windy ridge. Her roost mates did not fare as well. When after several weeks the timid Y-2 still appeared reluctant to fly, the biologists recaptured her. A few days later Y-3, who had accompanied Y-1 on many of her longest flights, ventured west into an area that was dotted with oil pumps and crossed by a major power line. She flew into a wire at full speed and was instantly killed.

Neither Wiley nor Michael Wallace, the principal investigator on the project from the Los Angeles Zoo, was surprised. Both Y-1 and Y-3 had seemed curious about the oil rigs, and they had shown little fear of the people who worked around them. Reluctantly, the researchers agreed that the pinnacle roost site was too close to power lines, mechanical equipment, and human activity to be suitable for future releases.

In the days after the collision, Y-1 spent most of her time in the areas she had explored with Y-3, as if searching for her companion. Finally she flew to the windy ridge where the second group of condors were learning to fly. After some aggressive snapping, the other birds accepted her.

As spring faded to summer, the first release of Andean condors appeared to be a success, although two birds had to be recaptured after they refused to socialize with the others. They were held in captivity for several weeks to give them time to mature and were rereleased. By summer's end the shapes of five condors could be seen among the peaks where years before Carl Koford began his study of North America's largest bird. They soared over the valley, stopping traffic; they hovered in thermals alongside hang gliders, curious about such strange, colorful birds. "People have come out to see them with tears in their eyes," Wiley said, ebullient.

In 1990, Y-1 and her nest mates all were recaptured and sent to Columbia for release there. Six more Andean condors were freed in remote portions of the Hopper Mountain refuge; they adjusted well and showed little inclination to stray into areas of human habitation. "They're curious about people, but they don't spend much time around oil rigs or industrial equipment," Wiley said. "They're acting the way we hoped they would—like wild birds."

The 1990 and 1991 breeding seasons also went well, so well that the Fish and Wildlife Service began laying plans to reintroduce California condors over the winter of 1991–92. Under the criterion adopted by the recovery team, releases were to begin whenever three breeding pairs had each produced six chicks. The first five offspring were to be held in the captive program; the sixth would be released.

In October 1992, amid much fanfare, two California condor chicks were moved to a roost site on a wind-beaten sandstone cliff within the Los Padres National Forest. With them were two Andean condor chicks that were to act as companions. One of the California condor chicks was a male, given the Indian name Chocuyens; the second, named Xewe, was a female. In January the four birds were quietly released.

As the condors began exploring their surroundings, they flew repeatedly to the shores of Lake Pyramid, a popular tourist resort. In October Chocuyens was found dead on a cliff near the lake. An autopsy showed that he had ingested antifreeze, perhaps from a spill or a discarded container. Xewe survived into 1993, although hunters shot at her one day while she was roosting near Lake Piru, south of the release site. After the incident service officials decided that condors could not be allowed to frequent populated areas. Field biologists began flushing Xewe and her Andean companions whenever they tried to land near human settlements.

In December 1992 the Andean condors were trapped, and six more California condors were freed. They formed a cohesive flock and, joined by Xewe, spent the spring slowly expanding their range. By the summer of 1993 biologists were making plans to release another eight condors, perhaps from a new site farther north in the forest. That spring Topatopa, the twenty-eight-year-old male brought into captivity in 1967, had finally sired two chicks. And Mololko, the first condor ever conceived in captivity, had laid a fertile egg at the tender age of five.

Given the commitment of money and personnel, given the popularity of the condor as a symbol for endangered wildlife, given, finally, the apparently malleable nature of the bird, the California condor recovery program seems likely to accomplish its goals. The return of the species to the wild will be a victory for the technical art of wildlife management. Beyond that, its merits will always be in question.

Noel Snyder believes the progress of the program so far has provided vindication for the years he spent arguing in favor of captive breeding. Snyder maintains that, despite the dissension, the recovery program accomplished a remarkable amount during the tumultuous years before the last condor was captured. "There certainly was a lot of political maneuvering," he says, "but there were enough contributors to the program that all proposals got to the floor. They had to be decided in an open forum. And the tactics that made the most sense were adopted."

Mark Palmer of the Sierra Club and David Phillips of the Friends of the Earth are less satisfied. Palmer wonders if the program will continue to be funded at adequate levels long enough for wild condors to become truly reestablished. "There's still the problem of lead poisoning," he says, "and there are even more roads into condor habitat than there were before. Why do they think they're going to do any better than before all the birds were brought in?"

Phillips, who now works for the Earth Island Institute in San Francisco, complains that condors will no longer fill the role they once filled in the California wilderness. "That ecological niche has been destroyed," he says. "If they're ever able to reestablish condors and have them act like condors, then sure, I'd have to consider the project somewhat of a success. But I don't think you can consider it successful if you just keep producing birds in captivity and throwing them out there. Eventually you're going to have to deal with hunter education, with condors getting shot, with off-road vehicles buzzing nests. You're going to have to do what we've been proposing all along."

Almost every day, Jesse Grantham of the National Audubon Society spends a few moments musing over the events that preceded the capture of AC-9. To him, the condor will never be the same, no matter what the future successes of the recovery program. "It's not so much the biological concerns that bother me," he says. "It's a question of ethics. The integrity of the species was destroyed with its traditions. We arbitrarily broke the condor's link with the past because we didn't want to deal with the problem of lead poisoning. I don't think those kinds of compromises will save wildlife, or the planet, in the long run.

"Our culture—where we live, what we eat, how we feed and raise our young—is all very important to people. Why shouldn't it be the same for the condor? Maybe in our society we don't think of tradition as important anymore. People will pull themselves up by their bootstraps under bad conditions; maybe we think the condor should be able to do the same.

"We're going to have a zoo species in the wild. That's all. Is *that* endangered species recovery? To me there's something missing. Was the intent of the Endangered Species Act to save animals by changing them to fit a man-altered habitat? I think that's a dangerous premise."

Toward the end of my conversation with him, Noel Snyder looked at me hard. "If you want to tell this story fairly," he said, "don't make it sound like there were good guys and guys in black hats. We were all human, all of us trying to do what we thought was right. We all made mistakes."

I left California with those words reverberating through my thoughts. Whatever mistakes were made may never be corrected. Whatever the condor has lost, it has lost irretrievably.

Having seen condors in flight, I can no longer imagine allowing the one species native to North America to perish simply because our society does not care to spend the time or the resources needed to save it. But I find it tragic that fifty years of recovery work has led us to a crossroads where we must choose between hand-feeding wild condors and letting them die.

It is comforting to think that the feeding program may be a temporary arrangement, a purgatory that will lead to the restoration of the truly wild condor. Perhaps before the end of the century condors will be reintroduced to sites outside California that are not under such intensive development pressure. There is talk of releasing condors in the Grand Canyon, where prehistoric nests have been found, or to the Gray Ranch, a five-hundred-square-mile preserve owned by a private landholder in southwest New Mexico. At least initially, clean meat would probably have to be provided for the birds at both sites. To move beyond the artificial feeding program will require, it seems, something more ambitious than simple reintroduction.

In a 1989 article for the journal *Current Ornithology*, Snyder and his collaborator and wife, Helen, note that for more than a century the condor has depended heavily on food provided by people, in the form of dead cattle and game killed by hunters. Its natural source of food, the carrion of large animals like elk and whales, is no longer widely available. The Snyders argue that the ultimate goal of condor recovery should be to reestablish the species in completely natural landscapes—in "megasanctuaries" large enough to support wolves, cougars, grizzlies, and other endangered species, as well as deer, antelope, bison, and elk. The sanctuaries would have to be well-defined by natural boundaries and vigilantly protected from poachers. As possible sites, the Snyders suggest the Gray Ranch and the Carrizo Plain of San Luis Obispo County, California, which could be connected through the Cuyama Valley to the Sisquoc Sanctuary of Santa Barbara County.

The creation of preserves large enough to contain entire natural systems is an intriguing, if expensive, proposal. These places would be more than homes for our vanishing plants and animals; they would be sources of nourishment for a national spirit badly battered by a legacy of thoughtlessness toward our native landscape. But it would be a tragedy to rebuild the intricate matrix of plant and animal life in a few remote megasanctuaries—and then to allow the rest of the American wilderness to deteriorate. If we restrict rare animals to one or two islands of prime habitat, if we do not provide them with the territory they need to live throughout their historical range, we rob them, and ourselves, of dignity.

By holding condors in captivity we have left ourselves a narrow footpath away from an abyss of utter failure. There is still a single hope that condors will one day live much as they did before the coming of Columbus. We must be careful with this fragile second chance. The odds are great that we will not receive a third.

XII

STEPPING BACK FROM EXTINCTION

The Black-footed Ferret

Late one evening in the fall of 1981, on the first step of land above the Greybull River of western Wyoming, a ranch dog encountered a strange animal in its back yard domain. From the house Lucille Hogg heard the dog's snarls but paid them little mind. The Hogg ranch backed up toward the dark Absaroka Mountains on the western edge of the Big Horn Basin; the dog, Shep, was part blue heeler and well accustomed to defending his turf against coyotes and badgers.

In the morning, Hogg found the remains of the intruder a short distance from Shep's food bowl. It was long and thin, a snaky mammal with short, black legs and soft fur. Its neck and belly were white, but mostly it was a light, buffy yellow. A black swath in the shape of a mask covered its eyes. Hogg had never seen anything like it. Neither had her husband, John, who shrugged and tossed it over the back fence.

The animal did not lay abandoned for long. The Hoggs were planning to run some errands in Cody that morning, and Lucille suggested that on their way they drop the carcass by a taxidermist in the nearby community of Meeteetse. It was pretty enough to have stuffed, and maybe someone in town could tell them what it was. Sure enough, when they stopped by his shop, the local taxidermist was astounded. "He told me, 'That's a black-footed ferret, and you'll get into a bunch of trouble if you mount it and even more if you display it'," Lucille said. "I told him, well, do what you think you have to, call the game warden or whatever."

For several years the black-footed ferret had been widely assumed to be extinct. The animal killed by Shep belonged to a population that had eluded detection by biologists—not a surprising feat, given that ferrets are nocturnal and live in some of the most isolated country in North America. Within a few days a team of scientists arrived to survey the land along the Greybull, including the Hogg property. At night spotlights scanned the pastures and hills in search of the chilling, emerald eye-shine characteristic of ferrets.

The Hoggs and their neighbors were surprised, and a little dismayed, that so much attention was focused on their small enclave of ranches, and so suddenly. And they never could have believed that the discovery of a sleek, secretive animal would lead to such acrimony and bitterness. From the beginning the ferret recovery project was plagued by political battles and personal jealousies. As in the case of the California condor, the last handful of animals were captured for breeding after a bitter struggle between key biologists.

If all proceeds as planned, by 1995 the black-footed ferret will be restored to the wild, at least in a few isolated locations. In 1991 and 1992 federal and state biologists reintroduced the first ferrets to Wyoming, and a release in Montana is scheduled for the fall of 1993. The only barriers to the species' recovery lie in the behavior of the animals and the attitudes of the people who will live around them. And scientists do not know for certain that either problem is surmountable.

At one time in the history of North America, the black-footed ferret, *Mustela nigripes*, may have been as well distributed in the West as the prairie dogs that make up the bulk of its food source. Although they spend most of their time below ground, ferrets do not dig dens of their own. Instead, they live in extensive burrow systems made by prairie dogs, emerging at night to hunt. They are unusually difficult to study. John James Audubon and John Bachman first described the species around 1849, but no other specimens were observed by naturalists for the ensuing twenty-five years. Until the late 1870s there was considerable skepticism as to whether the species existed.

Especially as kits, when they spend much of their time playfully wrestling, ferrets do not seem like ferocious predators. Their bodies, though unusually long, are so gracefully proportioned that they have the same elegant appearance as a mink or a weasel; they are, in fact, members of the weasel family. A similar European species of ferret has been domesticated as a pet. But the black-footed ferret, with its long canine teeth and strong jaws, is an unusually efficient hunter and much wilder in temperament than domestic species. It is not uncommon for a black-footed ferret to kill a prairie dog twice its size with a single bite to the neck.

Except during breeding season adult ferrets are loners. Like all mustelines, they have well-developed scent glands, and they make their presence known by rubbing their musky scent on plants, rocks, and the ground. Biologists estimate that each adult ferret requires about a hundred acres of prairie dog town to support itself, and a bit more if it is raising kits. It is this need for extensive territory that drew the species into its precipitous decline.

Two hundred years ago prairie dogs occupied millions of acres of short-grass prairie on the Great Plains and along the eastern flanks of the Rocky Mountains. In the late 1800s the naturalist and artist Ernest Thompson Seton estimated the North American population at five billion. Seton's estimate was based on crude census data, but it provides an image of how prolific the prairie dog was, and of how extensive were its communities. One prairie dog town in the Texas Panhandle, perhaps the largest on the continent, stretched for a hundred miles in width and two hundred fifty miles in breadth.

When prairie dogs dig their burrows, they may make obvious entrance mounds or else hidden entrances beneath bushes or clumps of sage. Either way, they move large quantities of dirt. As settlers migrated to the Great Plains and set up farms, they did not take kindly to prairie dogs disturbing their carefully sown fields. Nor did ranchers like the idea of rodents digging up their rangeland. But once populations of wolves and other predators had been reduced, and once the tall-grass prairies had been cropped by cattle, the prairie dog population exploded. By the turn of the twentieth century it was generally held that prairie dogs were a major nuisance to the nation's expanding beef industry. They ate the same grasses as cattle, and their burrows presented hidden hazards for stock animals. A cow or horse could break a leg by stepping in a prairie dog hole.

Around 1900 the federal government began hiring work crews to exterminate prairie dogs throughout the West. The crews used a variety of techniques—among them shooting, dynamiting, and gassing burrows—but the most common and effective method was poisoning with grain laced with strychnine. In 1920, the peak of the campaign, nearly seven million acres were poisoned in Montana alone. By the 1970s all but two percent of the ninety-eight million acres once inhabited by prairie dogs had been poisoned, much of it repeatedly. Prairie dog numbers had declined by more than ninety percent, and the Utah and Mexican black-tailed species had been added to the federal list of endangered species. Along with the prairie dogs disappeared the black-footed ferret.

Between 1946 and 1953 only about seventy ferrets, a third of them dead, were observed in the ten states of their former range. The federal government was debating whether the species should be declared extinct, when in 1964 a small population was discovered in southwest South Dakota. (The ferrets could not have chosen a less hospitable state in which to take a stand. Traditionally the government of South Dakota has run one of the most aggressive prairie dog control programs in the country; a law still on the books in 1990 required ranchers to kill prairie dogs on their land or pay the state to do so.) For several years biologists studied the population, spotlighting individuals at night in the summer and tracking their movements across the snow in winter. In all, more than ninety ferrets were seen, but they were too widely scattered to have much chance of survival.

In 1971 biologists from the Patuxent Wildlife Research Center captured six ferrets, with the hope of establishing a breeding program. Because

Mustela nigripes was known to be susceptible to canine distemper, the captured animals were given a distemper vaccine that had been effective in Siberian polecats, the species' closest relative. The polecats had not reacted to the live virus contained in the vaccine, but the ferrets proved less hardy. Within weeks, four of the animals contracted distemper and died. Although field researchers trapped another three ferrets, the animals never successfully raised young at the center. The South Dakota population dwindled, and after 1974 no more were seen.

In 1973 a biologist from Wyoming named Tim Clark attended a conference on the black-footed ferret in South Dakota. Clark was a specialist in prairie dog ecology and knowledgeable about small population biology. At the conference a researcher showed him two ferrets. Seeing the animals alive for the first time piqued his curiosity. "It just seemed to me that there was a high likelihood that there were other remnant populations out there," Clark said, "and it was obvious the species needed help." He decided to apply for some grant money to start searching for ferrets on his own.

Clark was something of a rambler. In the mid-1970s he was a research fellow at the University of Wisconsin and a faculty member at Idaho State University. Over the next fifteen years he would form his own company of field biologists, which he called Biota Research and Consulting, Inc. He would also hold teaching appointments at Yale University and Montana State University, and research appointments with two wildlife conservation organizations. Partly because he belonged to no single institution, he managed to spend three or four months a year studying prairie dogs and looking for ferrets. By 1981 he had searched unsuccessfully in eleven states, concentrating his efforts in Wyoming. Along the way he had started offering a $250 reward to anyone who could produce evidence of a wild ferret population. One September day he got a call from Lucille Hogg.

The residents of the western Big Horn Basin do not go out of their way to mix with strangers, but they are usually gracious to anyone who requests a favor. The Hoggs were no exception. When Clark asked if he could survey their property, they readily agreed. They also offered Clark and a small team of researchers a place to camp. "The first thing I did was hire a plane and do an aerial survey of the prairie dog population," Clark said. "It was quite large, easily big enough to support some ferrets."

Considering that Biota Research was working with one of the most endangered animals in the world, its associates enjoyed remarkably few restrictions during that first year. But Clark's interest in ferrets conveniently filled a bureaucratic void. The discovery of the Meeteetse ferrets had caught the U.S. Fish and Wildlife Service by surprise. The controversy over the preservation of the California condor was intensifying, and federal administrators were wary of becoming ensnared in another difficult endangered species project. President Reagan had just taken office; the political climate was not favorable for spending money on conservation. Biologists from the service came to Meeteetse that fall to help search for ferrets, and they fitted one animal with a radio collar. But in February 1982 federal officials appointed the Wyoming Game and Fish Department as the lead agency in the ferret recovery program; the state was to be responsible for managing the wild population and instituting a captive breeding program, if necessary.

Wyoming did not have much of a budget for endangered species work. Nor did state biologists have the contacts with wildlife conservation groups that Tim Clark had cultivated. After the discovery of the ferrets, Clark quickly pieced together enough grant money to keep his staff in the field studying the population for the next year. He, a business partner, and two Biota associates spent the autumn months searching for ferrets with spotlights, and the winter months tracking them in snow. The animal killed by Shep was a young male apparently moving into new territory. The core of the population lay several miles away, spread across about forty-five hundred acres. From what Clark could tell there were at least twenty animals.

It is difficult to determine the point where the feelings of distrust between Clark and state biologists first sprouted. Perhaps they grew from a simple problem of geography. Wyoming is a rural state and distant from the wildlife research centers where, especially in the early 1980s, the concepts of conservation biology and minimum viable populations were being shaped. The people who worked for the Wyoming Game and Fish Department tended to be hunters and fishermen—proud, independent, and partial to the dry, rugged landscape of their state. They were outdoorsmen; they were confident that they understood the best way to manage natural systems. They were also inclined to bristle at suggestions that their approach was too unrefined to save a critically endangered species. Or, perhaps the hard feelings drew their genesis from what state biologists perceived as arrogance on Clark's part. He knew more about black-footed ferrets and small population biology than anyone in state government, and was not afraid to say so.

At first, relations between the field biologists and the state staff were polite, if not cordial. The Biota team hunted ferrets through the spring and summer, counting the animals as carefully as possible. Occasionally they caught glimpses of ferrets in the daylight, of mothers with kits that bounded playfully over the sage or scampered and leaped between holes, their long bodies virtually turning loops. Most of the time, however, the researchers saw the animals only after dark.

Steve Forrest spotted his first ferret in the early hours of a cool spring night. He had come west to work with the Biota team during a break in classes at Yale University, where he was a graduate student. Panning the sagebrush pastures with a spotlight, he saw two small, emerald points of light, the eye-shine of a ferret. "The lighting and all made it seem like something from E.T.—this long, strange-looking animal peering out at me from a hole," he said. He had intended to work mostly on snow-tracking techniques during the trip, but he also went out with the spotlighting crew. It was haunting to spend nights in the shadow of ragged, snow-covered mountains searching for green-eyed phantoms. It was slow, difficult work to surprise the inhabitants of the prairie and try to distinguish the ferrets. The eyes of other animals—badger, long-tailed weasel, pronghorn antelope—shone green as well, but a different shade. The eyes of rabbits were red in the harsh light, while the eyes of coyotes shone yellow gold. Forrest found the work to be addictive. "The ferrets were kind of mysterious, and very challenging to study," he said. On Clark's recommendation he applied for some grant money and joined the Biota researchers full time.

From the start Clark advocated capturing some of the animals to protect them from outbreaks of disease and other environmental hazards. First, however, it was essential to find out how populous the ferrets were and whether they faced any immediate dangers. The biologists knew how to census and track the ferrets fairly well because of the studies conducted on the extinct South Dakota population. By August of 1982 they had found sixty-one ferrets. Twenty-one of the animals were adults; the rest had been born that spring. During the fall and winter Clark and his associates combed the prairies looking for juveniles that had scattered from their litters. Dean Biggins, a Fish and Wildlife Service biologist, arrived to try fitting several ferrets with radio collars. The researchers also surveyed the surrounding countryside thoroughly enough to be convinced that there were no other ferrets living nearby. The Meeteetse population had survived alone.

In August 1983 Clark and his staff estimated the ferret population to be at eighty-eight animals. Clark believed it was critical to capture some immediately for breeding, before the population suffered a setback of some sort. The Game and Fish Department wanted to move more cautiously. Many of the wildlife managers who worked for the state had a strong bias against holding wild animals in pens. (In 1985, Bill Morris, the director of the Game and Fish Department, would tell a local paper that he felt "no compunction about killing an animal, but I have a hell of a time putting one in a cage." The sentiment was shared by the coordinator of the ferret project and several other staff members.) Besides, Clark had a reputation among state biologists as being given to exaggeration. They were not inclined to let an outsider push them to move too hastily.

Technically, the state's decision on how to manage the ferrets could have been overridden at any time by the Fish and Wildlife Service. But once it had given control of the recovery project to the Game and Fish Department, the federal government could not have taken it back with any grace. The state biologists were sincerely committed to saving the ferret; their ideas of how to do so simply differed from those of Clark and other conservation biologists. And, as time passed, they grew more recalcitrant about bowing to outside pressure. In the spring of 1984 a meeting on ferrets was held in Cheyenne. It was attended by biologists from the state staff, the Fish and Wildlife Service, several wildlife conservation groups, and members of the Black-footed Ferret Advisory Team, an advisory board comprised of representatives from different government agencies and private interests. "Virtually everyone at that meeting thought some ferrets should be brought in as soon as possible," Forrest said, "except the people from Game and Fish." The state did appoint a committee to study possible sites for captive breeding and alternatives for funding; members were asked to present a report to the advisory team the following fall. The Biota researchers viewed the move as a stall tactic.

In retrospect, 1984 would have been the best year to take ferrets out of the wild for breeding, since the population produced twenty-five litters that year, bringing the late-summer count to one hundred twenty-eight. But with the ferrets healthy and reproducing well, the administrators and biologists of the Game and Fish Department saw little reason to worry about the species' well-being.

Harry Harju, the chief biologist for Wyoming Game and Fish, is a scrappy, outspoken man and a favorite among the state media. At one point in the debate over capturing ferrets, he told reporters the state had no intention of "losing control" of the animals by allowing them to be bred at facilities outside Wyoming. The comment brought him instant notoriety among the national and international wildlife organizations that helped fund Clark's field work.

When I met Harju in 1989, the question of whether or not to breed ferrets in captivity had long been settled. But the subject clearly still rankled him. "Even though we're isolated out here, we do read," he said. "If you look at the track record of how many captive breeding programs have actually put animals back out in the wild, you'll see it's not too shiny.

"The one thing we did to get into trouble with the black-footed ferret was that we refused to relinquish control. And if we had to do it over, I'd do it the same way. My agency has one goal—to breed these animals and put them back in the wild. If we hadn't kept control, there would have been absolutely no guarantee that any of the animals would ever go back out."

When the Game and Fish Department first raised the possibility of building a breeding facility for ferrets in Wyoming, no one was sure where the money for the facility would come from. Department administrators and members of the state Game and Fish Commission proposed that the cost be born largely by the Fish and Wildlife Service, with some money from the state. Clark objected, saying that adequate facilities already existed at several zoos. Partly at his urging, representatives from the World Wildlife

Fund and the New York Zoological Society contacted the state to offer financial assistance for a breeding program. "We had eastern conservation groups coming to us like crazy and telling us money would be no problem," Harju said. "So we said great, but we want to breed ferrets in Wyoming. All of a sudden they had damn little money they could spare."

Harju's attitude, and that of his staff, struck Clark as absurd. "There was simply never a cooperative climate created," he said. "Rather than go to meetings and try to solve problems, the people in charge had an agenda, and they pushed that agenda very hard. One of their key assumptions was that the ferrets belonged to the state of Wyoming. Period."

Privately, state biologists believed Clark had an agenda of his own—to make himself indispensible to the ferret recovery project. Relations between the Biota associates and the Game and Fish biologists deteriorated all through the summer of 1984. The staffing of the project was, at best, untenable. Since the state had only a few field researchers of its own, it depended heavily on the observations collected by the Biota scientists. But state biologists so distrusted Clark that they tended to disagree with virtually all his recommendations.

That September a conference on the ferret was held in Laramie. The Biota researchers were scheduled to present papers based on their field work over the previous three years. Biologists from the state and the Fish and Wildlife Service were to give reports on subjects ranging from scent-marking to efforts to radio-collar ferrets. The evening before the conference opened, the Black-footed Ferret Advisory Team met to hear the report prepared by the special committee on the feasibility of captive breeding. According to the Biota researchers, Don Dexter, who was then director of the Game and Fish Department, opened the session with the announcement that he had already decided it was too early to start capturing ferrets. Not enough was known about the species in the wild, he said, and no money was available for a breeding facility in Wyoming.

Later the Biota researchers speculated that Dexter wanted to proceed slowly because of fears that the state would have to bear the entire cost of building a breeding center, estimated at $180,000. While Fish and Wildlife Service administrators had spoken of providing funds for the facility, they had made no commitments. After the meeting, state biologists acknowledged that a breeding program was probably inevitable. But their reluctance to set up the program swiftly was roundly criticized by Clark, his associates, and a number of wildlife conservation groups. In refusing to be "hammered into submission," as Harju put it, the Game and Fish Department provided the fodder for a devisive test of wills during the long summer of 1985.

From Highway 120 north of Thermopolis, the mountains near Meeteetse look like waves rising in the distance, an ocean of chopped and jagged rock.

The countryside of western Wyoming is nearly skeletal in its hardness; a parched, sage-dotted soil stretches over the contorted hills like a nappy skin concealing the bones of the earth. Rows of toothy, decaying rock hang exposed in chalky bluffs. The land is dusty right down to the flood plain of the Greybull River, where vibrant cottonwoods replace the steely colors of the sage. Driving west along the river, the valley fills with hay and oats. The earth fills out too, as if more meat sits beneath the rocky hide. Brown hills fall to the river in smooth folds.

One autumn afternoon I sat in a jeep on a benchland high above the Greybull. The bench was on the Pitchfork Ranch, at eighty thousand acres one of the largest and wealthiest holdings on the west side of the Bighorn Basin. I had asked Jack Turnell, the manager, to show me something of ferret country. He obliged by driving up a steep rise that flattened to rolling pasture. Mountains surrounded us in three directions—the scalloped buttes of the Carters, the snowy blocks of the Absarokas. Pronghorn antelope cantered in long strings, flashing their moony, white rumps.

Turnell was a member of the Black-footed Ferret Advisory Team and a longstanding proponent of managing ranchland to benefit wildlife. The Pitchfork had not poisoned prairie dogs for more than twenty years. He parked to let me survey the core of a white-tailed prairie dog town, an expanse of pasture pocked with reddish mounds the size of small graves. I could not begin to guess the number of animals that lived there. Dozens of fat, brown rodents sat back on their haunches to study us, chewing ceaselessly, not threatened enough to duck into cover. Others busied themselves rambling from burrow to burrow or pushing egg-sized pebbles out of entrance holes. Two sentinels close to us clicked a staccato warning. There were hundreds of prairie dogs, maybe thousands. And not a single ferret.

In early 1985, when the state of Wyoming agreed finally to capture ferrets for breeding, no one could foresee that the dire predictions about the precariousness of small populations were on the verge of coming true. The Biota researchers and two state biologists had tracked ferrets through December on the benches of the Pitchfork and three surrounding ranches. The population appeared to be stable, although some of the kits born the previous spring had undoubtedly been killed by predators or perished in the subzero temperatures typical of the region. The yearly spotlighting counts were not scheduled to begin in earnest until late July, when new litters would be old enough to explore the terrain outside the burrows where they had been born.

The strain between the state biologists and the staff of Biota Research had been eased slightly by the announcement that some ferrets would be captured in the fall of 1985—provided that the population stood at roughly the same level as in 1984. In late spring, however, Harju told a Casper, Wyoming, newspaper there would probably be no breeding program started in 1985 because of a lack of money. The eastern conservation groups that had pledged support had lost interest, Harju said, once they realized the ferrets were to be bred in Wyoming. Federal funds would probably not cover the cost of building a facility in the state to house the animals.

That summer the remaining thin façade of cooperation between the Biota researchers, the Black-footed Ferret Advisory Team, and Harju's staff collapsed. The rift could not have happened at a worse time, for, unbeknownst to biologists, the ferret population was suffering a crisis. The first inkling of trouble appeared in June, when a graduate student who was studying prairie dogs at the Meeteetse site collected some fleas and sent them to a laboratory for analysis. The fleas were found to be carrying sylvatic plague, a disease that is fatal in prairie dogs. The following weekend Steve Forrest drove out to the benchlands to look around. As he hiked across the rocky pastures he saw a prairie dog crawl out of a burrow, spasm, and die. More fleas were collected, and a prairie dog carcass was tested for the presence of plague. The results were positive. "We immediately started worrying, because we knew a large decrease in the ferrets' prey base could wreak havoc with the population," Forrest said.

In early July, Dave Belitsky, the ferret coordinator for the state, ordered that prairie dog burrows within the range of the ferrets be dusted with the pesticide Sevin to control the spread of plague-infested fleas. To be effective, the pesticide needed to be applied widely and generously, and Belitsky enlisted the help of several dozen volunteers. Clark complained that the state had been too slow to act and that the volunteers were applying the chemical sloppily. He predicted that the ferrets would go extinct unless some were captured for breeding almost immediately. His criticisms, reported by local newspapers, angered Harju and Belitsky.

By then, however, the staff of Biota Research was too preoccupied to worry much about its relationship with the state. On July 8, Forrest took some spotlighting equipment into an area known as the East Core, which had been densely populated with ferrets the previous summer. He found none. Three nights later he tried again without success. A second team of state and Fish and Wildlife Service biologists surveyed a nearby area, also in vain. "By then we all had a feeling something was seriously wrong," Forrest said, "but Game and Fish wouldn't listen."

Harju, Belitsky, and several members of the Black-footed Ferret Advisory Team dismissed the concerns of the Biota researchers as exaggerated. In particular, they pointed to claims by Clark that the plague might be killing ferrets, for which there was no proof (and which proved to be false). "In our business we're used to seeing big fluctuations in wildlife populations," Harju said. "Tim Clark was an alarmist from the moment he set foot on that site. We just got tired of listening to him."

In mid-July the Biota researchers began conducting nightly spotlight surveys of prairie dog towns. The body of a ferret, apparently killed over the winter, was found to contain no trace of plague. Nevertheless, the disease had spread unevenly through the prairie dog colonies, and ferrets seemed most prevalent where it had not taken hold. Forrest worried that the carcasses of prairie dogs, which were mounting in number, might harbor bacteria or a secondary infection. "We worried about botulism, for example," he said. "At that point all we knew was that a lot of ferrets had disappeared." And some of the animals sighted by the survey crews seemed lethargic. In the West Core area, where thirty-six ferrets had been counted in 1984, the biologists found only four, a mother and three kits. "I should have known something was wrong," Forrest said, "because the mother was just lying on top of a little mound and the kits weren't jumping around like they normally do." After the first sighting they were never seen again.

When the spotlight surveys were completed in late summer, the count of ferrets stood at fifty-eight, with only twenty adults—a decline of sixty-eight percent over the previous year. Belitsky suggested that the animals might have dispersed to outlying areas. Frustrated, Clark and Forrest contacted Tom Thorne, a wildlife veterinarian in the Game and Fish Department, and asked him to draw a few blood samples from ferrets for analysis. Thorne refused. He was scheduled to examine ferrets in September, when field biologists conducted an annual trapping program to mark some of the animals with small ear tags. He saw no reason to stress the animals by capturing them sooner.

Clark and his colleagues, certain that the ferret population was crashing, could no longer bear what they viewed as gross negligence on the part of the state. On August 6 they submitted a report and a series of recommendations to Harju. They noted that the ferret population tended to peak in the summer, and that over the winter the population might be reduced by fifty to seventy-five percent from predation and cold. "It is obvious," they wrote, "that a small ferret population living on an eroding prairie dog base is more vulnerable to extinction than ever before." They recommended that blood samples be taken from a few ferrets and screened for disease, and that every litter be located each night to make sure the kits were still alive. If a litter was not seen for five days, it should be presumed dead and its burrow dug up to retrieve the bodies for autopsy. And although plans had already been made to trap six ferrets that fall for captive breeding, the Biota staff urged that some of the animals be taken from the wild immediately.

The report was forwarded to Harju and the Black-footed Ferret Advisory Team and simultaneously released to the Fish and Wildlife Service and half a dozen conservation groups, including the World Wildlife Fund, the New York Zoological Society, and the National Geographic Society. "If we made a mistake in all this, that was it—sending the reports to the conservation groups without giving Game and Fish a couple of days to mull it over," Forrest said. Several of the organizations issued harsh criticisms of the state's policies and demanded that ferrets be captured immediately. Harju responded by telling reporters that the conclusions of the Biota researchers were overblown and emotional. As far as anyone knew, he added, the ferrets were not in danger.

For another month nothing was done. On September 9, Tom Thorne arrived at the Pitchfork Ranch to assist with the annual mark and recapture program, a censusing technique in which animals were trapped randomly, tagged with ear clips, and released. Three weeks later the biologists trapped another sample group. By comparing the ratio of tagged animals captured during the second round of trapping to those that were not tagged, the researchers could estimate the size of the entire population. The technique had been used the two previous autumns as a way of verifying the summer spotlight counts.

Its results were horrifying. During the first round of trapping, the researchers spotted only about thirty ferrets. They caught twenty. By the second round the population seemed to have dropped even more. The biologists managed to capture only ten—less than a fifth the number of the previous fall.

The results of the census estimated that the ferret population had dropped from fifty-eight to thirty-one in six weeks. Six animals had been taken into captivity during the trapping program, and these were now housed in a state research station outside Laramie. Harju scheduled a meeting for late October to discuss what should be done. Those in attendance included the Biota researchers, some Fish and Wildlife Service biologists, and most of Harju's staff. "It was very hostile," Forrest said. "We were being nailed to the wall; the Game and Fish people were saying, 'How can you make these conclusions based on this little bit of data?' And then Tom came in."

The gathering knew immediately from Thorne's drawn face that he had bad news. One of the captive ferrets was dead. Another was gravely ill. The animals had canine distemper, and, based on the incubation period, they could have contracted it only in the wild. Which meant the entire population—what remained of it—was in mortal danger.

Meeteetse is a weathered western town fifteen miles and a steady down-stream drop from the Pitchfork Ranch. Like many rural Wyoming communities, it is more important to local residents than travelers might guess, for it serves as the business and social hub for dozens of outlying ranches. Its tiny business district is lined with squared-off western storefronts that stand shoulder-to-shoulder: a cowboy bar, the Meeteetse Mercantile, and, set a little apart, Lucille's café, owned by Lucille Hogg.

It is conventional wisdom in Meeteetse that biologists working for the state, the federal government, and Biota Research killed the local ferret population with meddlesome curiosity. A T-shirt for sale in the Mercantile bears a picture of a ferret with a defiantly raised front paw and the words, "Live Free or Die." A friendly woman named Edna Myers, waiting on me one Saturday morning, turned suddenly stern when I asked whether she agreed with the message on the shirt. "The people that came in here to study them killed them; we sure didn't," she said. "They wouldn't let them alone. Nothing can stand the kind of disturbance they put those animals through, following them around all night with lights."

Four years had passed since the outbreak of distemper among the ferrets, but the anger among local residents had diminished little. Once again the species was believed to be extinct, at least in the wild. The long-awaited breeding program had been established with eighteen animals rounded up over a year and a half. In February 1987, the last known ferret, a male the researchers named Scarface, was removed from the prairie dog towns above the Greybull River by federal biologist Dean Biggins.

Rumors of remnant ferret populations surfaced every few months, but none of the claims could be substantiated. Wildlife Conservation International, a division of the New York Zoological Society, was offering a $10,000 reward to anyone who could provide evidence of wild ferrets. In Meeteetse, though, no one seemed tempted by the size of the purse.

"If it was up to me to do over again," Lucille Hogg was saying, "I'd take that ferret Shep killed and throw it out back on my property and never tell a soul."

I had stopped for lunch in the café, which was simple and homey, with photos of ferrets hung around the room. But when I raised the subject, Hogg and most of her customers voiced disgust.

"I've got a buddy who says he knows where there are fifteen ferrets, and he's not telling," confided a cowboy drinking coffee.

I said that with $10,000 at stake, someone was bound to contact the Game and Fish Department if there were really ferrets in the wild.

"No ma'am," a waitress, passing by with a coffee pot, shot back at me.

"Not anyone from this town," said Hogg, "or most of the other towns around here, either."

Part of the local disillusionment grew from a common suspicion that biologists had infected the ferrets with distemper by taking dogs with them when they went out to do research. Some residents blamed Clark and his colleagues, some the Game and Fish Department, some the volunteers who dusted the prairie dog burrows with Sevin. Coinciding as it did with the outbreak of plague in prairie dogs, the timing of the distemper epidemic did seem curious. But canine distemper also infects skunks, raccoons, and, less frequently, coyotes and badgers, which prey on ferrets.

Three hundred miles from Meeteetse, in the college town of Laramie, Tom Thorne took a long draw from a cup of coffee and shook his head. "The distemper was probably introduced by wild animals," he said, "but we'll never be able to prove that, just like we'll never be able to change a lot of people's misconceptions about what happened. It's frustrating, but what counts is saving the species. And we're doing that."

Thorne is a ruddy-faced man with a crop of dark hair. Although he grew up in Oklahoma, he has lived in Wyoming for so long that he considers it his home. He unabashedly loves the state, its people, and especially its arid, broken landscape. On this morning we had met for breakfast before a tour of the new ferret breeding facility, built north of Laramie during the winter of 1985–86. Thorne had brought his wife, Beth Williams, who worked as a veterinarian at the University of Wyoming and frequently served as a consultant on the ferret project.

In late 1985 Williams had been assigned the task of caring for the first ferrets trapped from the wild. When they were initially brought into captivity, the animals were kept in cages in a small cinderblock room. After the first one died of distemper, Thorne and Williams knew it was only a matter of time before the others succumbed. One animal lingered for forty-nine days. "It was awful. Awful." Williams looked at me, and then at Thorne, as if for verification, as if she was still slightly numbed from the emotional strain. "All we could do was give them supportive therapy.

"We contacted some other vets to see if they knew of any treatments, anything at all that might help. They said we might as well go ahead and kill them. Of course we wanted to hold out on the slim chance that there would be a miracle. But it was one of the hardest things I've ever done, going in there twice a day and watching those ferrets die."

Times were happier now. The breeding facility, constructed with $188,000 in federal funds, held both black-footed ferrets and Siberian polecats, which could be used as surrogates in experiments. After a fruitless first year, the ferrets had started producing young with surprising ease. In June 1987, much later in the season than expected, the first two females became pregnant. They bore healthy litters, and a total of seven kits were weaned in late summer.

During 1986 and 1987, Fish and Wildlife Service administrators and scientists from the Captive Breeding Specialist Group had closely scrutinized the state's handling of the ferrets, at the invitation of Game and Fish administrators. *Mustela nigripes* had never produced young in captivity, and there was concern that the animals would not adapt well to living in cages. Scientists from the specialist group had traveled to Wyoming to inspect the Sybille facility and give advice on how the ferrets should be bred. But the day-to-day operation of the program was left up to the state. Thorne, who was in charge of captive breeding, hired a veterinarian named Don Kwiatkowski to live at the facility and care for the animals. In the spring of 1987, the two men spent night after night putting ferrets together in nest boxes and watching them through television monitors for signs of breeding.

"We knew it would take the animals a while to settle down, and we didn't want to manipulate them any more than necessary," Thorne said. "It got kind of uncomfortable, with people looking over our shoulders so much. Everyone had pretty much given up on getting any litters in eighty-seven except Don. He not only predicted that a couple of pairs would breed late; he predicted which ones."

The births marked the beginning of one of the most successful endangered species propagation programs ever undertaken. In 1988 twelve different females bore litters, and thirty-four kits were weaned. By the time of my visit in September 1989, the captive population had risen to a hundred and eighteen. Some of the animals had been moved to other facilities.

Thorne was clearly proud of the program's remarkable accomplishments. He acknowledged that the success was due largely to the recommendations made by members of the Captive Breeding Specialist Group. "But we also did some homework on our own," he said. "We talked to a few commercial ferret breeders who made some good suggestions about using photoperiod stimulation to control when the ferrets come into breeding. You just vary the amount of light by a few minutes every day. The animals need to go through a series of shortening days and lengthening nights and then have it switch. The females are real sensitive to it."

The technique had enabled Kwiatkowski to bring several females into heat at the right time to breed with the oldest males in the program. "The old wild-caught males tend to be ready to breed earlier in the season than the others, and we want to use them as much as we can, while we can," Thorne said. "Plus, it just makes it a lot easier on us if we can stagger the breeding and spread the work out a little."

The female kits first went into heat when they were slightly less than a year old. Most males could not impregnate females until their second year. Still, the young age at which the animals matured had made it possible for the population to expand very quickly. Mates for each animal were selected carefully to make sure all gene lines were equally represented in the offspring.

Scientists from the National Zoo and several universities were experimenting with cryopreservation, the technique of freezing sperm and eggs, to build up a bank of genetic material. Other researchers, including the Game and Fish staff, were designing experiments to see how well captive-bred ferrets could hunt and whether they would imprint on certain foods. With the breeding program going so well, biologists for the Game and Fish Department and the Fish and Wildlife Service talked optimistically of reintroducing ferrets to the wild in the autumn of 1991.

We had left Williams and were driving toward the breeding facility, following a winding road through the blackish-gray rock of the Laramie

Range, through the worn nubs of some of the oldest mountains on the continent. It was stark, wild country, full of subtle color, and Thorne remarked on its beauty. "We've got some land with a cabin out this way," he said. "It's where we come for sanity." He turned into a driveway by a sign for the Sybille Wildlife Research Unit of the Game and Fish Department.

The land held by the Sybille station spread across four square miles, most of it left undeveloped as forage for deer. An interconnecting series of split-rail corrals held moose, antelope, and bighorn sheep. Thorne was conducting a long-term study on brucellosis in elk, but recently he had spent scant time on anything besides the ferret project.

The ferrets were kept in a one-story aluminum building within a fenced compound. All who entered—and visitors were few—were required to shower completely as a precaution against distemper and other diseases. "We also wear surgical masks around the animals," Thorne said. "Domestic ferrets are very susceptible to human influenza, so we've got to assume the black-footeds are too."

Don Kwiatkowski waited for us in a laboratory inside the quarantine zone as we showered and dressed in blue coveralls. He was soft spoken and struck me as a patient, pleasant man. Thorne had described him as a talented animal handler. From a room next to the lab we could hear the muffled bangs and clatters of many ferret kits knocking around the food dishes in their cages. I peered through a plate-glass window but could see nothing of the animals, only rows of spacious cages set on long tables. The ends of the cages were screened with wood. Kwiatkowski smiled at a particularly loud thump. "They're rambunctious today," he said.

On one wall of the lab a chart marked with large block X's served as a record of which ferrets had been paired that year and which had born kits. Scientists believed that the ferrets were all descended from five or six unrelated "founder" animals—that is, five or six different genetic lineages. The more founders in a population, the greater its genetic diversity. Geneticists believe that small populations of animals must be based on about twenty founders to have a healthy measure of diversity. "We can't really tell how many founders we have in this population; there may be more," Thorne said. "Since the original animals were wild caught, it's impossible to know how they're all related. So we've tried to take a conservative approach in identifying founders." Still, the possibility of inbreeding was clearly a concern. So was the demographic make-up of the population. All the animals were either very old or fairly young.

"We're pretty sure some of the animals we have are much older than they would ever live to be in the wild," Kwiatkowski said. "We've had some problems with carcinomas, which you'd expect in a geriatric population." One female was suffering from a mouth cancer and was being held in an isolation unit. Some of the younger animals had contracted coccidiosis, a parasitic disease that had killed several kits that spring. "It's difficult to detect, because it builds up slowly in their guts and isn't always shed in their fecal samples," Kwiatkowski said. "They're healthy one day, nice and fat, and dead the next."

In a small room off the lab was an examining table, an assortment of veterinarian tools, and a small rectangular handling cage made of wire covered with vinyl. Several times a week during breeding season each female was flushed from its burrow into a handling cage and examined to determine if it was ready to breed. Ferrets come into heat once a year and are believed to be at peak fertility for only a few days. Thorne and Kwiatkowski had found they could tell where a female was in her cycle by taking vaginal washes and measuring the swelling of her vulva. "With the handling cage we can get a good look at the animal without ever putting our hands on it," Thorne said. At the right time, the female was placed in a cage overnight with a male.

As we talked, Kwiatkowski occasionally glanced toward a row of television monitors that showed the interiors of ferret nest boxes. One displayed a mother sleeping with several young around her. At birth ferrets are less than an inch long. These had grown nearly to the twenty-inch length of adults; their bodies curled around each other like sausages. "They look peaceful now," Thorne said, "but wait till they're disturbed." The comment carried particular weight. A study conducted by some Fish and Wildlife Service researchers had shown that Siberian polecats raised in captivity for many generations probably did not have enough instinctive fear of predators to escape being killed. The study suggested that polecats—and by inference black-footed ferrets—would have to be trained to avoid predators before they could ever be released to the wild. Thorne viewed the results as irrelevant. "The black-footeds are scared of everything from the time they're hotdog size," he said. "They're just a lot wilder."

The disagreement revealed a current of defensiveness, slackened but still running, on the part of both the Game and Fish staff and the federal biologists working on the project. Tim Clark was not involved in the breeding program, and he would not be invited to participate in any reintroduction conducted within Wyoming. But the Game and Fish Department was still wary of outsiders who tried to tell it how to run the program—especially outsiders who had never had any experience with captive black-footed ferrets.

It was time for a look at ferrets in the flesh. Thorne handed me a surgical mask and asked me not to talk while we were within earshot of the animals. "They're very sensitive to changes in routine or new voices," Kwiatkowski explained. "It can throw them off their feed for a day."

Moving with deliberate slowness, he opened the door to the adjacent room. All romping and banging stopped as dozens of ferret kits dashed for cover. Kwiatkowski smiled at me from behind his mask.

From the bottom of each cage hung lengths of beveled black irrigation tubing attached to a pair of wooden nest boxes. The tubing served as burrows and connected the nest boxes to the main part of the cage. Kwiatkowski opened a small trap door on the side of a nest box and reached in. "Look quick," he said. "This is about the only chance you're going to get to see an adult." The ferret flushed from the box, ran up the tubing, and ventured warily into the cage. He looked around for a few seconds, his body bent in a U. He was diminutive and long, almost comically long, but he moved with fluid grace. His rump and back haunches curled elegantly, reminding me of a long lock of hair. He stood still for only a moment, then crept into another nest box mounted on the side of the cage.

In surrounding cages I could see ferret kits peering from burrows, regarding us with intrepid curiosity. The masklike swaths across their faces were darker than in photographs, and their eyes and noses were jet black. As they strained upright to see over each other I could not tell where their thick necks ended and their torsos began; they were like tubes of flesh. "Watch this," Kwiatkowski said. He waved his arms suddenly toward a litter. The room exploded with shrill, chattering barks, and every ferret disappeared.

Bright fluorescent tubes were staggered overhead to provide as much light as possible. We made our way to a second large room, also brightly lit and airy, with windows running the length of two sides. Ferret kits crowded from burrows, ducking in as we approached but appearing again within a minute. I realized suddenly that I was grinning like a fool. The ferrets were funny and beautiful and immensely appealing. What a thrill it would be, I thought, to spot the eye-shine of one on the prairie in the middle of the night.

Thorne motioned me over to a cage. "I want you to see what happens when juvenile black-footeds are threatened," he said. Kwiatkowski pulled on a leather glove and opened the door on the side of a nest box to reveal a mother and six kits. All seven barked viciously; even expecting it, the shrillness made me jump. In an instant the ferrets had changed from appealing and inquisitive to aggressive and mean.

Kwiatkowski led me to a third room where smaller cages held Siberian polecats. Although they closely resembled the ferrets, the polecats were slightly larger, with pink noses and ears and long guard hairs on their backs. Kwiatkowski opened a nest box. A single black-footed ferret kit had been given to the mother to raise when its own mother stopped producing milk. The polecats looked at Kwiatkowski with curiosity, but he dug through them to find the ferret kit. It yelped loudly and tried to hide behind a litter mate. "You can see there's a difference," Thorne said. "The black-footeds just aren't going to need much training to behave like they're wild."

The ferrets were utterly silent as we made our way back to the lab. Thorne tiptoed to the end of a row of cages and peered cautiously around the wooden screen. He waved to me to follow. I arrived just in time to see a ferret kit standing in the open. At the sight of me it gave a shriek and disappeared.

The one achievement that remains for the complete recovery of the black-footed ferret is the most difficult: the task of returning it to the wild. Whether the species can be successfully reintroduced will depend on a combination of handling, technical skill, and luck—the kind of luck that will keep the animals out of the path of disease and environmental disaster. Whether the ferret can be truly restored as a wild species will depend on something outside the control of natural forces: a complete shift in the way wildlife managers and ranchers—and, over time, all Americans—regard our arid western rangelands.

I sat in a rickety straight-backed chair, concentrating on a small television screen. Beside me, Brian Miller leaned forward in his seat and hit the rewind button on a video recorder connected to the television. We were in a barn at the Conservation and Research Center near Front Royal, Virginia, watching tapes of experiments Miller had run the previous month on whether captive Siberian polecats could be trained to avoid predators.

A polecat kit appeared on the screen. It looked up in alarm and barked once before darting into a makeshift burrow. As it vanished, the head and front paws of a stuffed badger appeared from the left and jerked to a stop. The badger seemed to be swimming in midair. One paw was fully extended, the other flexed downward in a paddling motion. It was mounted on the chassis of a Radio Shack remote-control toy truck; a set of studded tires protruded from beneath its belly. The contraption was known, somewhat affectionately, as RoboBadger.

The scene changed. A younger kit slinked along a cinderblock wall, clawing occasionally at the dirt. RoboBadger rolled into view. The kit, startled by the sudden movement, spotted the makeshift burrow and dived in. A few seconds later its head popped back out. "Not a very good response," groaned Miller. Its appearance was greeted immediately with an onslaught of rubber bands shot from toy guns.

In the wild, confronted by a live badger, the polecat would have been dead. But at eight weeks of age it was young to be out of its burrow, and out of the company of its litter mates. In a month it might react to danger more quickly, though the whining, whirring RoboBadger was about as far removed from a stealthy, wild predator as it would ever encounter. "We'll test him again at twelve weeks and sixteen weeks," Miller said. "We've found so far that eight-week kits are just too young to have much of a clue about what they're supposed to do when they're threatened."

Another kit appeared on the screen, chased this time by a stuffed great horned owl suspended by fishing line from the ceiling. As the owl swooped over, the kit scrambled for the burrow and did not reappear.

These predator avoidance trials, carried out in a twelve-by-twelve-foot pen, were rather comical replications of the kinds of dangers young ferrets would face in the wild, but they were considered adequate for Miller's purposes. Miller was probing for clues, nothing more. Working in collaboration with Dean Biggins, he hoped to devise some way of teaching captive black-footed ferret kits to recognize danger and flee. As a first step he was exposing Siberian polecat surrogates to a facsimile of a wild predator—RoboBadger—to see how they would react.

There was a good chance, of course, that the ferrets would know innately to hide from predators, as Tom Thorne believed. "All we're trying to do is see whether training might be beneficial and, if so, at what age it works best," Miller said. "We're also trying to look at individual behavior. Are there certain things animals do that mean they shouldn't be reintroduced? That mean they're basically boneheads?"

Miller was a tall man, friendly and outgoing, with a scrubbed, earnest face and hair the color of straw. He had worked on the black-footed ferret project since 1984, first as a field researcher and, later, examining various aspects of the species' biology in captivity. Now he was studying polecat behavior through a grant with the Fish and Wildlife Service and the Smithsonian Institution. The barn where we sat was filled with cinderblock pens, most of which held polecat litters. The animals had been given cardboard boxes to play with, and these scraped noisily against the cement floor.

Miller had spent most of the summer of 1989 running polecats of various ages through tests. He chased them with the owl or RoboBadger on two consecutive days and timed how long it took them to duck into the nest box. He also released kits into an outdoor arena where he had dug several burrows and hidden a dead prairie dog. The kits were timed to see how long it took them to locate the prey. Most of the older kits had shown the right kinds of behavior; in the outdoor pen they had even run from burrow to burrow in a zigzag pattern typical of both wild polecats and black-footed ferrets. "They can do everything a wild animal can do," Miller said. "It's just a question of whether they can do it fast enough."

The study was the first attempt by the ferret recovery project to compensate for a major failing of captive breeding. It is believed that ferret mothers may give their young certain signals, beyond the obvious alarm barks, that help them learn to recognize danger. But when the mother herself has been raised in captivity, protected all her life from predators, she may not convey to her kits the importance of vigilant caution. After several generations, the behavior that captive ferret mothers teach their young may differ substantially from what wild mothers would teach. But there is no way for biologists to measure such subtle behavior.

The issue was especially important, since the plans for reintroduction called for releasing juvenile ferrets in autumn, when they would normally disperse from their litters. The released animals would have to learn to hunt and hide at the same time they adjusted to the strangeness of freedom. Miller and Biggins hoped to find a few ways to ease the transition.

The education of young ferrets would not be restricted to experiments by the Fish and Wildlife Service. In Wyoming a set of prairie dog pens was being constructed at the Sybille research center, where Game and Fish biologists planned to run trials of their own. After some prairie dogs had established a system of burrows in the deep dirt floor of the pens, young ferrets would be let in to see how well they could locate prey and kill. Thorne also planned to test the ferrets' hunting skill in winter. Because white-tailed prairie dogs hibernate in cold weather, ferrets must be able to sense their presence beneath the snow and dig them up.

Before the reintroduction, Miller and Biggins hoped to put the ferrets through additional experiments to measure both their hunting prowess and their ability to flee from predators. "It's possible that ferrets hunt differently in the presence of predators than when none are around," Miller said. "We think the whole range of behaviors may be interrelated. Animals respond differently in different situations. We may be able to teach captive ferrets how to hunt a little more efficiently, or recognize dangerous situations a little quicker, and give them a better survival rate." Thorne, however, was still skeptical that the ferrets would need to be taught to avoid predators. The difference of opinion had caused some strain between the Game and Fish Department and the federal researchers.

The issue of where the species would first be released was an equally delicate subject. The initial reintroduction would take place in Wyoming, at the request of the Game and Fish Department. Residents of the Meeteetse region were anxious for ferrets to be returned to the benchlands along the Greybull River, and Thorne agreed that the area was a logical first choice. But the prairie dog population at the Meeteetse site still suffered widely from sylvatic plague. Many biologists doubted that it was stable enough to support ferrets.

Selecting a site to reintroduce ferrets would have been easier if prairie dogs were still as abundant as in the nineteenth century. But the vast populations had been left in tatters by poisoning programs that were still in effect, both on private and public land. A number of the biologists involved in the ferret recovery program, including Miller, questioned the wisdom of continued prairie dog control. "One of the main things that's wrong with endangered species programs is there's too much single species management," Miller said. "With the black-footed ferret especially, there's a chance to save an entire ecosystem instead of concentrating on one animal."

Miller had recently written a paper with Christen Wemmer, the director of the Front Royal center, and two other biologists challenging the conventional stockmen's wisdom that prairie dogs strip the range of forage needed to support cattle. The paper argued that researchers had found little difference in the market weight of steers grazed on pasture where prairie dogs lived and those grazed on land where the rodents had been poisoned. Federal poisoning programs were expensive, benefitted ranchers exclusively, and had to be repeated frequently to be effective. Moreover, they had helped cause the near-extinction of the black-footed ferret.

The paper also noted that prairie dog communities attract other species of small mammals and an unusual diversity of birds. The burrows provide shelter not only for ferrets but for mice, voles, and two vanishing bird species: burrowing owls and mountain plovers. Many biologists believe prairie dogs may actually speed the growth of certain grasses by cropping them and by stirring up nutrients with their digging.

As part of the plan to restore black-footed ferrets, Fish and Wildlife Service administrators had requested that prairie dog poisoning be curtailed around potential reintroduction sites. As a concession to ranchers, however, poisoning restrictions were to be loosened in other areas. Miller and his colleagues believed the concession was unnecessary and unwise. Instead, they argued that the government should halt all poisoning on public land and offer subsidies to ranchers who do not kill prairie dogs on their own property. The subsidies could be drawn from funds that would otherwise be spent on poisoning programs, and they could offset any losses in steer weight—real or imagined—incurred by the ranchers. Over time the extensive prairie dog towns of the West would revive, providing shelter and food for the restored black-footed ferret and other animals.

The proposal seemed certain to be opposed by western ranchers and politicians. Nevertheless, the paper pointed out an important discrepancy in federal policy toward wildlife. A stable population of two hundred ferrets would need as many as twenty-five thousand acres of grasslands, all of it colonized by prairie dogs. Yet federal biologists were having difficulty finding prairie dog towns even as large as one thousand acres.

"There are only a handful of places left in the country where a population of black-footed ferrets could be reintroduced and not starve," Miller said. "The numbers are pretty depressing, especially when you consider that prairie dog towns once covered millions of acres. We've exterminated them, just like we exterminated the gray wolf. Now we've got a chance to bring them back, especially if we tie their recovery to an appealing endangered species like the ferret.

"It'll take a full-scale education program, the kind of effort the National Zoo has put into the golden lion tamarin project in Brazil. With the ranchers it'll be an uphill fight. But I do think it's possible to change attitudes and turn things around, if the government offers the right kind of economic incentives."

In the fall of 1989, only a few weeks after I visited the ferret breeding facility at Sybille, Dean Biggins, Brian Miller, and several other federal researchers released thirteen Siberian polecats to a wilderness area near Wheatland, Wyoming. The polecats had been given "mild aversive conditioning to predators," meaning that they had been briefly introduced to the stuffed owl and RoboBadger. They were freed wearing radio collars. Within two days, six of the animals were killed, at least four of them by predators. The remaining seven were killed within three weeks. "We knew some were going to get nailed," Miller said, "but at the least we expected to have a few left by Christmas."

The black-footed ferrets bred well enough in 1990 to prompt the Game and Fish Department to lay the final plans for a reintroduction in the autumn of 1991. Because of continued problems with sylvatic plague at Meeteetse, the ferrets would be reintroduced to a wilderness area in the Shirley Basin north of Laramie. The decision had been reached after much debate, but it satisfied all parties.

The Game and Fish staff, borrowing from techniques used in past reintroductions, decided to release the ferrets from a hacking cage built on stilts, much like the cages I had seen at the breeding center at Sybille. Beginning in September 1991, litters of ferrets were taken to the release site and held for ten days so they could imprint on the cages. On the tenth evening tube "burrows" were opened to the ground so the animals could leave. They could return to the cages whenever they wished.

Over a six-week period biologists freed 49 ferrets, all born the previous spring. Most immediately scampered to the ground and began to explore. By mid-November only nine could be accounted for. Six of the animals were known to have been killed by predators; the rest simply disappeared.

In early winter heavy snows and winds set in, making it impossible for researchers to track the animals. Biologists could do nothing but wait until spotlight surveys could be conducted in summer. "We just have to hope that a few of them have managed to hold on and are out there breeding," Biggins said.

With so many disappearances, it seemed impossible to believe that any of the animals had survived long enough to produce litters. But in July state biologists spotted at least two animals and possibly more. In September, a year after the first releases, Biggins and his colleagues captured four ferrets on the site. Two were females that had given birth to litters, and two were juveniles born in the wild. "We never did pick up the males that could have sired them," Biggins said.

The discovery of wild-born ferrets buoyed the spirit of the researchers, and in the autumn of 1992 another 90 animals were released in Shirley Basin. Half were freed from elevated pens, as before, and half from underground nest boxes. The method of release seemed not to affect the fate of the animals. Once again most quickly disappeared. Of the eleven still known to survive after a month, seven had been raised in an outdoor pen at the Sybille research facility and had been given a chance to kill prairie dogs before their release. "That apparently affected their ability to survive more than anything else," Biggins said.

In the interim Biggins and Brian Miller continued to experiment with conditioning Siberian polecats to avoid predators. They chased some of the animals around an arena with Miller's dog, a playful golden retriever named Rosa, and frightened others with a tethered owl. But when the conditioned polecats were released at two different sites, almost all were killed within days. Biggins decided to halt the experiments for awhile. He turned his attention to preparing for a release of black-footed ferrets at the Charles Russell National Wildlife Refuge, a prairie preserve in northern Montana, in the autumn of 1993.

Because of its ability to breed prolifically in captivity, the black-footed ferret could be reestablished in three or four locations, perhaps even in three or four different states, by the year 2000. Over time scientists are likely to invent ways to protect released animals, perhaps through predator control and supplemental feeding programs. Such support efforts could be discontinued slowly, until the ferrets learn to live on their own. Unless problems with inbreeding begin to appear in future generations, the final goal of the recovery project—to establish a total of fifteen hundred wild ferrets at ten different sites—appears to be within reach.

In a deeper sense, however, the black-footed ferret will not be truly recovered unless the landscape from which it was extirpated is allowed to recover as well. "If it's going to work, conservation has to draw not just from science but from history, from psychology," Brian Miller once said. "We need to look at the past, at why the ferret nearly became extinct. At why the situation developed as it has. And at what we can do now to keep it from happening again."

XIII

THE PANTHER VERSUS FLORIDA

The plane, a single-engine Cessna, looped neatly to the south and followed a swath of forest that gave way to grasslands. In the front seat, Deborah Jansen toyed with the radio switch, controlling the signal picked up by antennas mounted on each wing. Jansen, a wildlife biologist, was searching for five wild panthers in the swamps and thickets of Everglades National Park.

Through a set of headphones, I could hear her give directions to the pilot over the clear, steady beep of the radio signal. We leveled out over a plain of sawgrass with an uneven texture that reminded me of an animal's fur. Shallow pools of water caught the glint of the sun like mercury.

The pilot was the first to see the cat. "She's out in the open! Look over to the right!" he yelled, banking abruptly. We circled tightly, the right wing falling away toward earth. Pressed against the window, I scanned the grasses feverishly until I picked out a bright stripe of orange.

The panther was walking slowly across a broad prairie, undisturbed by the buzzing, blue plane overhead. The orange blaze of her radio collar made her easy to spot; otherwise she was exactly the color of the winter-killed grass. From three hundred feet she looked thin but well muscled. Her back was broad at the powerful front shoulders but narrow at the hips. The blood rushed from my head as the plane straightened and rose, and for a moment I was utterly disoriented. I was in south Florida, but from what I had just seen I could have been in Africa. The panther looked up as we circled over her a second time, her ears perked forward, her expression almost quizzical. Lower now, we could see white markings around her nose and mouth. Compared to the rest of her body her face seemed very round, her head very small. She stopped walking to watch us. "That's enough—let's get out of here," Jansen said.

Until then I did not have a full appreciation for why Jansen talked about the "lions" of south Florida. A panther is the same animal as a cougar or a mountain lion; in fact, the biologists who work on the Florida panther recovery project use the terms interchangeably. I had seen pictures of

Florida panthers, and I had visited animal breeders who kept western cougars. But until I spotted one in the wild—slinking through a damp savannah with its tail curled behind it like a whip—the species' resemblance to the great lions of Africa escaped me.

It was unusual to spot a panther from the air during radio tracking, but within an hour we had seen two. The other had trotted briskly into a patch of brush as the plane approached. We landed at the small airport at Homestead feeling a victorious glow, the glow of unexpected luck. As I started back to my campsite in the park, slowly the magnitude of what I had seen began to dawn.

The Florida panther is the only great predator that has managed to survive in the eastern United States. Another subspecies of cougar that ranged through northeastern North America is believed to be extinct in the wild, although there are unconfirmed reports of cougar populations in eastern Canada, New England, and parts of the Appalachian Mountains. The full name of the Florida panther is *Felis concolor coryi*, after Charles Cory, the biologist who first described it in 1896. (Thirty subspecies of *Felis concolor* have been identified in North and South America.) Florida panthers once lived throughout Georgia, Alabama, Mississippi, Louisiana, Arkansas, and parts of South Carolina and Tennessee, but for many years the subspecies has been restricted to Florida. Biologists believe there are only between thirty and fifty still alive.

I pulled into my campsite and fished a Coleman stove from my gear to begin supper. It was a windless evening in March, too cool for mosquitoes. As I lit the stove I waved to an elderly man, a park volunteer who was checking campsites for occupants. We exchanged greetings and I blurted, "I just saw two panthers."

The man was clearly taken aback. "Where?"

"On a flight with a biologist."

He looked thoughtful. "That's one more panther than I've ever seen, and I've lived here all my life," he said.

The glow from the afternoon dimmed. I told him briefly about the cats and let him continue on his rounds. It occurred to me that Jansen, the pilot, and I had cheated nature by using technology to make her reveal what she would only rarely reveal by choice.

I did not volunteer to anyone else that I had seen panthers from the air. During the rest of my travels through south Florida, I discovered a reticent uneasiness about the way the last populations of panthers are being managed. The uneasiness grows partly from misunderstanding and misinformation. Though it is not shared by all residents of the state, it is felt widely enough to raise the question of whether it is seeded in something deeper than reason— perhaps in a paralyzing sadness that the last vestiges of North America's subtropical wilderness are succumbing to development.

During the 1990s the Florida panther is likely to die out unless scientists can overcome formidable odds against inbreeding, habitat loss, and hybridization. Recent genetic research has produced evidence that the gene pool of *Felis concolor coryi* may already have been diluted by other subspecies. Although biologists are attempting to establish a captive breeding program, the Florida panther is so beset with infertility problems that it may not be possible to breed enough animals to prevent a serious loss of diversity. And although steps are slowly being taken to preserve panther habitat, Florida is besieged by environmental problems. The swamps on the southern tip of the Florida Peninsula, where most land is publicly owned and undeveloped, have been badly degraded by pollution and by alterations in the traditional flow of water through the Everglades. Exotic vegetation has crowded out native plants on which deer, the primary prey of panthers, depend for forage. Especially in the southern reaches of Florida, development runs rampant, fueled by a constant flow of new people (an average of a thousand a day) moving into the state.

It is fair to say that the Florida panther is among the sickest of America's ailing wildlife, and that it will not recover without visionary assistance. It is fair to say that the panther offers the sorcerers of wildlife science a chance to work miracles, if indeed they can.

Of the many sensitive issues involved in the panther recovery project, few are as pervasive or troublesome as the simple question of how many of the animals remain.

Panthers are loners, and among the most secretive animals on earth. Especially in vegetation as dense as the palmetto thickets and slash pine and cypress swamps of south Florida, their presence is difficult to document. Every year the Florida Panther Records Clearinghouse, a service set up by the state Game and Fresh Water Fish Commission to investigate alleged sightings, receives hundreds of reports of "panthers" prowling backwaters and back yards. The overwhelming majority turn out to be dogs—or bobcats, which, contrary to popular belief, may have tails as long as eighteen inches.

With so many cases of mistaken identity, state and federal biologists have learned to discount sightings without some form of physical evidence—a print, a scat sample, or a photograph of the animal. "We've found that even with the most credible stories, the best descriptions, the most trained observers, we still get a lot of false sightings," sighed Chris Belden, a state biologist who has worked on the panther project since 1976. "I had one guy, a cop, really nice, who swore he saw a panther outside Jacksonville. The description sounded right, and he had a track to show me; he had put a tin can over it to make sure nothing messed it up. So I drove over there one Sunday morning. It was a raccoon track."

During the winter biologists attempt to capture panthers within certain study areas in south Florida, including Everglades National Park and the Big Cypress National Preserve. When evidence of a panther is found, professional hunters are dispatched to search the woods and swamps with trained hounds. The dogs sniff out the cat and chase it until it takes refuge in a tree. Then they guard it until a team of biologists, driving swamp buggies and all-terrain cycles, can invade the scene. A veterinarian decides whether the animal looks healthy enough to be handled and, if so, subdues it with a tranquilizer dart. Within minutes the panther grows so docile that a biologist can climb the tree and loop a rope around its chest. If all goes well, it is lowered to a crash bag, a large, air-filled cushion. On the ground it is examined and fitted with a radio collar. For the rest of its life it will be tracked and recaptured every year or two to have its collar replaced.

Despite biologists' reassurances that collared panthers suffer no lasting hardship, critics of the radio tracking program claim that the animals are placed under severe stress. Some people charge that at least two panthers have been killed by injuries suffered during the collaring process; wildlife officials accept responsibility for only one death. There are financial reasons as well for not reporting evidence of panthers. Frequently, landowners worry that federal officials will place restrictions on how they may use their property if an endangered species is found to reside on it.

Although the official estimate of the panther population is only a few dozen, many residents believe the number to be much higher. In my travels through Florida I talked to people from virtually every region—including the suburbs of such major cities as Tampa and Tallahassee—who insisted they knew, or had friends who knew, of panthers that lived near their homes. This is a curious phenomenon in so populous a state. It speaks, perhaps, of a need to believe in a few last shreds of mystery and romance, to believe that humans have not tamed the woods to the point that all the phantoms have been driven off.

Panthers and cougars are adaptable animals, and they have been known to live near urban areas. In parts of the West, where their numbers have increased recently, cougars sometimes are sighted on the outskirts of thickly settled communities. They occasionally venture into the east suburbs of Los Angeles to prey on dogs and cats. But within Florida, biologists are not inclined to believe reports of panther populations near cities, or in the northern half of the state. "Where would they all be? There's just not the habitat," said Oron Bass, a wildlife research biologist at Everglades National Park. "Males have home ranges of two hundred and fifty square miles or larger. There are panthers in the Panhandle? Then why don't we ever hear of any getting hit by cars up there? Either there aren't any, or someone's doing an awfully good job of keeping them secret."

To the early settlers of the southeastern United States, the Florida panther was no less dangerous or diabolical than the wolf, and it was persecuted with equal vigor. Hunters killed it without restriction until 1950, when the state of Florida declared it a game animal. It was not granted full protection until 1958, and it continues to be shot illegally on occasion. In addition, it is gradually being crowded out of its preferred habitat, the higher, drier tracts most suited for economic development. The fragments of land left to it are some of the most remote, hostile environments in all the South—the Everglades, Big Cypress Swamp, the Fakahatchee Strand.

Until the mid-1970s, no one paid much attention to whether the Florida panther population was expanding or quietly going extinct. The panther had been on the endangered and threatened species list since 1967, but the federal government had taken no steps to preserve it. In 1976 the Florida Audubon Society sponsored a conference on the subspecies. "It became pretty clear at that meeting that no one knew much about the panther," said Chris Belden, who attended as a representative of the Florida Game and Fresh Water Fish Commission. "We didn't even know whether or not it still existed."

At the time of the conference Belden was working as a deer biologist for the commission. Within a few months, however, he was put in charge of a recovery team for the panther and asked to do field studies on whatever animals he could find. The task was staggering. "We weren't sure where to begin," he recalled. "We started encouraging people to report sightings, and I spent a lot of time checking those." He and another state biologist searched for panther sign in the Big Cypress National Preserve and the Fakahatchee Strand, a long, thin limestone depression that forms the major drainage channel for the Big Cypress Swamp. Walking the abandoned tram roads that cut through the humid forests of the strand, the men came across panther scrapes, prints, and scat on a regular basis. "It was an incredible high to find panther sign," Belden said. "Back then I could go from the highest highs to the lowest lows—in days."

For several years Belden studied panthers in the woods of south Florida, commuting on weekends to his home in Gainesville. Although he was in charge of the program to save the subspecies, he was also expected to keep up with other duties, including a time-consuming study of wild hogs. "I didn't have a life outside work," he said. "I thought that if the Florida panther was going to get saved, I'd have to save it myself."

During 1981 Belden and a professional hunter treed two panthers in the Fakahatchee Strand. The captures marked the first time Belden had ever gotten close to wild cougars. More than anything, he was surprised by their size. "You know how your perspective and sense of size can be all out of whack in the woods? In thick vegetation a panther looks as big as a cow." Both animals had sharp crooks in the ends of their tails and whorls of hair in the middle of their backs. Belden fitted the animals with radio collars, released them, and began tracking them. A short time later he visited a state museum to examine some skins of panthers killed in the 1940s and 1950s. The skins had cowlick whorls, and their necks were flecked with white. These characteristics, along with the crooked tail, seemed peculiar to the Florida panther. In 1982 Belden put collars on another five panthers in the strand and the Big Cypress preserve. The cats tended to be slightly lankier than western cougars, and more docile. Where a western cougar's nose was rather flat, the Florida panther's was arched, like a Roman nose. The differences were subtle, but they offered evidence that the unique Florida subspecies still existed.

Unfortunately, Belden could not tell whether the panthers he caught were pure *Felis concolor coryi* or hybrids. Florida has an unusual number of residents who keep cougars as pets; wildlife officials estimate that there are as many as a thousand captive cougars living in the state at any time. Captive cougars tend to be a mix of different subspecies. As kittens they are fluffy, spotted, and cute, but they become difficult to handle as they mature. An owner, tired of caring for a cantankerous cat, might be tempted to turn it loose in the woods to intermix with indigenous panthers.

Belden believed passionately that the animals he was collaring and tracking were pure Florida panthers, but he had no proof, little funding, and little time for the detailed studies he thought were needed. Even then, south Florida was developing with frightening speed. He knew it was imperative to learn as much about the range and food needs and social nature of the animals before their habitat became more fragmented. And then, in January 1983, he shot a tranquilizer dart into the leg of a female that had been treed by hounds. Underweight and anemic, the panther reacted badly to the drug, slipped quickly into unconsciousness, and died.

The incident drew unprecedented criticism of the radio tracking study. But it also forced a reevaluation of the amount of money and staff dedicated to panther preservation. If wild cats were going to be handled, state officials decided, they should be handled only in the presence of a veterinarian. Several months after the panther's death, the Game and Fresh Water Fish Commission established a residency position in wildlife medicine at the University of Florida College of Veterinary Medicine to train a vet to work with panthers. The person chosen to fill the residency was Melody Roelke, a veterinarian who had extensive experience in caring for cheetahs and other great cats. Roelke had worked at a private wildlife park in Oregon, and she had collaborated with scientists at the National Cancer Institute and the National Zoo on a groundbreaking study on inbreeding in cheetahs. Her experience in genetic research would greatly affect the thrust of the panther project as it matured.

Her principal responsibility, however, was to help state biologists collar panthers. She began in the winter of 1983–84 by accompanying the capture team on hunts through the swamps and forests where Belden had previously found panther sign. The membership of the team had been expanded; in addition to Roelke, the game commission had hired a full-time panther biologist to work in south Florida. Through aerial radio tracking, the first sketchy pictures were being drawn of where panthers moved and how they interacted. Roelke's presence added another dimension to the research. "We wanted to look at panthers in different habitats and see if we could figure out why some were healthier than others," she said. "We had a hypothesis that we'd find bigger cats on the private land north of Big Cypress. And boy, did we."

The most robust panthers were consistently captured north of Interstate 75, the east-west corridor between Miami and Naples known as Alligator Alley. Much of the land north of the highway was owned by large farming and real estate concerns and had been left in slash pine and palmetto, swamp forest, or sawgrass prairie with scattered hardwood hammocks—in other words, the mixed woodland habitat panthers prefer. The property was drier than the swamps south of Alligator Alley, and it held more deer. In the spring of 1987 the panther team collared two unusually large male cats on private property in Hendry County. One weighed a hundred and forty-eight pounds, then the biggest Florida panther on record.

"They were just gorgeous," Roelke recalled, "big and healthy, with packed cell volumes of forty-five. That's not just good, that's great." The percentage of packed cells in a blood sample can be a measure of anemia. In contrast, some of the female panthers collared to the south on the Fakahatchee Strand and the Big Cypress preserve had packed cell volumes as low as twenty-five, a sign of a serious deficiency. They also tended to be underweight, haggard, and in consistently worse health than the males. One female, treed by hounds in the preserve, was so sickly looking that Roelke refused to work on her. "We just backed off and let her go," she said. "I was afraid I'd kill her with the drug."

The condition of the panthers south of Alligator Alley could not be explained by any single factor, but several biologists speculated that it might be caused in part by insufficient prey. Although deer is the panther's primary prey, scat samples showed that panthers south of the highway were feeding heavily on smaller animals, such as raccoons. The findings led some biologists and representatives of several conservation groups to question whether the deer population of the Big Cypress National Preserve had been depleted by hunters using off-road vehicles and dogs.

In 1987 administrators at Everglades National Park agreed for the first time to let biologists radio-collar panthers on park property. Although panthers were frequently reported in the Everglades, no one knew how many might reside there. That first spring the capture team collared two females, each with two kittens, on the east side of the Shark River Slough. To the biologists' surprise, none of the animals had kinked tails, and only one had the whorl of hair on its back. The Everglades cats were separated from those of the Big Cypress region by the slough, a seasonal river that bisects the park, flowing through a rich sawgrass marsh. Depending on rainfall, the slough can be several inches or several feet deep; during the dry season panthers occasionally crossed it.

Although the Everglades panthers had the lanky legs typical of the Florida subspecies, the absence of other characteristics seemed peculiar, as if the population had been isolated from the Big Cypress panthers for many generations. Park biologists did not believe the lack of a kinked tail or a cowlick whorl meant anything significant. Over time, however, genetic studies would raise a number of questions about how, and why, the Everglades cats had evolved differently from those west of the slough.

East of the Gulf Coast town of Naples, Alligator Alley is a crowded road through a landscape of heat, humidity, and swarming bugs. Although listed on maps as Interstate 75, only a narrow, two-lane strip was open to traffic until 1990. A feeling of dread settled in my stomach one March afternoon as I took my place in a single line of cars traveling east at sixty miles an hour, close enough together to touch shadows. Just to the north was a newly graded roadbed where eventually westbound traffic would flow. Gritty, white dust caught in the warm spring winds and coated the pickups, sedans, and Airstream trailers sliding like beads along an asphalt string. Every few miles bright yellow signs warned of Florida panther crossings. "Only 30 left," one sign read, but a vandal had crossed out the "30" and painted "28" below it.

Occasionally the newer sections of the highway crossed what looked like dry creek beds. These were wildlife underpasses being built, at considerable cost, so panthers could cross safely. When the road was completed, high cyclone fences would extend along either side, blocking animals from the right-of-way and funneling them into the underpasses. Vehicular traffic is one of the largest known causes of panther mortality. From 1980 to 1990 sixteen panthers were killed by cars in south Florida. Four others survived being hit, but two were too badly injured to be returned to the wild.

I had driven south by way of Hendry County and the Corkscrew Swamp, a bald cypress sanctuary reputed to have a resident panther. The countryside of the region varies between tracts of dark forest, cleared pasture, and new, treeless subdivisions. Where vegetation still stands, it is thick-leafed and prickly—brier, palm, saw palmetto. A panther concealed in a patch of stirring fronds and splintered light would be virtually impossible to detect.

Farther south, in the moist limestone gulley of the Fakahatchee Strand, royal palms grow in graceful clusters and hardwoods crowd out the sun with spindly limbs. The strand is renowned for having an unusually diverse collection of orchids and other epiphytic plants; the state has set aside most of it as a preserve. I turned off State Highway 29 and followed the flat, gravel road described on a map of the preserve as a scenic drive. It was hard to see into the vegetation that massed in the open sun of the roadside, so I parked and walked down an old tram road, part of a network of roads built through the strand by logging companies in the early part of this century.

In the midst of the swampy woods, the tree trunks were covered with toothy ferns, waxy bromeliads, and pale-green lichens as thick as carpets. Pockets of warm, stagnant air and cool breezes buffered each other. In the mud around a puddle lay a collage of spindly handprints left by raccoons. The tram road was too overgrown for use by cars, but it probably served as a highway for wildlife. In all likelihood it also served as routes of transmission for parasites and diseases. Roelke had found a high incidence of exposure to feline panleukopenia virus, or feline distemper, among the panthers captured in the strand. The virus is fatal to all species of cat and is extremely virulent; it has been known to live for a year at room temperature. Ultraviolet light can inactivate it, but in the deep shade of the strand it might fester for months in a pile of scat. Biologists had never documented a fatal case of feline distemper in panthers. Nonetheless, Roelke feared it might be killing large numbers of kittens that were already weakened by hookworms and other parasites.

East of the strand lies the Big Cypress National Preserve, a wilderness only in the sense that logging and residential development have been prohibited. Along its edges billboards advertise airboat cruises, alligator wrestling, and swamp buggy rides. In certain sections of the preserve hunters are allowed to keep houses as base camps; other sections are dotted with gas and oil rigs. Several conservation groups believe that regulations controlling hunting and the use of off-road vehicles in the preserve are far too lax. Hunters, on the other hand, complain bitterly that their rights to use the preserve are being unfairly curtailed. Regardless of how severe the regulations are, many people ignore them. The preserve is large enough, and its swamps labyrinthine enough, that game wardens catch trespassers and poachers only rarely.

Down a narrow corridor known as the Loop Road I could see into cool, green groves where broad cypress trees gripped the dark soil with ropy roots. Bromeliads sprouted from their sides, and poison ivy crept up their trunks, as lush and ornate as garden vines. The vegetation opened suddenly to a sawgrass prairie, where a thin line of hardwoods followed a narrow ridge. Naked cypresses lined the horizon, skeletal and contorted. A swamp buggy with five-foot tires pulled onto the road from a muddy trail. The driver grinned and saluted me with a beer.

At the beginning of the panther capture season, usually in January, the scientists who make up the capture team may hunt the public lands of southwest Florida and, where they have permission, the private tracts just to the north. Or they may first work the Everglades east of Shark River Slough. Their days consist mostly of biding their time while a houndsman, Roy McBride, searches for panthers.

McBride is lanky and unusually deft at picking his way through poison-wood and pinnacle rock behind his team of pied Walker hounds. For most of his life he has specialized in hunting rare cats—not only panthers in Florida but jaguars in Bolivia and leopards and jackals in southern Africa. He learned his trade stalking cougars outside Big Bend National Park, near his Texas home. In the early 1970s he also trapped red wolves in southeast Texas and Louisiana.

In south Florida McBride often hunts with one string of dogs while his two sons hunt in different areas. The men release their animals around 4:00 A.M. to run silently through the sawgrass marshes and slash pine forests. Sometimes the dogs will trail a panther for which the McBrides have found sign, but more often they simply stumble across a line of scent. The chase takes less than a half hour. On most days the dogs find no panthers. The rest of the capture team, at work by six, waits by a radio for a call from McBride or his son. During a week in 1988 when I visited the Everglades, the weather was windy and dry enough to dissolve any scent trail within an hour. "I'm just out here wearing grass down," McBride complained over the radio.

It was only the second year of hunting in the Everglades, and the houndsmen were searching hard for a male. Five of the collared panthers on the east side of Shark River Slough were females; the sixth was a year-old male still traveling with his mother. McBride and Oron Bass, the project leader of the Everglades team, speculated that an uncollared adult male was roaming the park, along with at least one other female. After nearly a week, though, the hunters had found no trace of either.

One cool, clear morning Bass and I followed the main road through the park to a field a few miles south of an observation tower known as the Pa-hay-okee Overlook. The McBrides were hunting nearby—fruitlessly—and Bass had suggested that the rest of the capture team look for sign on their own.

Bass grew up just outside the Everglades in the town of Homestead. A direct, friendly man, he had worked for the park since 1976, and he had thought a good deal about what should be done to preserve the Florida panther. Driving toward the overlook shortly after 6:00 A.M., he said he thought it was essential for scientists to build a foundation of knowledge about wild panthers through radio tracking and field studies. "People say we ought to leave them alone, or just increase their prey base and save their habitat and they'll be okay," he said. "That's not the way ecosystems work; that's too simple an equation. Suppose there are other elements, factors we haven't even realized yet, that play some pivotal role."

By monitoring their movements, Bass hoped to learn about the behavior and social interactions of the Everglades panthers, including the dispersal patterns of kittens. "We're looking at the whole ecology of the panther," he said, "including population dynamics, so we'll know what kind of room they need, what kind of habitat they prefer. We're trying to manage the subspecies to save it; that's the goal."

I asked whether his studies of social interactions were considered as critical to saving panthers as, say, his research on what kind of prey base the animals needed.

"I know where you're leading," Bass replied. "Are we studying these animals to death? I'll answer that with a rhetorical question: How much do we need to know to save the animal? Even though it's a solitary animal, the presence of other panthers affects it. So how can you save it, manage for it wisely, without knowing something about it?

"Believe me, I'm as concerned about harassing the animals as anyone. It's something I turn over in my mind all the time. How much data do we need? When is enough enough?"

We pulled up in back of a van where Roelke, her assistant and husband, Steve Parker, and Deborah Jansen were readying themselves for a hike across a sawgrass meadow. Bass, Roelke, and Parker started for a distant hammock. I drove another mile south with Jansen, a slight, hardy woman with steady, blue eyes.

We got out and made our way through the sawgrass, scanning for prints or scat or trampled swaths. "The lions we collared last year used to travel through this area some," Jansen said, "but they haven't been here in a while. That's a sign to me that they may be avoiding it because another cat has moved in." She had a way of talking about panthers as if she was intrigued by, and worried about, them.

I could see it would not be easy to find tracks. The Everglades formed on an ancient reef of sand, shell, and the skeletons of marine animals. The white pinnacle rock, as it was called, was as rough and uneven as cement dribbled randomly over the earth. In places, water had dissolved the limy stone, forming craters three feet deep. "It reminds me of the surface of the moon," Jansen said. Much of the ground was obscured beneath a paste of periphyton, a mixture of microscopic plants and animals that forms on the surface of standing water during the wet season, then dries. The periphyton was as resilient as a sponge; it sprang back when we stepped on it, leaving no trace of our passage. "Not an easy place to look for prints," Jansen said, "but if we're lucky we might come across an animal trail with some scat or a scrape."

In the distance gray cypress trees grew in messy stands, dwarfed from a lack of nutrition. We pushed through the grass toward a thick hardwood hammock, the kind of hiding place frequently used by panthers. "Aha,"

Jansen said, stooping over a patch of stiff, brown soil. "I don't know that we've got something here, but maybe. Do these look like toe marks?" She pointed to two small depressions.

I had not a clue, and said so.

The hammock was ringed by young cypresses with wispy needles. Myrtles, buttonwoods, and an occasional poisonwood crowded its cool interior. The packed ground, a pale butterscotch, was refreshing to study after the bumpy, confusing surface of the pinnacle rock. Jansen, combing patiently among the bushes, found a set of bobcat prints. "That's probably what left the other print," she said, "but look." She pointed to a small mat of fur made by the cat as it plucked its prey. "Panthers do that too," she said. "There's not much left; they even eat the bones."

Over the past year and a half Jansen had examined the remains of about two dozen panther kills, most of them deer. "They'll kill out in the open," she said, "and then drag the prey to a hammock to eat it. There'll be a real distinct drag mark through the grass. Sometimes you find a trampled area where the animal bedded down for a while after the kill."

There were signs of rabbit and opossum, but no panthers. We moved west to a smaller hammock, chatting in spurts. Jansen told me about caring for a yearling panther in 1987 that had been abandoned by her mother. "We named her Annie, as in Little Orphan," she said. The kitten was separated from her mother and sister twice, the first time after all three had been fitted with radio collars. They were recaptured and reunited. A few months later Roy McBride treed the sister to make sure her collar fit correctly. "We hadn't had much experience putting collars on juveniles, and we wanted to make sure they weren't too tight," Jansen said. "Afterwards the mom and Annie's sister took off. Annie got left. We captured them again and tried putting them back together, but that time it didn't work."

All told, the mother was treed nine times that season and tranquilized four times. Her kittens were also handled frequently. "We all had qualms about handling them so much, but I didn't think we had any choice," Jansen said. Finally Annie was taken to White Oak Plantation, an exotic wildlife ranch outside Yulee, Florida, where the state hoped to start a breeding program for panthers. She was scheduled to be freed again in the Everglades in a few months.

"We kept Annie here for about a month before she went to White Oak," Jansen said, "and I took care of her. One day I was trying to clean some scat out of her cage with a stick; it was right by the edge and I figured I could reach it from outside. I thought Annie was on the other side of the pen. But I looked up and her face was only about two feet from me. It really surprised me. She was just curious, wondering what I was doing."

We checked three more hammocks without success. The other searchers also had found nothing. I was scheduled to leave the next afternoon. Bass looked at me, smiled, and shrugged. "Sorry things haven't been busier," he said, "but at least you're getting a taste of what it's like to hunt panthers."

The capture team did not handle a panther for another three weeks, and then the McBrides treed two at once. One was a female with a radio collar that needed to have its battery replaced. The other was a young female never caught before. There was no evidence of the male that had sired the two litters born in 1987—and that was badly needed to sire more.

"Look at this," Melody Roelke said excitedly, pulling a chart with rows of neat figures from a bulging file. "These are the results of more than two years' worth of samples and genetic testing. They show that Florida panthers have a lot less diversity than other subspecies. It means they could be having all sorts of problems we don't know about."

Roelke spoke quickly, and with an air of anxious concern. It was the spring of 1989, and I was visiting her and Steve Parker at their temporary home, a ramshackle house on the edge of the Big Cypress, where the McBrides were hunting. The Florida panther, it seemed, was growing more beset by trouble each year. Of the seven panthers Chris Belden had studied in the Fakahatchee Strand during the early 1980s, all had died or been removed from the wild, and none had replaced them. A number of cats had also disappeared from the Raccoon Point area of the Big Cypress preserve. Once again the capture team had found no sign of a breeding male in the east Everglades. The team continued to capture robust panthers north of Alligator Alley, but there were now signs that they too were being squeezed for space. The previous August a young male had been killed, probably by an older male that was unwilling to share his turf. "South of the alley, in the Big Cypress and the Fakahatchee, I worry about nutrition," Roelke said. "North of the alley prey isn't a problem, especially on the private land. There are deer and wild hogs you can trip over. But sooner or later, probably sooner, a lot of that property's going to be developed."

Recently the U.S. Fish and Wildlife Service had arranged to purchase thirty thousand acres just north of Alligator Alley to form a preserve called the Florida Panther National Wildlife Refuge. Another one hundred forty-six thousand acres were to be added to the north section of the Big Cypress preserve. Outside of this protected island of forest and swamp, the region was being overrun by change. In addition to habitat loss from expanding residential development, woodlands in Hendry and Collier counties were rapidly being converted to citrus agriculture.

Although cougars tend to be adaptable to change, Roelke's research suggested that the Florida panther had so little genetic diversity left that sudden alterations in its habitat might place the species under severe stress. A year before, she had told me that diminishing prey and habitat loss seemed to be the most critical problems to be solved if the panther population was to recover.

Now she was equally worried about inbreeding. It had been known for several years that panthers produced very low concentrations of sperm, and that most of their sperm was abnormal—a common phenomenon in inbred species. In the past year Roelke had also found an increased incidence of cryptorchidism, where only one testicle descends properly during sexual development. "Nearly half the males I've handled have it," she said. "If you compare the birth dates of the cats, you see that the older ones tend not to have it, but the younger ones do. We're not absolutely sure it's a genetically determined trait. But if it is, we're seeing a deleterious gene being bred into the population right before our eyes. The scary thing is, what about the things we *can't* see, the physiological or immunological changes we can't measure, or the reproductive problems the females may be having?"

Since 1986 Roelke had collected more than eighty blood samples from Florida panthers, free-ranging western cougars, and captive cougars. The samples from caged animals included ten "Piper cats," a strain of Florida panther kept at the Everglades Wonder Gardens, a tourist park in Bonita Springs. (In the 1950s and 1960s, seven Piper cats were released in the east Everglades. Biologists speculate that the addition of their bloodline might account for the lack of kinked tails and cowlicks in the Everglades population.) Roelke processed the blood samples to isolate enzymes that could be tested for slight biochemical variations. "These enzymes are the workhorses of the body," she said. "They have housekeeping responsibilities, but you can't see any obvious physical evidence of them." The synthesis of each enzyme in the bloodstream is genetically controlled, just as certainly as eye color and other physical characteristics. Each gene is located on a chromosome and has a mate, or allele, found on a paired chromosome. The animal inherits one chromosome from its mother and the other from its father. If the two alleles are the same, or homozygous, only one version of the enzyme is produced. But if they are different, or heterozygous, two versions of the enzyme are produced.

In the blood samples from captive and wild western cougars, Roelke found an unusual number of heterogenous alleles—nearly as many, in fact, as geneticists had found in domestic cats. The captive cougars, which were a mix of different subspecies, were especially rich in diversity. In comparison, the wild Florida panthers were genetic paupers. Of eleven possible heterogenous alleles, the Florida panthers shared only three, and those only part of the time. "It's too early to say for sure what all this means," she cautioned. "We don't have a good profile of the entire species yet; I'd like to get some samples from cougars in South America and Canada to see what they look like." Nevertheless, her findings raised doubts about the ability of the Florida panther to breed, and to withstand environmental disturbances and outbreaks of disease. And her results added weight to the argument, voiced by some scientists, that a few western cougars should be introduced to Florida to bring new genetic material into the population.

This suggestion was a radical departure from traditional endangered species management, and both the state Game and Fresh Water Fish Commission and the federal Fish and Wildlife Service opposed it. Still, the idea had caught the attention of the Florida Panther Interagency Coordinating Committee, a group of state and federal wildlife managers responsible for directing the recovery program. Roelke's research had been structured to evaluate whether Florida panthers could continue to reproduce on their own. "I'm opposed to bringing western cats into the population," she said, "but things may get to the point where we don't have any choice."

Roelke's study was only one of several major investigations into panther genetics and reproduction, all of which would affect the management of the subspecies in the wild. Other tests were being conducted at the National Zoo and by Stephen O'Brien, a geneticist at the National Cancer Institute. O'Brien, in fact, was directing Roelke's genetic work.

Roelke first met O'Brien while working at Wildlife Safari, the private wildlife park in Winston, Oregon, where she had cared for a variety of animals, including forty cheetahs. O'Brien had developed the protocol for teasing out genetic variation in felines, using domestic cats as models. In the early 1980s he analyzed fifty blood samples collected from cheetahs at the DeWildt Cheetah Breeding and Research Center near Pretoria, South Africa. He was astounded to find that every enzyme in every sample contained identical alleles. The cheetahs seemed to have no measurable genetic variation at all. Subsequent tests also suggested that, as a species, the cheetah was remarkably inbred. Finally, O'Brien and two colleagues, Mitchell Bush and David Wildt of the National Zoo, decided to see whether they could transplant patches of skin from one cheetah to another.

The rejection of transplanted skin and organs is controlled by a genetic locus known as the major histocompatibility complex. In most animals the possible gene combinations in the MHC are quite numerous; the chance of two humans having identical gene combinations is less than one in ten thousand. Scientists have learned how to test the genetic similarities of individuals by seeing how quickly one will reject a skin graft from the other.

Over a period of months, small patches of skin were exchanged among seven pairs of cheetahs, three in South Africa and four in Oregon. Two of the animals were siblings; the others were unrelated. All the grafts were accepted. "The significance of that was just tremendous," Roelke said. "We can't say that every single cheetah is one hundred percent identical, genetically, but we can say they're close. At some point the species must have gone through an incredible bottleneck. To get mice that are that inbred, you'd have to breed brothers and sisters for twenty-two generations."

Soon after the experiment, Roelke discovered just how vulnerable the cheetah was to catastrophe. A pair of cheetahs sent to Wildlife Safari on breeding loan died shortly after their arrival of feline infectious peritonitis, a viral infection that is fatal in only ten percent of domestic cats. The infection spread to every cheetah at the park and killed half of them. "It was emotionally wrenching," Roelke said, "but it showed us how quickly disease can destroy an inbred population."

After the epidemic Roelke moved to Florida to work with panthers. The experience had both sobered her and piqued her interest in the genetics of endangered felines. At the time, O'Brien and other scientists were discovering that many of the world's great cats shared the same genetic and reproductive problems. David Wildt had found more than seventy percent of cheetah sperm to be abnormal. The sperm Roelke collected from wild Florida panthers was even worse. Each ejaculation contained only between one million and fifteen million sperm, a small fraction of the number contained in domestic cat samples. And more than ninety percent of the panther sperm showed some form of abnormality. By most medical standards panther males would be considered infertile; yet kittens continued to be born in the wild. "We don't know exactly what the pregnancy rate is compared to other subspecies of cougar," Roelke said, "but at least they're producing something."

With such dismally low sperm counts it seemed doubtful that panthers would breed well in the artificial conditions of captivity. Although members of the panther coordinating committee advocated captive breeding as the best way to compensate for genetic shortcomings, Florida panthers had never produced young in confinement. A panther named Big Guy, the only male in captivity, had been held at White Oak Plantation since 1984, when he was hit by a car, but for four straight years he ignored the female western cougars his keepers put in his pen. When he finally copulated in 1989, his mate developed a false pregnancy. The only other Florida panthers in captivity were an aging, anemic female captured from the Fakahatchee Strand and a young female from the Everglades that had been badly injured by a car.

It seemed that the only way to breed panthers in captivity might be through the delicate wizardry of infertility medicine. In the spring of 1988, Roelke and three animal reproduction specialists from David Wildt's laboratory at the National Zoo had met at the University of Florida College of Veterinary Medicine in Gainesville for an *in vitro* fertilization experiment on western cougars and Florida panthers. The principal investigator, Annie Miller, was attempting to conduct *in vitro* fertilizations in several rare species of feline. Another researcher, Karen Goodrowe, had already worked out a technique for doing *in vitro* fertilizations in domestic cats. Miller hoped to adapt the procedure for use in panthers and tigers, among other species. First, however, she and Goodrowe needed to see whether their method of harvesting eggs would work in a common species of great cat. They selected western cougars as models. The third scientist, JoGayle Howard of the National Zoo, had studied sperm in endangered mammals and would help the researchers test whether cougar sperm could fertilize eggs in a culture dish.

Infertility medicine generally has been used much more successfully in humans than in animals. Procedures such as artificial insemination, *in vitro* fertilization, and cryopreservation (the freezing of sperm and eggs) are expensive and difficult to perfect. Nevertheless, many wildlife scientists believe their development is essential for preserving critically endangered species. In addition to solving fertility problems, such techniques could be used to boost the number of possible pairings in a small population, and so increase its diversity. Through artificial insemination, for example, animals that are behaviorally or sexually incompatible could produce offspring without ever coming in contact. A female red wolf, mated to one male continually for many years, could be inseminated with another male's sperm simply to mix her genes with a new bloodline. After birth, her mate would help her raise the pups, oblivious to the fact that they are not his. The possibilities offered by *in vitro* fertilization were especially intriguing to Roelke, since embryos created in a petri dish could be implanted in more common western cougar females. By using western cougars as surrogate mothers for Florida panthers, researchers might greatly increase the number of panther kittens born each year.

But in 1988, when Miller and her colleagues gathered in Gainesville for the first attempt at *in vitro* fertilization in cougars, dreams of surrogate motherhood were only glimmers in a distant future. Progress in infertility medicine comes in small increments as doctors and scientists learn to harvest sperm and eggs without stressing the patients, to keep sperm and eggs alive in cultures outside the body, to coax embryos to grow in petri dishes. Howard had developed techniques for isolating the most viable sperm in each sample. Miller and Goodrowe knew how to treat domestic cats with hormones so they would develop mature eggs, virtually on command, and they could harvest eggs with a fiber optic device called a laparoscope. There, however, their experience ended.

As a preliminary step, Howard, Miller, Goodrowe, and Roelke injected seven female cougars and one Florida panther with a dose of hormones they thought would be adequate to stimulate the growth of follicles in the animals' ovaries. Precisely eighty-four hours later, the cats were given another hormone that caused the eggs within the follicles to mature. "We were guessing at the timing and hormone levels to use," Miller said later, "but it was a pretty well-informed guess because of all the work we had done on domestic cats." Six of the animals had been wild, and were caught in Texas. A seventh was a captive Florida panther–western cougar hybrid. The researchers also decided to try harvesting eggs from the old female caught in the Fakahatchee Strand, then the sole Florida panther female in captivity. She was the only animal that did not respond to the hormones.

A day later, just before the follicles containing the ripe eggs ruptured, the researchers carefully pulled a hundred and forty eggs from the cougars' ovaries, using laparoscopes and tiny syringes. "It worked beautifully; the hormone levels were just right," Roelke said. At that point, however, trouble appeared. Miller and Howard had planned to collect sperm from three wild, western cougars through a simple procedure known as electroejaculation. An electric prod, inserted into the cat's rectum, delivers a mild charge that causes him to ejaculate. Unfortunately, the quality of the sperm from two of the cougars was too poor to use. "No one had anticipated that," Howard said, "but it threw us for a loop because we didn't have a lot of time. Melody ended up going out at eight at night and scrounging up a pet panther from someone she knew."

The researchers managed to collect adequate sperm samples from three males, including Big Guy, the captive Florida panther. Then they selected seventy-five mature eggs and exposed them to the sperm. Nearly half—many more than predicted—became fertilized. To the scientists' surprise, even Big Guy's sperm fertilized several eggs. "When I first looked at Florida panther sperm under a microscope, I was shocked at how bad it was," Howard said. "I'm still shocked. The fact that it was successful in an *in vitro* fertilization is a real encouraging sign."

Ten of the fertilized eggs cleaved and began forming embryos, which then were implanted in two of the western cougar females. The embryos died, but even so, the investigators considered the experiment a great success. "All we had hoped to do was to get some fertilization and cleavage," Miller said, "and we managed that better than we thought we would. If we had gotten panther kittens too, it would have been icing. It would have been outrageous."

A year later, sitting at home on the fringe of the Big Cypress, Roelke eagerly looked forward to a time when the experiment could be repeated. "Nothing's planned right now, and we don't know when it will be," she said. "But with all we learned, and all the work Annie's done since, the technology's there to make *in vitro* fertilization a workable tool for preserving whatever genetic diversity the panther still has. And right now the panther needs some help."

In a small plane above a north Florida pineland, Chris Belden turned on a Loran receiver and set a course toward the north. It was a clear spring morning in 1989, and once again we were tracking "lions." This time, however, we were flying the countryside fifty miles west of Jacksonville, and our quarries were western cougars. Nine months before, Belden had released five sterilized Texas cougars on land just outside the Osceola National Forest. If the cougars adapted well to the pressures of the area— if they stayed out of settlements, did not prey on domestic animals, and avoided automobiles—captive-bred Florida panthers might be released there in the future.

Belden leveled the plane and headed toward the eastern side of the Okefenokee Swamp, where a male had lingered for several weeks. He had not expected any of the cats to range much into Georgia, and this male's preference for territory so far to the north was causing logistical problems. "It's just a long way to come to check on one animal," he said over the plane intercom. The straight rows of pine plantations, dull and monotonous, slowly gave way to grasses, and then to the green, soupy waters of the Okefenokee. "It's probably too wet in there for cougars," Belden said. "He's made trips into the swamp, but he hasn't stayed in it long."

The test release was one of the most important experiments ever run by the Florida panther recovery program, since it had the potential to open two thousand square miles of habitat outside of south Florida. So far, the project was not going well. Belden had designed the experiment to evaluate not only the behavior of transplanted cougars but the attitudes of the local populace toward large predators. Farmers and ranchers worried about panthers attacking their livestock, and hunters grumbled that panthers would deplete stocks of deer. The Game and Fresh Water Fish Commission had tried to alleviate the stockmen's concerns by promising to pay for any domestic animals destroyed by released cougars. The hunters had been more difficult to placate. "We're hoping to show that there's plenty of deer for both panthers and hunters," Belden said. "Most people support the idea of putting panthers back here. But we're having a hard time convincing a few groups that it's not going to cause all sorts of consequences."

Problems with hunters already had exacted a high toll on the experiment. Within six months of their release, two of the cougars were killed by poachers. The first disappeared in October. Belden believed it had been shot near a hunting stand, where he had last located it. The animal's radio collar had been cut off and discarded in a small pond several miles away. Its body was never recovered. Then, in December, another cougar died from a secondary infection after being shot in the leg. (A third cougar had been found floating in the Suwannee River only a month after her release, dead from unknown causes.)

The poaching incidents clearly troubled Belden. "If we can't keep panthers in an area without having to worry about most of them getting shot illegally, then we'd have to say the area's not a good reintroduction site," he said. "But I just can't reach that conclusion yet; it's too early." He was also worried about a change he had observed in the behavior of the cougars after the opening of deer season in the fall. In November two of the cougars abandoned their home ranges, or territories, when hunters began using the areas, as if they could not tolerate the presence of so many people and so many dogs. One of the animals, the male that moved to the Okefenokee, killed five goats on a local farm. "He'd been in pastures with cattle dozens of times and hadn't touched them, but all of a sudden he got a taste for goats," Belden said in a bemused tone. The cougar had not preyed on any stock since settling into a new home range.

Belden's original plan had called for also freeing young Florida panther-western cougar hybrids, sired by Big Guy and raised at White Oak Plantation, to see whether they behaved any differently than animals raised in the wild. Since Big Guy had not bred, however, there were no hybrid animals available. "Eventually I think we'll get some," he said. "We're just having to take things a lot slower than most of us would like."

The signal for the Okefenokee male increased in strength a short distance south of Woodbine, Georgia. Belden circled over a small creek and marked the cat's position on the Loran. "He's been there several days," he said. "Probably on a kill." He turned the plane south toward a tract of timber where the only other surviving cougar, a female, had been the previous day. He fell silent, leaving me to gaze at the flat piney woodlands. To the east, the Atlantic Ocean glinted gold in the early light. The country we were flying over was some of the least developed in the state. According to a survey Belden had conducted, it was the most suitable area outside south Florida for a population of panthers. The thought that it might be unusable because of hunting pressure was discouraging.

Belden located the female, noted her position, and circled back toward the airport at Lake City where we had taken off. It occurred to me that, as a panther biologist working for the Game and Fresh Water Fish Commission, he was in somewhat of an untenable position. Like every other state wildlife agency in the country, the commission was operated for the benefit of hunters. It did not seem likely that its board members would agree to close some gamelands for the protection of panthers. Yet if hunting really proved too disruptive, was there any place left outside south Florida where panthers could live?

The translocation experiment continued to go poorly. In late March Belden released two additional cougars to replace the animals that had been killed by poachers the previous fall. A short time later one of them strayed into the suburbs of Jacksonville and was treed by dogs. Then, within days of the opening of turkey season in April, the female cougar left her home range and discovered a private exotic game ranch north of Lake City. She killed five African antelope before she could be recaptured. Belden also trapped the Okefenokee male and rereleased him on the western side of the Osceola National Forest. He immediately moved northeast toward the farm where, a few months before, he had killed the goats. The commission ordered that he be recaptured and not freed again.

Belden had planned to run the experiment through June. But with only one animal still loose, there seemed little point in continuing. Discouraged, he removed the last western cougar from the wild in late April.

Over the next year, the thrust of the panther recovery project was to become more controversial than it had ever been. Belden's study was supposed to have laid the groundwork for the release of Florida panthers bred in captivity. Up until 1989, though, the coordinating committee had placed little emphasis on expanding the fledgling breeding program at White Oak Plantation. Panthers were taken into permanent custody only when they were too weak or too badly injured to survive in the wild. Yet the results of Roelke's genetic research indicated that the population might be on the verge of becoming badly inbred.

In the fall of 1989 the coordinating committee invited Ulysses Seal and the Captive Breeding Specialist Group to Florida to write a viability analysis and species survival plan for the panther. By then it had become clear that some panthers would be taken from the wild; the only questions to be resolved were when and how many. In January 1989 Seal's group had conducted a preliminary population viability analysis and concluded that the panther population could well be extinct in twenty-five years. Members noted that the subspecies was extremely vulnerable to inbreeding, disease, and environmental catastrophe. Soon after the meeting, a form of immuno-deficiency virus—a feline AIDS virus—was discovered in blood samples taken from a quarter of the panthers handled by Roelke. And in the summer of 1989 a panther was found dead in the Everglades with a shocking concentration of mercury in her liver, raising the possibility that panthers were being poisoned by toxic contaminants.

The conclusions to emerge from the autumn meeting of the Captive Breeding Specialist Group were more detailed. According to computer simulations, the panther population was suffering from such high mortality that it could not be expected to sustain itself for long. Although several field biologists doubted that the panther population was declining so quickly— a few even believed it might be growing—Seal's group recommended that a number of animals be taken into captivity to reduce mortality and for breeding. To compensate for genetic problems, group members concluded that the population needed to be expanded as rapidly as possible to three hundred, and eventually five hundred animals, through captive breeding and careful management of the wild stock. The group's final report noted that a number of zoos had offered to make pen space available for panthers at no cost to the government. Captures should begin immediately, with four adults and six kittens taken in the spring of 1990. Another six kittens and a pair of yearlings or adults should be taken in both 1991 and 1992.

For the previous several years the panther project had enjoyed a period of relative grace in terms of its public image. No panthers had been killed during captures. Fears that cats with radio collars would not breed, or that they would not use the wildlife underpasses beneath Interstate 75, were proving untrue. The reservations voiced to me by Florida residents had been expressed in the quiet tone of private misgivings, not the stridency of public complaint. With the announcement of plans to capture twenty-six panthers, however, the years of public acquiescence drew to a sudden close.

The coordinating committee based its decision to capture six kittens a year on assurances from field biologists that, as long as a panther's kittens were taken at the right time, she would soon go back into heat and breed again. It was possible, then, that the captures would cause little loss to the wild population. And the biologists noted that kittens left in the wild had only a fifty percent chance of surviving. The Captive Breeding Specialist Group recommended that some adults be taken from the wild as well, so the breeding program would be guaranteed of obtaining a wide sample of available genes.

In January 1990 the committee began applying for the federal permits needed to remove the first ten panthers from the wild. The action drew a barrage of protests from Florida residents and animal rights activists. To an extent, the objections were prompted by misinformation; after the committee's announcement, several newspapers reported, erroneously, that the number of animals removed over three years would total between half and all of the wild population. Many of the critics, however, were better informed. They complained that the agencies in charge of the recovery project had done little to preserve prime panther habitat, to improve prey base in marginal habitat, or to limit hunting in areas where panthers were known to live. They feared that Florida panthers would never breed in pens, except through *in vitro* fertilization or perhaps artificial insemination. They also predicted that removing panthers from the wild would intensify development pressures on the private land north of Alligator Alley.

The breeding proposal drew fire as well from the scientific community. Biologists on the capture team worried that removing too many panthers, especially females, would seriously disrupt the population. (The coordinating committee promised that the capture program would be halted if it had significant detrimental effects on the wild population.) Researchers outside the program questioned whether *Felis concolor coryi* still existed in its pure form and, more to the point, whether a subspecies was worth such fuss and bother. Subspecies are thought to be temporal forms, evolutionary half steps between one species and the next. With so many species disappearing, critics wondered why the federal government would consider devoting several million dollars to building a population of five hundred Florida panthers. Wouldn't it be more practical simply to mix new genetic information into the subspecies by adding a few western cougars?

Before the coordinating committee could obtain federal permits to capture panthers, the Fish and Wildlife Service was required to evaluate the consequences of leaving panthers untouched in the wild, capturing a portion of them, or capturing them all. In February 1990 the agency agreed to delay its final environmental assessment for several months to give the public time to comment on the breeding proposal. The decision meant that no panthers would be captured for at least a year.

Politically, the delay was essential to avoid the appearance that panthers were being captured frivolously. It also gave proponents of the breeding program time to rally support. During the public comment period, the Fish and Wildlife Service received more than a hundred letters and statements from people who strongly opposed bringing panthers into captivity, but two hundred and fifty from people who strongly supported the proposal. It was decided that captures of kittens would begin in 1991. Unfortunately, the delay had unanticipated costs. When the Captive Breeding Specialist Group issued its report in November 1989, it identified sixteen genetic founders—animals that were unrelated, and so represented different lineages. During 1990 five of the founder animals died. Their deaths left the population with a greatly diminished gene pool. And unknown to either federal wildlife officials or the coordinating committee, the story of the panther was about to take a strange twist.

In October 1990 Stephen O'Brien of the National Cancer Institute, Melody Roelke, and several colleagues published the results of a study on mitochondrial DNA samples taken from nine subspecies of cougar, including one from Chile, seven from western North America, and the Florida panther. Mitochondrial DNA is inherited from the mother only and does not vary from generation to generation. If a male breeds with a female from another subspecies, their offspring will pass the female's form of mitochondrial DNA on to subsequent generations forever, even if no other hybridization ever occurs.

In analyzing the mitochondrial DNA, O'Brien found that all the cougars from western North America shared certain similarities; they belonged to a clade, or group related by a common evolutionary heritage. Most of the panthers living in the Big Cypress region and on the private lands of southwest Florida shared the same heritage, with one slight variation, as if they had been isolated from other North American cougars for many generations. But the east Everglades panthers—those without kinked tails and whorls of hair on their backs—appeared to be more closely related to South American cougars than to their western counterparts.

The results of the study were not conclusive, since O'Brien and Roelke had analyzed samples only from the southern tip of South America. Nevertheless, the study introduced the possibility that the Everglades cats were not pure Florida panthers. And apparently their bloodline was spreading. In the late 1980s young panthers began to appear in the Big Cypress Swamp without kinked tails or cowlick whorls.

It was possible that the South American blood within the Everglades population had come from the Piper cats released in the national park in the 1950s, or from a single captive cougar that had been abandoned by its owner. It was possible, in fact, that the Everglades animals were ninety-five percent

---

Florida panther. O'Brien and his colleagues hoped to draw more detailed conclusions from refined mitochondrial DNA work and from DNA finger-printing, a technique that enables researchers to discern close familial relationships and to differentiate between distinct populations. But by early 1991 only thirteen of the twenty-three radio-collared panthers in south Florida could be shown to have mitochondrial DNA with no trace of South American influence. The others, to one extent or another, were hybrids.

On my first visit to the panther recovery project in 1988, Oron Bass had advised me: "Don't put much stock into minor morphological distinctions like kinked tails. Eventually the question we're going to have to answer is, do we want to have the Florida panther, or do we want to have panthers in Florida?" He couldn't have known how prophetic his words would be.

From the whole cauldron of problems that face the Florida panther two facts emerge that are at once illuminating and troubling. First, without intervention the genetic deficiencies that plague the subspecies are likely to worsen. In 1991–93 increasing numbers of wild kittens had congenital heart defects, probably caused by inbreeding. In many respects the recovery project was going well. Six kittens were successfully taken from the wild in 1991, four more in 1992. In February 1993, in a repeat of Chris Belden's reintroduction experiment, ten western cougars were released in the Osceola National Forest. Nevertheless, federal officials were worried about the subspecies' health. By 1993 they were talking of cross-breeding Florida panthers with western cougars to augment the gene pool.

Second, it is alarming that by 1993 all the panthers living in Everglades National Park had either died or moved west of Shark River Slough. Only four animals were known to be using the southern portion of Big Cypress National Preserve and other public lands south of Alligator Alley. The other eighteen radio-collared panthers tended to stay on the million acres to the north, mostly on the private uplands that are likely to be developed in the next decade.

Something may be wrong with the nature of the territory we have bequeathed to the Florida panther. As recently as the early 1980s, the half million acres in the southern Big Cypress region alone supported at least five or six adult panthers and three kittens. Panther sign was common, where now it is rare. Perhaps the decrease is the kind of cyclical swing that characterizes certain animal species. Perhaps it is indicative of too little prey, or too much human activity, or too much toxic runoff from the agricultural lands and urban centers to the north. No one can say for sure.

It is intriguing to think that *Felis concolor coryi* can be preserved in its unadulterated form through captive breeding or various forms of manipulation in the wild. Yet it seems important to preserve not just the pure panther but the habitat where it once thrived.

In *Conservation Biology* in 1990, David Maehr, a panther biologist with the Game and Fresh Water Fish Commission, noted that by acquiring the rights to privately held acreage within the panther's range, the state and federal governments could preserve diverse forest ecosystems. Maehr was not talking about pristine wilderness. The private woodlands where the largest, healthiest panthers live abut agricultural land and cattle pasture and subdivisions. At the very least they have been timbered once. But in comparison to the land in public ownership in south Florida, the private tracts are drier and more fertile. They contain abundant wild hogs, large herds of deer, and a variety of plants and animals that have disappeared through most of the state. They also offer ample cover for great cats.

Maehr noted that public property in south Florida is generally swamp and wetland. He urged government agencies to find creative ways of conserving upland panther habitat, for instance by purchasing strategically placed parcels that would make it difficult for surrounding tracts of forest to be developed. The cost might be high, and many landowners are hostile to suggestions that they forfeit economic gain to help save the panther. But in a conversation, Maehr suggested that landowners be paid to leave their property in forest. "I don't know what it would cost, because no one's asked them," he said. "But to me that's one of the key biological questions. What needs to be done, how much might it cost, to save this prime habitat? There have to be economic incentives the landowners will accept."

In 1990 and 1991 the Fish and Wildlife Service began to inventory panther habitat in Hendry, Glades, and Collier counties, based on radio telemetry data compiled over several years. Through the survey, administrators hoped to identify the most important parcels of land and to open discussions with their owners about keeping the property in forest. In 1990 federal officials worked out an agreement with the Game and Fresh Water Fish Commission banning the use of dogs for deer and hog hunting in the Big Cypress preserve.

Sadly, it is only now, with the panther possibly a few generations from extinction, that officials have given serious thought to saving room for it—not just swampy park land set aside as a recreational playground but prime habitat where it will not be frightened off its home range by hunters. Because we have waited so long, the panther's story is following the same bitter trail as the red wolf, the California condor, the black-footed ferret, and hundreds of other species. The animals must be subjected to radical manipulation that might have been avoided if the settlement of America had unfolded with more compassion, and less greed.

On the day I went flying with Chris Belden over the piney forests of north Florida, I asked him why he thought it was so important to save a subspecies of cougar in a state so weighted with environmental problems. "If you can't save the panther in Florida," he replied, "then what that means is there's no wilderness left in the East, not even in small pieces. It means we've not only overhunted them, we've completely destroyed the last places where they survived. This is the front line for the mountain lion in this part of the country. If we lose them here, the front line becomes the West. Where do you draw the line then? Where?"

EPILOGUE

On a mild, sunny morning in January 1991, a hundred and fifty people gathered around a pickup truck parked on a road in the Great Smoky Mountains National Park. The crowd's attention was focused on two vinyl shipping kennels, in each of which was a disoriented red wolf. The wolves, which had just been flown in from a breeding center, were to be held in a pen near Cades Cove, a small valley deep in the mountains on the Tennessee side of the park. With luck, they would be freed with pups the following autumn.

A little more than three years after the release of wolves in the inhospitable marshes of eastern North Carolina, *Canis rufus* was being reintroduced to the heart of the Southeast. Most of the people present that day were reporters. They had come to welcome the wolves, but also to celebrate a new attitude of tolerance toward predators—a key reason that Fish and Wildlife Service biologists wanted to try restoring wolves in country where, a century earlier, they had been systematically killed off.

Comparisons to the reintroduction at the Alligator River National Wildlife Refuge were unavoidable. As one of the first major wildlife restoration projects that involved only captive animals, the Alligator River reintroduction had laid a critical foundation. Its success had ensured that other species would be reintroduced, and that red wolves would be freed at additional sites. But the project in the Smokies had the potential to become even more important, simply because of its scope. With a half million acres, the Great Smoky Mountains National Park is one of the largest pieces of federal property in the Southeast, and it is surrounded by a million acres of national forest. If all went well, the Smokies might eventually be home to several dozen red wolves, perhaps even to a population large enough to sustain itself.

Residents around the park had reacted to the planned reintroduction with a mix of enthusiasm and doubts similar to those expressed by the people of coastal North Carolina. All possible care had been taken to win their trust. A local resident ran a replica of an eighteenth-century cattle farm in Cades Cove, and there were a few small livestock farms on the outskirts of the park. The Fish and Wildlife Service had promised to reimburse the owners for any livestock killed by wolves and to release the wolves with radio capture collars, despite their experience with the devices at Alligator River.

Once freed, the wolves would wander up rocky ridges and down dark hollows forested with oaks, poplars, pines, and rhododendrons. They would feed on deer, rabbits, raccoons, and groundhogs. As they traveled they would probably encounter another species much like themselves, but smaller. For years the coyote, the adaptable animal that had driven the red wolf from its last stronghold, had been steadily invading the eastern United States. Although relatively few coyotes lived in the park, an extended family group inhabited Cades Cove. Their presence would provide the wolves, and their human guardians, with a significant challenge. As long as biologists could keep enough wolves in the wild so they could find suitable mates, Warren Parker did not expect them to breed with coyotes. He recognized, however, that the recovery of the red wolf would suffer a major setback if he was wrong. Coyotes were now too pervasive, and uninhabited land too scarce, for the wolves to have much acreage to themselves.

With the groundwork laid for the Smokies reintroduction, Parker had decided to retire from the service. Chris Lucash was reassigned to Cades Cove to oversee the release program there; it was he who greeted reporters and answered questions the day the wolves arrived. Mike Phillips remained in charge of the Alligator River project. Pairs of wolves had been established on three barrier islands in the Southeast—Bulls Island, South Carolina; Horn Island, Mississippi; and St. Vincent Island off the Florida Panhandle—and were raising pups in the wild every spring. Plans were also being made for wolves to be freed in southwest Mississippi, on an Army munitions disposal site bordering the Pearl River, and there was talk of eventually releasing more in northern Arkansas.

The Alligator River program, meanwhile, was expanding quickly. The corporate owner of three large farms adjoining the refuge had signed an agreement allowing wolves to hunt and breed on its property. The agreement had left Phillips and Parker jubilant, for it proved that local landholders were willing to accept the presence of wolves. The refuge had recently been expanded; now, with the addition of forty thousand acres of farmland, the wolves were free to roam over two hundred and twenty-five thousand acres. And it was likely that more territory would soon be made available nearby. The Pocosin Lakes National Wildlife Refuge, a ninety-three-thousand-acre preserve, had been established west of the Alligator River. By 1992 the wolves had begun traveling to the new refuge, circumnavigating the river by moving south through private farm fields and then turning west. Phillips planned to release a new group of wolves at Pocosin Lakes in the summer of 1993. With luck, in coming years more animals would be freed on the surrounding farms.

Canis rufus had been returned to the East and had been accepted, if somewhat begrudgingly, by the people most affected by its presence. Mortality was high among the released animals; Phillips had come to expect about half of them to die within a year. Yet enough survived and bred for biologists and the conservation groups that had followed the reintroduction closely to consider it a success. By early 1993 at least thirty wolves, and perhaps as many as five uncollared pups, were thriving in the boggy forests of Alligator River.

Twelve wolves, two pairs of adults and eight pups, were living in the Smokies. As with the first release Alligator River, Lucash spent much of his time hazing the animals with cracker shells to scare them away from roads and human settlements. "It's kind of like living in a time warp," he said. "The pups especially are way too comfortable around people." But one family group had settled into a home range in the Cade's Cove area of the park. Their easy adjustment to wild life would be successful.

Roland Smith, the curator of the breeding program in Washington State, had found homes for red wolves in zoos and breeding facilities across the country. The species' population had not expanded as quickly as biologists had hoped, and with a total of only one hundred and fifty animals the recovery program would continue to concentrate many of its resources on captive propagation, at least for the immediate future.

Ironically, just as the red wolf was becoming solidly reestablished in the wild, geneticists were again raising questions about its validity as a species. In 1988 and 1989, Fish and Wildlife Service administrators had commissioned a study on the composition of the mitochondrial DNA of the wolves in the breeding program. (The genetic molecules known as mitochondrial DNA are inherited only from the mother, and their mark is indelible. Once a female reproduces, evidence of her parentage is passed on to future generations forever through mitochondria molecules. As a result, mitochondria can be used to trace the evolution of species over hundreds, and in some cases thousands, of years.)

The population geneticists commissioned to conduct the study were Robert Wayne of the University of California at Los Angeles and Susan Jenks of the University of California at Berkeley. When they embarked on the study, Wayne and Jenks expected to find clear evidence of a unique red wolf gene line. To their surprise, they failed to uncover any consistent genetic difference between red wolves and coyote-gray wolf hybrids.

The findings left service administrators in an untenable position, for under federal law it was illegal to use endangered species funds to preserve hybrid animals and plants. Was the red wolf a true species? If it was not, should federal officials continue to spend money on its preservation?

Wayne believed the recovery project should be continued, since the animals being released at Alligator River and in the Smokies might well be the closest living relatives of an extinct species. Mike Phillips agreed. "I think you can make a case that humans trapped and poisoned the red wolf nearly to death, and that we have a responsibility to save what's left of it, even if it's not in a pure form," he said. "This isn't a black- and-white issue. We need to give the red wolf some middle ground." In mid-1993, Fish and Wildlife Service administrators were reviewing a new policy on hybrid species that, they hoped, would do just that.

However *Canis rufus* was defined, it was slowly regaining its rightful place in the American landscape. One warm day shortly before his retirement, Parker and several biologists were on the Alligator River refuge settling a new family of wolves into a pen. He and I sat down in the shade of a swamp willow to eat lunch. Parker's hair was a bit grayer than when the first wolves had arrived in 1986, and the lines on his face cut a bit deeper. He was looking forward to having time to himself. The work and worry of reconstructing the wild red wolf had exacted its toll.

I asked if he were gratified by the way the recovery project had unfolded.

"Oh yes," he said. "It's gone very, very smoothly, all things considered. Largely because I've had such a competent staff here."

"Did you ever think it would come as far as it has, with wolves in the Smokies and all?"

Parker grinned. It was not a playful smile but a candid one, half serious, half shy. It belied, for just a moment, the façade of self-confidence he preferred to display. "No," he replied at last. "I have to say no. I wasn't even sure, at first, that we would ever get this Alligator River project off the ground."

Even with all the animal and plant species lost since the European settlement of North America, there remains a great richness of life on this continent—in the bromeliads and ferns that erupt from the trunks of cypress trees in the Florida swamps; in the autumn passage of warblers along the barrier islands of the Atlantic; in the wildflowers of tall grass prairies, the black bear and grizzlies of the northern Rockies, the raptors and hummingbirds of the Southwest, the butterflies that congregate by the hundreds of thousands to breed in California, the anemones and urchins that crowd tidepools along the northern Pacific Coast. We still have much to preserve.

Yet simply maintaining the status quo will require us to make difficult choices. Many species can no longer survive without some form of good will or charity from humans. We have entered an age where populations of wild animals, especially predators, need to be carefully managed; on that most wildlife scientists agree. But we must still decide how intensely the lives of rare species will be controlled. Will the animals range, at least to an extent, freely? Or will they all be restricted to carefully managed preserves, refuges, and zoos?

In examining the recent history of endangered species recovery projects, several lessons become clear. First, the largest single cause of extinction is the destruction or alteration of habitat. Unfortunately, habitat preservation is an expensive task, and generally opposed by economic interests such as the timber, building, and agriculture industries that depend on the easy availability of land. Rather than pursue an aggressive policy of preserving or restoring wilderness, the government has chosen to confine endangered animals to small tracts of federal property. This is most easily apparent in the recovery programs for the black-footed ferret, the California condor, and the Florida panther. But if allowed to continue, the insidious loss of territory can be expected to undermine the health of hundreds of species that are still common, as well as the nearly eight hundred North American species already listed as threatened or endangered.

Second, federal wildlife managers frequently put off taking the needed preservation measures until species become highly endangered—that is, well past the point at which the Endangered Species Act could be invoked to come to their aid. By then it is too late to learn much about the species in the wild, and biologists must turn to tactics such as captive breeding that may seriously disrupt what remains of the free-ranging population. Perhaps it is naïve to think that the business of saving unique forms of life might be conducted in something more orderly than a string of last-ditch, crisis-driven rescue missions. Nevertheless, scientists have compiled compelling evidence to show that a more thoughtful approach is badly needed.

In 1989 Brad Griffith of the Fish and Wildlife Service's research unit in Orono, Maine, and three other wildlife researchers published the results of an analysis of nearly two hundred wildlife translocations conducted between 1973 and 1986 in the United States, Canada, Australia, and New Zealand. A translocation was defined as a release of animals to augment or restore a wild population. Some of the projects involved animals bred in captivity; others involved animals caught from the wild and moved to new locations. About ninety percent of the releases involved game species. The game animals proved relatively simple to restore; eighty-six percent of the projects conducted with them were successful. In contrast, only forty-six percent of the translocations that involved endangered, threatened, or so-called sensitive species—those that were affected by slight environmental disturbances—were considered successful. Apparently, the authors wrote, the chances of saving an animal through translocation are greatly reduced once it becomes endangered. "It is clear," they concluded, "that translocation must be considered long before it becomes a last resort for these species—before density becomes low and populations are in decline."

Griffith and his colleagues documented a strong correlation between the quality of habitat at the release sites and the success of the projects; the better the habitat, the more likely it was that the animals would survive. They also found that animals captured from the wild and moved to other areas were more likely to survive than those bred domestically and released. Apparently the best conditions for translocations occur when species populations are at least moderately high and are increasing in number—the very circumstances, the scientists noted, that tend "to make endangered species biologists relax." The easiest way to preserve wild animal species is to take action well before they become endangered. To delay is to invite defeat.

Third, rare animals should be reintroduced to the wild before they have endured too many generations in captivity. Regardless of how many precautions breeding specialists take, any wild animal that is confined to a pen is bound to undergo changes, many of which may be extremely subtle. Often the alterations are psychological, but scientists are finding increased evidence of small but potentially damaging physical changes as well. A study of red grouse, for example, showed that birds kept in captivity for as few as six generations had intestines measurably shorter than those of wild birds—a consequence, apparently, of eating domestic feed. Wild turkeys confined to captivity for several generations have been shown to have smaller adrenal glands than free-ranging birds, which may help explain several behavioral traits that make captive turkeys difficult to release. Whether the process of domestication affects an animal's behavior or its physical attributes, many species kept in captivity for prolonged periods lose the ability to make their way in the wild.

Sadly, this problem greatly reduces the potential for zoos to serve as modern arks, carrying endangered species through long and troubled times. Although scientists may devise means to return even thoroughly domesticated species to the wild, it seems unwise to depend too heavily on captive breeding as a solution to the problem of disappearing biological diversity. This is especially true when one considers the large number of species that face extinction and the limited resources available for preserving them.

A fourth lesson to be drawn from recent experience is that when endangered species recovery projects come under the direction of several different government and private agencies, they tend to become mired in administrative bickering. This is, perhaps, simply because it is trickier to reach consensus among many policymakers than few. It is also one of the drawbacks to working with animals that are highly endangered. With little time to experiment with solutions, and with virtually no room for error, the pressure placed on wildlife managers becomes extreme. As the situation grows more desperate, proponents of different options become strident and unwilling to compromise. In the worst cases recovery programs degenerate into political chess games, and key players come to distrust anyone who questions their views.

     Meant to Be Wild

What can be done to rectify all this? How can we avoid the political and philosophical snares that have kept us from developing a wiser national policy toward endangered wildlife? There are no ready answers. However, there are ways that answers might be sought.

Clearly, more attention must be given to preserving animal populations before they begin to fail. Just as important, stronger efforts must be made to save habitat for wild animals, not only for species that are already in decline but for others that may become threatened in the future. There are two types of habitat to consider, wilderness and land that has already been settled. Many species of wildlife have proved to be tremendously tolerant of human activity. Falcons now nest on skyscrapers, and foxes haunt the edges of suburban back yards. Small patches of forest can shelter a surprising variety of wildlife—as long as they are not completely surrounded by shopping malls, parking lots, and treeless subdivisions. One of the simplest things we can do to help conserve nature is to preserve tree cover, both in our front yards and in parkland throughout our cities.

This is not to say that large tracts of open forest are unimportant. As David Wilcove's 1983 study of songbird predation showed, deep forest is critical to the survival of certain animals. If we are to preserve species that cannot adapt easily to environmental change, we must leave them ample territory, whether they live in the cool deciduous woods of Vermont or the stark salt pans of Nevada. Here, however, a word of caution is in order. Much has been written recently about the need for entire ecosystems to be held intact, and possibly linked by natural corridors through which animals could travel. These expansive showcase preserves could provide habitat for many kinds of animals, and also limit human encounters with rare or dangerous species. Yet there is a danger that, once several large preserves have been created, political pressure will increase to loosen environmental regulations elsewhere. The establishment of biopreserves should not be used as an excuse to destroy the natural integrity of the rest of the continent.

Beyond this the course becomes less clear. Despite their dedication, many of the conservation scientists I have spoken with believe their efforts are doomed to fail. The world is preoccupied with accumulating wealth; only a small segment of the population truly cares about holding onto what remains of the natural world. Modern culture is not likely to dismantle its prevailing system of values, not even if an environmental catastrophe of some sort shakes us momentarily to our senses. The one path to profound change appears to be through some transformation, religious or otherwise, that would alter the view we hold of ourselves in relation to nature. We are neither separate from nature nor above it. We must somehow learn to participate in, but not control, its evolution.

As bleak as the current situation appears, there is some reason for hope. Since the mid-1980s astounding political changes have occurred in eastern

Europe and the former Soviet Union, and the institution of apartheid in South Africa has crumbled. These revolutions have not furthered the cause of conservation; I mention them only because they provide evidence that a small group of people, fighting against all odds, can sometimes triumph. If broad political change is possible, perhaps broad social change is too. Nevertheless, the social and attitudinal changes needed to avert or end a mass extinction will have to be much more widesweeping than any isolated political struggle. They will have to transcend borders; they will have to shake modern civilization to the same degree that the Industrial Revolution and the discovery of the Americas shook it.

It is impossible to foresee how such radical alterations might be brought about. For the present all we can do is concentrate on improving our own national affairs. Many countries look to the United States to set world standards for the conservation of nature, partly because of the Endangered Species Act, and partly because of our scientific knowledge and material wealth. The example we set should indeed be exemplary.

There are a number of ways we might move toward a wiser, more visionary conservation policy. For one, we must abandon our crisis mentality and begin to combat the loss of natural diversity in a more intelligent, efficient manner. This will require wholescale, creative solutions instead of simple, piecemeal ones. Captive breeding, though an important component of species preservation, ultimately will not cure the extinction epidemic. Neither will relatively painless steps, such as creating wildlife corridors between key parcels of habitat. We must rethink our emphasis on preserving species one at a time. And we must learn to choose more carefully which strategies to pursue.

In recent years some conservationists have begun to speak more and more readily of using a triage approach to protecting wild animals, of sorting out species and spending precious resources only on those with the greatest chances of survival. In a system of triage, wildlife managers must consider the possibility that the most critically endangered animals, those at the very edge of extinction, may not be worth trying to save. They must select which will be rescued and which will be allowed to die out. Triage is anthropocentrism at its worst; it is based on the assumption that humans have the right to make life-and-death decisions for other species. Unfortunately, wildlife managers inevitably are forced into such choices when they attempt to rescue species one at a time with limited resources.

If we are to avoid sentencing animal species to extinction either consciously or through neglect, we must greatly increase the amount of money available for their preservation, both through the federal government and through private conservation groups. At the same time, we must strengthen our commitment to using resources wisely and carefully, so as many species as possible will reap the benefit.

One step in this direction would be to increase public and private support for biological field studies, especially studies of animals whose populations are dwindling but not yet endangered. Field study is tedious, and often its value is not immediately apparent. Occasionally, however, it yields sudden insights that enable biologists to increase a species' breeding success or rate of survival. It was this kind of realization that made Noel Snyder and his colleagues begin experimenting with artificial nest boxes for pearly-eyed thrashers. By situating the boxes within a few yards of active Puerto Rican parrot nests, the biologists reduced competition between the two species and so enabled more parrots to raise young. Such solutions provide quicker results and are much less expensive than dramatic measures like captive breeding. Just as important, through field studies scientists can maintain an ongoing connection with the wild nature of animals.

Many field biologists who work on endangered species preservation complain that federal wildlife managers have abandoned classical natural history as a conservation tool. Under the current crisis-oriented mode of operation, there can be little time given over to painstaking study. Yet a thorough knowledge of species' behavioral traits, as well as their food and habitat requirements, can be vital to the success of recovery projects, especially in cases where biologists must create new wild populations using only captive stock. Field researchers are especially fearful of what they regard as a dangerous trend toward captive breeding at the expense of maintaining wild populations. "Plain old nineteenth-century biology has gone out of vogue, but it's always been the most important thing in determining what the animals need," the parrot biologist Jim Wiley once told me. "Call it ecology or behavioral biology or whatever you want, but it's the natural history that continues to be important."

If field biology is to be an integral part of species preservation, however, it must be plied with a gentle hand. Cartesian science functions only as a narrow lens; to perceive the vast intricacy of the natural world, scientists must broaden their vision to include a sense of the great mysteries inherent in the wild spirit. They must anthropomorphize a bit, and accept the possibility that animals have their own cultures. They must realize that animals may have unique ways of perceiving their landscapes and moving through them, ways that are not easily understood. And they must recognize that intrusive procedures such as radio tagging may affect the behavior of animals, and try to minimize those effects.

Even if scientists learn how to regard their subjects with more compassion, the struggle to save America's most endangered species cannot be won without significant shifts in how we perceive wild animals and wild lands. In how hunters view wolves and panthers, how ranchers think of prairie dogs and wild predators, how loggers look at old-growth forests, how urban Americans regard the need for open space. The Golden Lion Tamarin

Conservation Program in Brazil proved that public thinking can be changed, quickly and profoundly, through careful, creative education programs. We need to build on that lesson at home.

During the 1990s two proposed wildlife reintroductions will give us the opportunity to reverse the destructive tide that has washed across the continent during the past four hundred years. Both projects involve wolves, and both will require us to overcome the fear and hatred of predators that has dictated national policy in the past. One entails restoring the Mexican wolf, a subspecies of gray wolf, to the Southwest, perhaps by releasing the species on national forest land in Arizona or on the White Sands Missile Range in New Mexico. Except for a few animals that may still roam the mountains of northern Mexico, the Mexican wolf is believed to be extinct in the wild. As of early 1993 only fifty-three were alive in captivity, scattered among a half dozen zoos in the United States and Mexico. Until recently it seemed unlikely that the subspecies would ever be returned to the wild, because of extreme opposition by ranchers, politicians in Southwestern states, and Army officials at the missile range. In 1990, however, the Army dropped its objections to a reintroduction on the range, and in 1992 Arizona officials agreed to study the possible repercussions of a release.

The other project involves the return of the northern gray wolf to Yellowstone National Park. Given the groundswell of support for wolf restoration among wildlife scientists and conservation groups, and considering the public's more lenient attitude toward wolves, the reintroduction of the species to Yellowstone appears likely. It would be undertaken with wild wolves trapped in Canada and transferred to the park. The proposal, of course, has been vigorously opposed by stockmen's associations in Montana, Wyoming, and Idaho. However, biologists for the National Park Service, the Fish and Wildlife Service, and wildlife conservation groups believe that if the reintroduction is conducted carefully, wolves will be able to range freely through Yellowstone without causing significant problems for ranchers. The released wolves would be classified not as endangered animals but as an experimental population. If they began to kill stock, they could be trapped or shot.

Some ranchers have conceded that they stand to gain from supporting a wolf reintroduction conducted under such provisions. In the past decade, wolves have begun to recolonize northern Montana, and biologists believe they will eventually move farther south. If the animals return to Yellowstone on their own, their standing as an endangered species will give ranchers and wildlife managers few options for dealing with individual wolves that attack stock.

As recently as ten years ago, the possibility that wolves would be returned to Yellowstone was considered unlikely. The increased willingness of many

Americans to grant a place in their landscape to potentially dangerous animals is encouraging, though the pendulum can reverse direction without warning. All ground gained in the past decade can be quickly lost. Even if it is not, the wolves freed at Yellowstone or in the Southwest may be monitored so closely, and their movements restrained so severely, that they become only facsimiles of wild animals. I wish things were different. Yet I continue to remember what I learned from the wolves of Alligator River—that there is no way to predict how any wildlife restoration project will unfold, or how it may succeed. To be a conservationist is to be an eternal optimist. Pessimism serves no purpose, not when you are trying to change the world.

It is conceivable that human attitudes toward nature will be radically improved through education, or creative financial incentives, or (who can say?) an environmental catastrophe that may actually shake us so deeply that we alter our lifestyles. Despite the terrible odds against us, the primary work of conservationists—the fight to redefine the relationship of our own species to nature—must continue. We owe nothing less to the animals that are living out their lives in cages, the refugees that were meant to be wild.

BIBLIOGRAPHY

Over the past three decades a great many scientific studies have been conducted on animals kept in captivity, and on wild populations of endangered species. In addition, several good books have been written for the general public on extinction and the loss of natural biological diversity.

My purpose here is to note the sources that were most helpful to me in my research, and to give direction to readers who would like to delve more deeply into the issues raised in the text.

THE WOLVES OF ALLIGATOR RIVER

Chapters I to V

Most of the historical information in this section was drawn from interviews with federal biologists and from technical reports issued by the Fish and Wildlife Service. In Chapter II, I relied on several books for information about worldwide extinction rates, including *The Last Extinction*, edited by Les Kaufman and Kenneth Mallory (Cambridge: MIT Press, 1986), *Biodiversity*, edited by E. O. Wilson (Washington: National Academy Press, 1988), and *Animal Extinctions: What Everyone Should Know*, edited by R. J. Hoage (Washington: Smithsonian Institution Press, 1985).

Information on the natural history of red wolves and the early years of the recovery project can be found in the following sources.

Carley, Curtis J. *Activities and Findings of the Red Wolf Field Recovery Program from Late 1973 to 1 July 1975.* U.S. Fish and Wildlife Service, 1975.

_____. *Status Summary: The Red Wolf.* U.S. Fish and Wildlife Service, December 1979.

Hoagland, Edward. "Lament the Red Wolf." In *Red Wolves and Black Bears.* New York: Random House, 1976.

McCarley, Howard. "The Taxonomic Status of Wild Canis (Canidae) in the South Central United States." *The Southwestern Naturalist* 7 (December 10, 1962): 227–235.

_____, and Curtis J. Carley. *Recent Changes in Distribution and Status of Wild Red Wolves*. Endangered Species Report No. 4. U.S. Fish and Wildlife Service, 1979.

Nowak, Ronald M. *North American Quarternary Canis*. Museum of Natural History Monograph No. 6. Lawrence: The University of Kansas, 1979.

Parker, Warren. *Red Wolf Recovery Plan* (updated version). U.S. Fish and Wildlife Service, 1984.

_____. *A Technical Proposal to Reestablish the Red Wolf on Alligator River National Wildlife Refuge, N.C.* U.S. Fish and Wildlife Service, August 1986.

_____. *A Historical Perspective of Canis rufus and Its Recovery Potential*. Red Wolf Management Series Technical Report No. 3. U.S. Fish and Wildlife Service, May 1988.

Phillips, Michael K. *Reestablishment of Red Wolves in the Alligator River National Wildlife Refuge, North Carolina*. A series of progress reports on the reintroduction that are issued annually by the U.S. Fish and Wildlife Service.

BIRDS IN THE HAND

Chapter VI: A Ruinous Legacy

One of the best sources on the history of American settlement and its consequences for animal species is Peter Matthiessen's *Wildlife in America* (New York: Viking, 1959). Barry Lopez presents a fascinating account of the destruction of North American predator populations in his book *Of Wolves and Men* (New York: Scribners, 1978). A disturbing discussion of Christian attitudes toward the wilderness can be found in Frederick Turner's *Beyond Geography: The Western Spirit Against the Wilderness* (New Brunswick: Rutgers University, 1983).

The Defenders of Wildlife, a nonprofit wildlife conservation group based in Washington, publishes an excellent annual report on the status of endangered species within the United States. It can be obtained through the organization's headquarters in Washington.

American Fisheries Society Endangered Species Committee. "Fishes of North America Endangered, Threatened, or of Special Concern: 1989." *Fisheries*, Vol. 14, No. 6 (November–December 1989): 2–19.

Bartram, William. *The Travels of William Bartram*, Mark van Doren, ed. New York: Dover Publishers, 1928.

Lawson, John. *A New Voyage to Carolina*. Chapel Hill: The University of North Carolina Press, 1967.

Lewis, Meriwether, and William Clark. *The Journals of Lewis and Clark*, Bernard DeVoto, ed. Boston: Houghton Mifflin, 1953.

Newmark, William D. "A Land-bridge Island Perspective on Mammalian Extinctions in Western North American Parks." *Nature*, Vol. 325 (January 29, 1987): 430–432.

United States Department of the Interior. *Audit Report: The Endangered Species Program, U.S. Fish and Wildlife Service*. Report 90–98. Washington: September 1990.

United States General Accounting Office. *Endangered Species: Management Improvements Could Enhance Recovery Program*. Washington: December 1988.

_____. *National Wildlife Refuges: Continuing Problems with Incompatible Uses Call for Bold Action*. Washington: September 1989.

White, Lynn. "The Historical Roots of Our Ecological Crisis." *Science*, Vol. 155 (March 10, 1967): 1203–1207.

Wilcove, David. "Empty Skies." *The Nature Conservancy Magazine* (January–February 1990): 4–13.

Chapter VII: THE NEED FOR COMPASSION

The best resource on the early history of whooping crane recovery efforts, including the lengthy controversy surrounding capture of the cranes, is Faith McNulty's book *The Whooping Crane* (New York: E. P. Dutton, 1966).

Dozens of scientific articles have been published on whooping crane breeding and behavioral biology and on the history of the recovery program. I drew extensively from the following works.

Derrickson, S. R. "Captive Propagation of Whooping Cranes, 1982–84." In J. C. Lewis, ed., *Proceedings of the 1985 Crane Workshop*. Platte River Whooping Crane Maintenance Trust and U.S. Fish and Wildlife Service, 1987.

Drewien, R. C., and E. G. Bizeau. "Cross Fostering Whooping Cranes to Sandhill Crane Foster Parents." In S. A. Temple, ed., *Endangered Birds: Management Techniques for Preserving Threatened Species*. Madison: The University of Wisconsin Press, 1977.

_____, and E. Kuyt. "Teamwork Helps the Whooping Crane." *National Geographic*, Vol. 155 (May 1979): 680–693.

Erickson, R. C., "A Federal Research Program for Endangered Wildlife." *Transcript, North American Wildlife Conference* 33 (1968).

_____. "Captive Breeding of Whooping Cranes at the Patuxent Wildlife Research Center." In R. D. Martin, ed., _Breeding Endangered Species in Captivity._ New York: Academic Press, 1975.

_____, and S. R. Derrickson, 1981. "The Whooping Crane." In J. C. Lewis and H. Maasatomi, eds. _Crane Research Around the World._ Baraboo, Wisconsin: International Crane Foundation, 1981.

Kepler, C. B. "Captive Propagation of Whooping Cranes: A Behavioral Approach." In S. A. Temple, ed., _Endangered Birds: Management Techniques for Preserving Threatened Species._ Madison: The University of Wisconsin Press, 1977.

The brief history of The Peregrine Fund was compiled from interviews and the annual newsletters published by the fund, especially those from 1973 through 1976, and from the following articles.

Cade, T. J. "What Makes Peregrine Falcons Breed in Captivity?" In S. A. Temple, ed., _Management Techniques for Preserving Threatened Species._ Madison: The University of Wisconsin Press, 1977.

_____. "Reasons for Using Non-Indigenous and Exotic Peregrines for Release and Establishment in the Eastern United States." Unpublished memorandum, 1978.

Gilroy, M. J., and J. H. Barclay. "Recovery of the Peregrine Falcon in the Eastern United States." In S. K. Majumdar, F. J. Brenner, and A. F. Rhoads, eds., _Endangered and Threatened Species Programs in Pennsylvania and Other States: Causes, Issues, and Management._ Philadelphia: The Pennsylvania Academy of Sciences, 1986.

Wade, Nicholas. "Bird Lovers and Bureaucrats at Loggerheads over Peregrine Falcon." _Science,_ Vol. 199 (March 10, 1978): 1053–1055.

Chapter VIII: LEAVING THE ARK

The best single source on the natural history of the Arabian oryx and the first releases of captive oryx to Oman is Mark R. Stanley Price's book, _Animal Re-introductions: The Arabian Oryx in Oman_ (Cambridge, Great Britain: Cambridge University Press, 1989). Deborah Forester and Timothy Tear wrote an enjoyable account of tracking reintroduced oryx that appeared in _International Wildlife_ (Vol. 20 [November–December 1990]: 18–24). The early efforts to breed oryx in captivity are detailed in an article by Richard Fitter, "Arabian Oryx Returns to the Wild," in _Oryx,_ (Vol. 16, 1982: 406–410).

In 1987 the Smithsonian Institution issued a copyrighted paper entitled "An Overview of the Golden Lion Tamarin Conservation Program" that is informative, if somewhat dated. In addition, a number of books on animal reintroductions contain chapters on the golden lion tamarin reintroduction project. I found the following to be especially helpful.

Beck, Benjamin B., and four coauthors. "Preparation of Captive-Born Golden Lion Tamarins for Release into the Wild." In D. G. Kleiman, ed. *A Case Study in Conservation Biology: The Golden Lion Tamarin* (in press).

Kleiman, Devra G., and five coauthors. "Conservation Program for the Golden Lion Tamarin: Captive Research and Management, Ecological Studies, Educational Strategies, and Reintroduction." In Kurt Benirschke, ed. *Primates: The Road to Self-Sustaining Populations.* New York: Springer-Verlag, 1986.

Kleiman, Devra G., and three coauthors. "Costs of a Reintroduction and Criteria for Success: Accounting and Accountability in the Golden Lion Tamarin Conservation Program." In J. H. W. Gipps, ed. *Beyond Captive Breeding: Reintroducing Endangered Species to the Wild.* Oxford: Oxford University Press (in press).

Chapter IX: The Essence of Wildness

A number of popular books contain informative discussions of the genetic management of endangered species. One of the best primers on the subject has been issued privately by the Captive Breeding Specialist Group. It has been included, in slightly different forms, in the Population Viability Analyses for the Puerto Rican parrot, the Florida panther, and other highly endangered species. All are available from the group's headquarters in Apple Valley, Minnesota.

Hoage, R. J., ed. *Animal Extinctions: What Everyone Should Know.* Washington: Smithsonian Institution Press, 1985.

Lande, Russell. "Genetics and Demography in Biological Conservation." *Science*, Vol. 241 (September 16, 1988): 1455–1460.

Lovejoy, Thomas E. "Genetic Aspects of Dwindling Populations: A Review." In S. A. Temple, ed. *Endangered Birds: Management Techniques for Preserving Threatened Species.* Madison: The University of Wisconsin Press, 1977.

Seal, Ulysses S. "The Noah's Ark Problem: Multigenerational Management of Wild Species in Captivity." In S. A. Temple, ed. *Endangered Birds: Management Techniques for Preserving Threatened Species.* Madison: The University of Wisconsin Press, 1977.

Simberloff, Daniel. "The Contribution of Population and Community Biology to Conservation Science." *Annual Review of Ecological Systems 19* (1988): 473–511.

Wilson, E. O., ed. *Biodiversity.* (Washington: National Academy Press, 1988).

Chapter X: A SINGLE STRUGGLING FLOCK

Much of the historical information was taken from *The Parrots of Luquillo: Natural History and Conservation of the Puerto Rican Parrot* by N. F. R. Snyder, J. W. Wiley, and C. B. Kepler (Berkeley: Western Foundation of Vertebrate Zoology, 1987).

I also referred extensively to the annual performance reports for the project for 1987 and 1988 and to memorandums written between the Fish and Wildlife Service southeast regional office, the Patuxent Wildlife Research Center, the Puerto Rico Department of Natural Resources, and the Captive Breeding Specialist Group during 1984, 1985, 1986, 1989, and 1990.

Captive Breeding Specialist Group. *Population Viability Analysis for the Puerto Rican Parrot*. Published privately by the group, July 30, 1989.
Puerto Rican Parrot Working Group. *Puerto Rican Parrot Recovery Plan* (updated version). Atlanta: U.S. Fish and Wildlife Service, April 1987.
Wiley, James W. "Bird Conservation in the United States Caribbean." *Bird Conservation* 2 (1985): 107–159.

Chapter XI: THE SOUL OF THE CONDOR

The amount of material written on the California condor is the most voluminous of any of the species that I investigated. I include here only a small sampling of the articles and books that are available. Most contain additional sources for readers who wish to research the subject more thoroughly.

Brower, Kenneth. "The Naked Vulture and the Thinking Ape." *The Atlantic Monthly* (October 1983): 70–88.
California Condor Recovery Team. *Revised California Condor Recovery Plan*. Portland: U.S. Fish and Wildlife Service, July 1984.
Koford, Carl B. *The California Condor*. Research Report No. 4. New York: National Audubon Society, 1953.
_____. "California Condors, Forever Free?" *Audubon Imprint* (Newsletter of the Santa Monica Bay Audubon Society), Vol. 3 (April 1979): 1–7.
Miller, Alden H., Ian I. McMillan, and Eban McMillan. *The Current Status and Welfare of the California Condor*. Research Report No. 6. New York: National Audubon Society, 1965.
Snyder, Noel F. R. "California Condor Recovery Program." In S. E. Senner, C. M. White, and J. P. Parrish, eds. *Raptor Conservation in the Next Fifty Years*. Research Report No. 5. Hastings, Minnesota: Raptor Research Foundation, 1986.

_____, and Helen A. Snyder. "Biology and Conservation of the California Condor." In Dennis M. Power, ed. *Current Ornithology*, Vol. 6 (1989). New York: Plenum Press.

Wilbur, Sanford R. *The California Condor, 1966–76: A Look at Its Past and Future.* North American Fauna No. 72. Washington: U.S. Fish and Wildlife Service, 1978.

Chapter XII: STEPPING BACK FROM EXTINCTION

For the history of the recovery project, I depended on interviews with the parties involved, on popular articles, on memos from staff members for the Wyoming Game and Fish Department, and on the annual job performance reports published by the state. A great many studies have been conducted on ferrets in captivity—far more than on the species in the wild. The publications listed below provide only a small sampling of available material.

Biggins, Dean E., and three coauthors. "Trial Release of Siberian Polecats (*Mustela eversmanni*)." Field Study Progress Report, February 28, 1990.

Forrest, Steve C., and three coauthors. "1985 Black-Footed Ferret Litter Survey at Meeteetse, Wyoming." Interim Field Report, August 5, 1985.

Miller, Brian, Christen Wemmer, and two coauthors. "A Proposal to Conserve Black-Footed Ferrets and the Prairie Dog Ecosystem." *Environmental Management*, Vol. 14 (1990): 763–69.

_____, and six coauthors. "Development of Survival Skills in Captive-raised Siberian Ferrets (*Mustela eversmanni*)." *Journal of Ethology*, Vol. 8 (1990): 89–104.

Seal, Ulysses S., ed. *Conservation Biology and the Black-Footed Ferret.* New Haven: Yale University Press, 1989.

Thorne, E. Tom, and Elizabeth S. Williams. "Disease and Endangered Species: The Black-Footed Ferret as a Recent Example." *Conservation Biology*, Vol. 2 (March 1988): 66–74.

U.S. Fish and Wildlife Service. *Black-Footed Ferret Recovery Plan.* Denver: 1988.

Weinberg, David. "Decline and Fall of the Black-Footed Ferret." *Natural History*, Vol. 95 (February 1986): 62–69.

Wyoming Game and Fish Department. *A Cooperative Management Plan for Black-Footed Ferrets at Meeteetse.* Draft, August 4, 1989; updated June 22, 1989.

Chapter XIII: THE PANTHER VERSUS FLORIDA

At the end of each fiscal year biologists employed by the state of Florida write detailed performance reports on the studies they have in progress.

These reports often include informative data tables and graphs, as well as the biologists' own conclusions and recommendations. They are available from the state on request.

Anderson, Allen E. *A Critical Review of Literature on Puma Felis* concolor. Colorado Division of Wildlife Special Report No. 54, February 1983.

Captive Breeding Specialist Group. *Florida Panther Viability Analysis and Species Survival Plan.* Apple Valley, Minnesota: Captive Breeding Specialist Group, December 15, 1989.

Florida Panther Interagency Committee. *Florida Panther Revised Recovery Plan.* Atlanta: U.S. Fish and Wildlife Service, 1987.

Maehr, David S. "The Florida Panther and Private Lands." *Conservation Biology,* Vol. 4 (June 1990): 167–170.

O'Brien, Stephen J., David E. Wildt, and Mitchell Bush. "The Cheetah in Genetic Peril." *Scientific American,* Vol. 254 (May 1986): 84–92.

_____, and nine coauthors. "Genetic Introgression within the Florida Panther *Felis concolor coryi.*" *National Geographic Research,* Vol. 6 (October 1990): 485–494.

Chapter XIV: EPILOGUE

Many scientists and conservationists have published papers containing suggestions for improving the management of endangered species. I drew ideas from far too many sources to list. A number cautioned that the science of reintroduction is in its infancy, and that it may be many generations before we know how well we have succeeded in our attempts to preserve the rarest animals on earth.

Derrickson, Scott R., and Noel F. R. Snyder. "Potentials and Limits of Captive Breeding in Parrot Conservation." In S. R. Bessinger and N. F. R. Snyder, eds. *New World Parrots in Crisis: Solutions from Conservation Biology.* Washington: Smithsonian Institution Press (in press).

Griffith, Brad J., and three coauthors. "Translocation as a Species Conservation Tool: Status and Strategy." *Science,* Vol. 245 (August 4, 1989): 477–480.

Grumbine, R. Edward. "Viable Populations, Reserve Size, and Federal Lands: A Critique." *Conservation Biology,* Vol. 4 (June 1990): 127–134.

Kaufman, Les, and Kenneth Mallory, eds. *The Last Extinction.* Cambridge, Massachusetts: MIT Press, 1986.

Stanley Price, Mark. *Animal Re-introductions: The Arabian Oryx in Oman.* Cambridge, Great Britain: Cambridge University Press, 1989.

Wilson, E. O., ed. *Biodiversity.* Washington: National Academy Press, 1988.

INDEX

Aboriginal cultures and nature, 100, 103
Allen, Robert Porter, 107–8
Alligator River National Wildlife Refuge, North Carolina
  description of, 5, 11–12, 15, 19, 23, 57, 77, 80, 91
  selection as reintroduction site for red wolves, 10, 50
American Association of Zoological Parks and Aquaria, 154, 158
American attitudes toward wildlife and nature, 4, 5, 8, 9, 21–22, 26, 27, 99–100, 102–3, 163, 172, 176, 243, 275, 283, 285
American fauna, extinct
  Carolina parakeet, 93
  ivory-billed woodpecker, 92–93
  Labrador duck, 92
  passenger pigeon, 92, 153
  sea mink, 92
American landscape before European settlement, 90–91, 106
American Society of Mammalogists, 33
Anahuac National Wildlife Refuge, Texas, 33, 36
Anderson, Betsy, 180–81
Anthropocentrism, 100–101, 102, 282
Arabian oryx, 105, 129–38
  ability to withstand heat and drought, 130, 135
  acclimation of captive animals to desert, 134–35, 136
  in Arabic culture, 130
  captive breeding program for, 131–32
  extinction from overhunting, 129, 130–31, 132, 133
  funding for recovery project, 133–34
  genetic diversity in, 136, 137
  original range, 130
  physical characteristics, 129, 130, 134
  recovery project as model for other programs, 129, 137

release of captive animals in Oman, 129, 134, 135–36, 137
  reproduction in wild, 135, 136, 137
  social behavior, 134, 135–36, 137
Aransas National Wildlife Refuge, Texas, 106, 107, 119
Archibald, George, 116, 117, 119
Audubon, John James, 226

Bachman, John, 226
Baker, Andy, 147
Bass, Oron, 252, 258–59, 260–61
Bears
  black, 162
  giant panda, 161
  grizzly, 95, 152
  polar, 149, 156
Beasley, Jim, 70, 71, 72, 73, 76
Beck, Benjamin, 143, 145–46, 147
Behrns, Sue, 13–14, 27, 44, 47–51, 87
Belden, Chris, 251, 253–54, 255, 261, 266–68, 269, 273
Belitsky, Dave, 234, 235
Big Cypress National Preserve, 252, 253, 254, 255, 257, 261, 272, 273
Biggins, Dean, 230, 237, 244, 245, 247, 248
Biota Research and Consulting, Inc., 228, 230, 232, 233, 234–36
Bird species
  disappearing songbirds, 98–99
  increasing generalist species, 99
Bison, 92, 93
Black-footed ferret, 7, 15–16, 225–48
  breeding biology, 239, 241
  causes of near-extinction, 226, 227
  discovery of Meeteetse population, 225–26, 228
  distemper in, 227–28, 236–37, 238
  extinction of Meeteetse population, 233–36
  feeding and social behavior, 226, 229, 244–45

original range, 226
physical characteristics, 225, 226, 242
South Dakota population, 226–27
temperament of wild versus captive
animals, 241, 242
Black-footed Ferret Advisory Team, 231,
232, 233, 234, 235
Black-footed ferret recovery project
breeding program established, 230, 231,
232, 234, 236, 238, 239
cages and nest box design, 242, 247
capture of entire population, 226, 237
censusing of wild population, 227, 229,
230, 233, 234–35, 236
discord among biologists and conservationists, 226, 229, 230–32, 233–36
genetic management of population, 240
illnesses in breeding population, 240–41
lack of habitat for reintroduction, 245–46,
248
manipulation of breeding by photo-
stimulation, 239
predator avoidance skills, 241, 242,
244–45, 247–48
reintroduction to wild, 226, 239, 245,
247, 248
tests of hunting skills in captive kits, 239,
245, 247, 248
Blumig, Cathy, 167–69, 182
Bobwhite, masked, 111–13
Boeker, Erwin, 110
Bosque del Apache National Wildlife
Refuge, New Mexico, 114–15
Bossert, William, 33, 38
Buffalo. See Bison
Bureau of Land Management, U. S., 96
Bureau of Sports Fisheries and Wildlife,
U. S., 37, 108, 109–10, 111–12
Butler, Harold, 73–74

Cade, Tom, 120–24, 125, 126–27
Cage stereotypes in confined animals,
149–50, 162
California condor, 17, 191–224
breeding biology, 196, 198, 207
causes of near-extinction, 196, 207
deaths of wild birds, 200, 208, 211
flight, 193, 194, 203, 210, 217–18, 219, 220
genetic inbreeding in, 197, 200, 208–9,
220
and hunters, 201, 202, 209, 222
lead poisoning in, 194, 207, 208, 210–11,
222
original range, 194–95
physical characteristics, 193–94, 215
social behaviors and traditions, 199, 209,
215, 222–23
California condor recovery project
alteration of species in captivity, 197,

199, 204, 220
breeding program established, 197,
198–99, 205–7, 219
capture of entire population, 194, 208–9,
211–12
censusing of wild population, 197, 198,
204
discord among biologists and conservationists, 194, 197–98, 199, 200–202,
203–11
funding for, 95, 195–96
habitat preservation, 198, 201, 204–5, 209,
222, 223–24
manipulation of condor foraging habits, 17,
191–92, 208, 210, 213–14, 216, 223–24
radio telemetry research, 201, 202, 203–4,
205, 207, 208, 210
raising of condor chicks with puppets,
215, 219
reintroduction of captive chicks to wild,
207–8, 210
reintroductions outside California, 224
removal of eggs and young from wild,
199, 206–7, 219
role of recovery team in, 198
role of zoos in, 192, 197, 206, 211, 212,
218–20
study of wild Andean condors in Peru,
203
California Fish and Game Commission, 197,
200, 202, 205, 206, 209, 210
Canadian Wildlife Service, 106, 112, 114,
115, 125
Captive breeding
limitations of, 7–8, 9, 18, 102–3, 108,
116–18, 122, 156, 157, 199, 244, 279,
280, 282
need for special handling of animals,
105–6, 113–14, 160, 161–62
philosophy behind, 6–7, 17, 18, 102–3,
108, 150, 153–54, 157, 159, 163
Captive Breeding Specialist Group
and black-footed ferret recovery project,
238, 239
and Florida panther recovery project,
269, 270, 271
formation of, 154–55
and Puerto Rican parrot recovery project,
184, 185, 186
Caribbean National Forest
crime in, 175
description of, 167, 170, 180
logging in, 172
Carley, Curtis, 38–43, 44, 45, 46, 49, 50
Carlstead, Kathy, 162
Carson, Rachel, 100, 111
Castro, Inez, 145, 146
Cattle ranching. See Livestock ranching
Cheetah, genetic inbreeding in, 153, 263–64

Christian attitudes toward nature, 100–101, 281

Clark, Tim, 228, 229, 230, 232, 234, 235, 237, 241

Clendenen, Dave, 210, 211, 214, 215, 216, 217–18

Coimbra–Fihlo, Adelmar, 138, 139, 140, 143

Collins, Larry, 160–62, 163

Condor Research Center, 194, 201, 202, 204, 205, 206, 209, 212

Condors. *See also* California condor

Andean condors used in release experiment in California, 191–93, 212, 213–18, 220–21

Conservation and Research Center, Front Royal, 118, 157–58, 243

Cooper, Larry, 54–55, 56, 63, 78, 79

Cory, Charles, 250

Cougar, western

comparisons with Florida panther, 249, 254

increase in population, 252

used in pilot release program in Florida, 266–68

Cranes. *See also* Whooping crane

Mississippi sandhill crane, 117, 119

sandhill crane as surrogate parent for whooping crane, 109, 110, 112

sandhill cranes first released in Florida, 158

Curlew, Eskimo, 93

Daly, Ralph, 132, 133, 134

Deep ecology, 101–2, 103

Defenders of Wildlife, 96, 288

Derrickson, Scott, 115–18, 119, 157–60, 163

Dexter, Don, 232

Diaz–Soltero, Hilda, 184, 187

Dietz, James, 140–42, 143–44, 147, 148

Dietz, Lou Ann, 140, 142–43, 147–48

Drewien, Roderick, 115, 119

eagle, bald, restoration program for, 95, 125, 127

Earth First!, 102, 220

Ehrlich, Paul, 100

Elder, William, 32–33, 46

Emerson, Ralph Waldo, 100

Endangered Species Act, 6, 32, 38, 42, 44, 94, 95, 126, 156, 172, 223, 278, 282

Erickson, Ray, 108–12

Everglades National Park, 252, 255–56, 258, 259, 273

Extinction

causes of, 16, 17, 105, 137, 152

world rates of, 17

Fakahatchee Strand, Florida, 253, 255, 256–57, 261

Falcon, peregrine. *See* Peregrine falcon

Falconry, 120, 122, 123

Fauna Preservation Society, 131, 133

Ferrets. *See also* Black-footed ferret

domestic, 226, 240

Siberian polecat as surrogate for black-footed ferret, 228, 243–45, 247–48

Field research, importance of, 174–75, 179–80, 282–83

Fish, disappearing species, 98–99

Fish and Wildlife Service, U.S.

guiding policy behind endangered species programs, 15–16, 95–97, 195

mismanagement of endangered species programs, 95, 96–97, 112

progress toward preserving endangered species, 94–95

recovery plans for endangered species, 40, 155

southeast regional office, 10–11, 61–62, 179, 181, 182, 188, 189

Florida Game and Fresh Water Fish Commission, 119, 251, 253, 254, 263, 267, 268, 273

Florida panther, 152, 249–73

breeding and social behavior, 255, 259, 261, 270

causes of near-extinction, 251, 253, 273

condition of panthers on public versus private land, 255, 261

deaths of wild panthers, 252, 254, 256, 261

differences between panthers in southeast and southwest Florida, 256, 262, 271–72

disease in, 257, 269

feeding habits, 260

hybridization with exotic species, 251, 254, 262, 271–72

inbreeding in, 152, 251, 262–63, 269

original range, 250

persecution by settlers, 253

physical characteristics, 249–50, 253–54

sperm deficiencies in, 152, 262, 264, 266

Florida Panther Interagency Coordinating Committee, 263, 264, 269, 270

Florida Panther National Wildlife Refuge, 261

Florida Panther Records Clearinghouse, 251

Florida panther recovery project

captures of animals for radio tracking studies, 252, 253–56, 258, 260, 261

controversy over captive breeding program, 269–71

difficulties of locating and censusing animals, 251–52, 258, 259–60, 261

genetic research, 254, 261–63, 271–72

habitat preservation, 251, 258, 261, 270, 272–73

and hunting pressures, 255, 257, 267–68, 270, 273
in vitro fertilization experiments, 264–66, 270
introduction of new bloodlines to gene pool, 262–63
population viability analysis, 269, 270
public attitudes toward, 250, 252, 258, 267, 269–71
release of western cougars in north Florida, 266–68
wildlife underpasses on Interstate 75, 256, 269
Foott, Jeff, 199–200
Forest Service, U.S., 96, 172, 173, 174, 178, 179, 182
Forrest, Steve, 229–30, 234, 235, 236
Francis of Assisi, Saint, 101
Friends of the Earth, 198, 200, 201, 202, 222

Geese, Aleutian Canada, 111, 113
General Accounting Office, reports on endangered species, 95, 96
Genetic diversity
importance in animal populations, 151–52
size of population needed for, 153–54, 155–56, 240
tests for determining, 153, 262, 263, 271, 272, 277
Genetic inbreeding
causes, 151–54
and demographic trends, 152
and disease resistance, 105, 151, 152, 262–64
and environmental change, 105, 151, 153, 161, 262
reproductive failure caused by, 105, 151, 152, 153, 262, 264
Golden lion tamarin, 105, 129, 137–48
breeding by zoos, 138–40
causes of near-extinction, 138, 140
effects of habitat destruction on, 138, 140, 141, 148
feeding habits, 139, 140, 144, 145
genetic diversity in, 139, 143, 145
movement through jungle, 141, 144
original range, 138
as pets, 138, 143
physical characteristics, 137–38
predation on, 144, 158
social behavior, 139–40, 141, 144, 147
vocalizations, 141, 144, 147
Golden Lion Tamarin Conservation Program
and employment for local residents, 138, 146–48
and forest fire, 141, 143, 148
and forest regeneration, 138, 142
funding for, 142, 147

and local land owners, 143, 146–47
as model for other endangered species programs, 105–6, 138, 148
public education program, 138, 140, 142–43, 146–47, 246, 283
releases of captive animals, 143, 144, 145, 146
research on wild population, 141, 147
survival of released animals, 144, 147
training programs for released animals, 138, 143–46
Goodrowe, Karen, 264, 265
Grantham, Jesse, 205–6, 207, 208, 209, 210, 222
Grays Lake National Wildlife Refuge, New Mexico, 114–15, 116, 118–19
Great Smoky Mountains National Park, 85, 98, 275–76
Greens political party, 102
Griffith, Brad, 279
Grimwood, Major I.R., 131

Habitat degradation and loss, effects on animal species, 16, 30, 94, 96, 97–99, 106, 115, 119, 138, 140, 141, 148, 224, 251, 253, 261, 272–73, 278–79
Harasis nomads
assistance given to Arabian oryx project, 133, 135
culture of, 129, 132, 136
as rangers for oryx project, 129, 132, 134–35, 136, 137
Harju, Harry, 231, 232, 233–34, 235, 236
Hoagland, Edward, 34, 37
Hogg, John, 225–26
Hogg, Lucille, 225–26, 228, 237
Hopper Mountain National Wildlife Refuge, 191, 212–13, 214, 216, 217–18, 220–21
Howard, JoGayle, 264–65, 266
Hunters
attitudes toward wildlife, 11, 21–23, 25–26, 61–62, 257, 267, 283
influence with state and federal wildlife officials, 16–17, 62, 96, 268
Hurricane Hugo, 186–88, 189

Infertility medicine and endangered species, 239, 264–66
Ingram, Jack, 216, 217–18
International Crane Foundation, 116, 119
International Species Inventory System, 154
International Union for the Conservation of Nature and Natural Resources, 154–55, 184

Jansen, Deborah, 249, 250, 259–60
Janzen, Daniel, 108, 110
Jiddat-al-Harasis desert, 132, 133–37

Johnson, Murray, 32, 37, 42, 44
Jones, Mike, 44, 45
Joslin, Paul, 33
Judaism and the natural world, 100–101
Juneamann, Greg, 58, 59, 61, 62

Kennedy, Beth, 24–26
Kepler, Cameron
    and Puerto Rican parrot, 172–73
    and whooping crane, 113–14, 115
Kleiman, Devra, 138–40, 143, 145, 147, 148
Koford, Carl, 196–99, 201, 203, 204, 221
Kwiatkowski, Don, 238, 239, 240–41, 242, 247

Lawrence, Barbara, 33, 38
Lawson, John, 91
Lehman, Bill, 199–200
Leopold, Aldo, 100
Livestock ranching
    effects on grasslands, 16
    ranchers' and cowboys' attitudes toward wildlife, 16, 35, 36, 227, 246, 283, 284
Long, Buddy, 35, 37, 42, 44, 45
Los Angeles Zoo, 131, 194, 206, 212, 213, 220
Lost Colony, the, 91
Loxahatchee National Wildlife Refuge, Florida, 158
Lucash, Chris, 19–20, 58, 63, 64, 65–66, 67–69, 70, 71–72, 74, 75–77, 78–79, 80, 82–84, 276
Luquillo Mountains. See Caribbean National Forest
Lynch, John, 108, 109

Macaw, extinct Caribbean species, 171
Maehr, David, 272–73
Makey, Dale, 154
Malheur National Wildlife Refuge, 108–9
Mammal species
    difficulties of preserving, 129–30
    extinction in fragmented habitat, 97
Manns Harbor, North Carolina, 5, 11, 20–21, 70, 71–74
Matthiessen, Peter, 91, 98
Maytag Zoo, Phoenix, Arizona, 131
McBride, Roy and Rocky, 258, 260, 261
McCarley, Howard, 31–32, 33, 42, 45, 46
Mech, L. David, 33–34
Meeteetse, Wyoming
    ranches inhabited by ferret population, 225–26, 232–33
    residents' attitude toward ferret recovery project, 236–37
Meng, Heinz, 123
Miller, Annie, 264, 265–66
Miller, Brian, 243, 244, 245–46, 247, 248
Monte Vista National Wildlife Refuge, Colorado, 110

Montezuma National Wildlife Refuge, New York, 125, 127
Morris, Bill, 230
Morse, Mike, 86
Mountain lion. See Florida panther and Western cougar
Muir, John, 100
Mundt, Senator Karl, 111
Myers, Edna, 237

Naess, Arne, 101–2
National Audubon Society, 107, 108, 109, 112, 126, 168, 194, 196, 197, 198, 199, 202, 210, 211, 222
National Marine Fisheries Service, 95
National Zoological Park, Washington, D.C., 7, 137, 138–40, 142, 143–44, 146, 156, 157, 161, 162, 239, 246, 254, 263
New York Zoological Society, 7, 232, 235, 237
Newmark, William, 97
Noegel, Ramon, 177, 178, 180, 181, 182
Nowak, Ronald, 37–38, 41

O'Brien, Stephen, 263, 271–72
Ogden, John, 199, 200, 201–4, 205–8
Owl, northern spotted, 95

Paleudis, George, 67–69, 70, 71–73, 74, 80
Palmer, Mark, 202, 204, 222
Panther, Florida. See Florida panther
Paradiso, John, 37–38
Parakeets
    Caribbean species, 171
    Carolina parakeet. See American fauna, extinct species
Parker, Warren, 3, 8, 10–11, 15, 20, 27, 50, 53, 54, 57, 58–61, 62–66, 70–73, 74, 82, 83, 85, 86, 87, 276, 277–78
Parrots. See also Puerto Rican parrot
    Caribbean species, 171
    Hispaniolan parrots as surrogates for Puerto Rican parrots 169, 175, 176
    thick-billed parrot, 184
Patuxent Wildlife Research Center, 109, 110–12, 114, 115, 119, 170–71, 178, 181, 182, 192, 199, 202, 206, 209–10, 211, 212, 214, 226–27
Pedersen, Dale, 44–45, 47
Pelican, brown, 94–95
Peregrine falcon, 119–27
    artificial insemination in, 122
    captive breeding program established, 121–22
    endangerment by pesticides, 120–21
    extinction of eastern North American population, 120–21
    flight, 120, 124
    fostering into wild nests, 122, 126

hacking techniques, 123, 124, 125
normal reproduction, 122, 126
releases of captive-bred chicks, 122, 123, 124, 125, 126, 127
releases of exotic subspecies, 122–23, 126, 127
Peregrine Fund, the, 121, 122–27, 247
Pesticides, effects on wildlife, 94, 120–21, 122, 202
Petrochemical development and wildlife, 16, 201, 205, 213, 221
Phillips, David, 202, 203, 205, 206, 207, 222
Phillips, Mike, 23–26, 53, 54, 55, 56, 57, 58–61, 62–63, 64, 65, 66, 67, 69, 70–72, 74, 75, 77, 78–79, 80–81, 82–83, 84, 85, 86, 87, 276, 277
Pimlott, Douglas, 33, 46
Plover, golden, 93
Poço das Antas Reserve, Brazil, 140–42, 143–44, 148
Point Defiance Zoo and Aquarium, Tacoma, 9, 36–37, 38, 42, 44, 45, 46
Poisoning of wild animals, 16, 30, 39, 196, 199, 201, 202, 207, 227
Population Viability Analysis
for black-footed ferret, 184
for Florida panther, 184, 269
general concept, 155–56
for Puerto Rican parrot, 184–86
Prairie dogs
burrows used by other species, 226, 246
effects on grasslands, 16, 227, 246
extermination of, 16, 227, 245
original range, 226–27
as prey for black-footed ferret, 16, 226
proposal by biologists to eliminate poisoning programs, 245–46
and sylvatic plague, 234–35, 245, 247
Puerto Rican Department of Natural Resources, 170, 173. 175, 177, 181, 182, 189
Puerto Rican parrot, 167–90
causes of near-extinction, 171–72
competition with pearly-eyed thrashers, 173–74, 282–83
destruction of nest trees by charcoal industry, 172
domestication in captivity, 170, 171, 175
and hurricanes, 171–72, 185, 186–88, 189
mortality among juveniles, 177, 183, 184
nesting behavior, 169, 170, 173, 178, 179
original range, 171
as pets and food, 172
physical characteristics, 169, 180, 189–90
predation on, 173, 178, 183, 187
seasonal movements, 172, 188
vocalizations, 169–79
Puerto Rican parrot recovery project
artificial insemination used in, 176, 179, 182
artificial nests used in, 168, 173, 174, 178, 282–83
breeding program established, 170, 173, 175
cage design, 177, 180
censusing of wild population, 172–73, 183, 184, 187–88
and commercial aviculture, 176–77, 180, 181
discord among biologists, administrators, and breeding specialists, 170–71, 177–80, 181–82
fostering of eggs and chicks into wild nests, 170, 177, 181, 183, 185, 188
funding for, 176, 178, 181
genetic management of wild flock, 170, 185
guarding of wild nests, 167–69, 176, 180, 183
infertility in captive birds, 175–76, 178, 182, 183
Luquillo aviary, 175, 180, 186, 187, 188
mainland aviary, 189
population viability analysis, 184–86
preservation of wild flock, 172–75, 176, 177, 178, 179–80, 181, 182, 183, 184, 185–86, 188
Rio Abajo aviary, 177–78, 181, 182, 183, 189
survey of nest trees, 173, 174

Qaboos bin Said, H.M Sultan, 132

Radio capture collars, used on red wolves, 10, 53, 57–64, 70, 71–72, 75, 77, 81, 276
Reagan, President Ronald, 228–29
Red wolf
classification as separate species, 6, 33–34, 37–38, 277
effects of parasites on, 9, 34, 36, 48, 56, 57, 80
extermination in Southeast, 6, 30, 275
hybridization with coyotes, 9, 31–32, 33–34, 35–36, 38–43, 276, 277
natural history, 8–9, 29, 34, 35
original range, 5, 30
physical characteristics, 5, 6, 45, 53, 87
predation on livestock, 29–30, 34, 35, 275–76
and predator control programs, 30, 31, 32, 33, 34
reluctance of Texas officials to preserve species, 36, 37
subspecies of, 30–31
taxonomy of, 31–32, 33, 41–42
vocalizations, 33, 34, 43, 48, 51, 85
Red wolf recovery project
Bulls Island, South Carolina, releases, 13, 82, 276

captive breeding program established, 6, 9, 36–37, 40–43, 46

delay of first release in North Carolina, 53, 57–62

established, 6, 34–43, 46–47

euthanasia of captive hybrids, 42, 45, 46, 49

extermination of wild coyotes and hybrids, 35, 39–40, 42

genetic management of species, 81, 83–84, 277

handling of wolves in captivity, 13–14, 45, 47–48, 53, 54–56, 62–63

mortality in captive pups, 47, 48, 56

preparation of wolves for release, 13, 20, 23–24

problem of distinguishing red wolves from hybrids, 31, 32, 35–36, 41–43, 45–46, 48, 49, 50, 277

proposed reintroduction in Kentucky and Tennessee, 10–11

recovery team, 38, 40, 42, 43, 54, 82

reintroduction in Great Smoky Mountains National Park, 98, 275–76

releases of captive wolves, 64–65, 67, 79, 85

Red wolves, released animals

aggression between, 68, 77–78

behavior, 4, 67–69, 70, 71–74, 77, 78–79

capture of wild-born pups, 82–83, 84, 85

deaths, 78, 79, 80, 84, 276–77

encounters with humans, 68–69, 70, 71–74, 277

and private property, 70, 71–74, 75–77, 80, 85, 276

recaptures, 70, 71–74, 75–77, 84–85

reproduction of, 79–80

Restoration ecology, 97

Riley, Glynn, 34–38, 39, 40, 42

Ripley, S. Dillon, 107–8

Rippons, Benny, 22–23

Roelke, Melody, 254–55, 259, 261–64, 265, 266, 271

Salyer, J. Clark, 108–9

San Diego Wild Animal Park, 7, 132, 134, 151, 193, 194, 206, 219

San Diego Zoo, 197, 212

Santa Cruz Predatory Bird Research Group, 127

Seal, Ulysses, 46, 154, 155, 156, 184, 188, 269

Seton, Ernest Thompson, 226–27

Shaefer, Donald and Donna, 41

Sierra Club, 197, 198, 200, 201, 202, 222

Smith, Roland, 54–56, 58–60, 63, 64, 277

Smithsonian Institution, 142, 157

Snyder, Helen, 223

Snyder, Noel

and California condor, 184, 199–210, 211, 212, 213, 222, 223

and Puerto Rican parrot, 174–75, 179, 184–85, 188, 189, 282–83

Sontag, Hugh, 58–59, 60–61, 62

Sorenson, Teri, 186, 187

Species

generalist and specialist, 99

number of endangered and threatened in U.S., 94–95, 96

reproductive barrier between, 31

r-selected and K-selected, 152–53

triage approach to preservation, 282

world extinction rates of, 17

Species Survival Plan

for Florida panther, 269

general concept, 155

for red wolf, 83–84, 155

Stanley Price, Mark, 133–36

Taapken, John, 174

tamarins, Brazilian species, 138, 147

Taylor, John, 14, 21, 23, 54–56, 57, 58–60, 61, 62, 64, 69, 70, 71, 72, 74

Tear, Timothy, 136, 137

Thoreau, Henry David, 100

Thorne, Tom, 235, 236, 237–39, 240, 241, 242, 244, 245, 247

Timbering

effects on natural systems, 16, 278

in Pacific Northwest, 16, 96

Toone, William, 151, 158

Turnell, Jack, 233

Turner, Frederick, 101

Vivaldi, Jose, 171, 177–78, 179, 181–82, 184

Wallace, Michael, 221

Washington Park Zoo, Portland, 149

Wayne, Robert, 277

Wemmer, Christen, 158, 245–46

White, Lynn, Jr., 100–101

White Oak Plantation, 260, 264, 268, 269

White Sands Missile Range, New Mexico, 283

Whooping crane, 106–10, 112–19

artificial incubation of eggs, 115–16

artificial insemination in, 114, 115, 116–17, 119, 159

captive breeding program established for, 107–8, 109, 112, 113–14

copulation problems, 114, 117–18, 119

equine encephalitis in, 118

failure of captive birds to tend young, 159–60

Grays Lake reintroduction project, 114–15, 116, 118–19

growth of wild population, 109, 112, 119

incubation by sandhill cranes, 115–16

migration, 106–7

original range, 106

physical characteristics, 106
release at Kissimmee River prairie,
  Florida, 119
social behavior, 107, 113–14, 116
Whooping Crane Advisory Group, 108, 110
Wilbur, Sanford, 198
Wilcove, David, 98–99, 281
Wild Animal Propagation Trust, 138
Wild animals
  effects of stress on, 161
  instinctive behavior versus learned be-
    havior, 158, 161
  physical changes in captivity, 157, 280
  psychological and behavioral changes
    in captivity, 7, 8, 18, 150, 151, 156,
    158–62, 280
  restrictions on movements, 8, 17, 70,
    71–74, 75–77, 278, 284
  windows of learning in, 159
Wilderness Society, the, 96, 98
Wildlife biopreserves, 223, 224, 281
Wildlife Conservation International, 237
Wildlife Preservation Trust International,
  142
Wildlife Safari, Winston, Oregon, 263
Wildlife translocations, analysis of, 279
Wildt, David, 263, 264
Wiley, Beth, 175, 176, 177
Wiley, James
  and California condor, 179, 212, 213–14,
    216–18, 221
  and Puerto Rican parrot, 175–80, 184,
    283
Williams, Beth, 238

Wilson, Alexander, 92, 93
Wilson, Marcia, 182–83, 184, 186–87, 188
Windley, John, 82–83, 85
Wolves. See also Red wolf
  gray, 5, 6, 8, 16, 33–34, 95, 161, 246,
    283–85
  Mexican, 283–84
Wood Buffalo National Park, Alberta, 106,
  107, 109, 112, 114, 118
World Center for Birds of Prey, 127
World Wildlife Fund, 131, 133, 142, 172,
  231–32, 235
Wyoming Game and Fish Department
  breeding project for black-footed fer-
    rets, 238–39, 240–42
  and ferret reintroduction, 247, 248
  as lead agency for ferret recovery project,
    229–32, 233–36, 238–39, 248

Yellowstone National Park, wolf releases
  in, 3, 16, 283–85

Zerda Ordáñez, Enrique, 216–17
Zoos
  cage design and landscaping, 149–50,
    156
  management of genetic diversity in en-
    dangered species, 139, 154–56, 269
  and reintroduction of animals to the wild,
    160
  and species preservation, 102, 150,
    154–57, 160–62, 189, 206, 211, 218–20,
    269, 276, 280
  and stress in animals, 161